Complex Analysis

This user-friendly textbook introduces complex analysis at the beginning graduate or advanced undergraduate level. Unlike other textbooks, it follows Weierstrass's approach, stressing the importance of power series expansions instead of starting with the Cauchy integral formula, an approach that illuminates many important concepts. This view allows readers to quickly obtain and understand many fundamental results of complex analysis, such as the maximum principle, Liouville's theorem and Schwarz's lemma.

The book covers all the essential material on complex analysis, and includes several elegant proofs that were recently discovered. It includes the zipper algorithm for computing conformal maps, a constructive proof of the Riemann mapping theorem, and culminates in a complete proof of the uniformization theorem. Aimed at students with some undergraduate background in real analysis, though not Lebesgue integration, this classroom-tested textbook will teach the skills and intuition necessary to understand this important area of mathematics.

Donald E. Marshall is Professor of Mathematics at the University of Washington. He received his PhD from UCLA in 1976. Professor Marshall is a leading complex analyst with a very strong research record that has been continuously funded throughout his career. He has given invited lectures in over a dozen countries. He is coauthor of the research-level monograph *Harmonic Measure*, published by Cambridge University Press.

CAMBRIDGE MATHEMATICAL TEXTBOOKS

Cambridge Mathematical Textbooks is a program of undergraduate and beginning graduate-level textbooks for core courses, new courses, and interdisciplinary courses in pure and applied mathematics. These texts provide motivation with plenty of exercises of varying difficulty, interesting examples, modern applications, and unique approaches to the material.

A complete list of books in the series can be found at www.cambridge.org/mathematics

Recent titles include the following:

Chance, Strategy, and Choice: An Introduction to the Mathematics of Games and Elections, S. B. Smith
Set Theory: A First Course, D. W. Cunningham
Chaotic Dynamics: Fractals, Tilings, and Substitutions, G. R. Goodson
A Second Course in Linear Algebra, S. R. Garcia & R. A. Horn
Introduction to Experimental Mathematics, S. Eilers & R. Johansen
Exploring Mathematics: An Engaging Introduction to Proof, J. Meier & D. Smith
A First Course in Analysis, J. B. Conway
Introduction to Probability, D. F. Anderson, T. Seppäläinen & B. Valkó
Linear Algebra, E. S. Meckes & M. W. Meckes
A Short Course in Differential Topology, B. I. Dundas
Abstract Algebra with Applications, A. Terras

Complex Analysis

Donald E. Marshall
University of Washington, Seattle, WA, USA

Shaftesbury Road, Cambridge CB2 8EA, United Kingdom

One Liberty Plaza, 20th Floor, New York, NY 10006, USA

477 Williamstown Road, Port Melbourne, VIC 3207, Australia

314–321, 3rd Floor, Plot 3, Splendor Forum, Jasola District Centre, New Delhi – 110025, India

103 Penang Road, #05–06/07, Visioncrest Commercial, Singapore 238467

Cambridge University Press is part of Cambridge University Press & Assessment, a department of the University of Cambridge.

We share the University's mission to contribute to society through the pursuit of education, learning and research at the highest international levels of excellence.

www.cambridge.org
Information on this title: www.cambridge.org/9781107134829

First published 2019

A catalogue record for this publication is available from the British Library

Library of Congress Cataloging-in-Publication data
Names: Marshall, Donald E. (Donald Eddy), 1947– author.
Title: Complex analysis / Donald E. Marshall.
Description: Cambridge, United Kingdom ; New York, NY :
Cambridge University Press, 2019.
Identifiers: LCCN 2018029851 | ISBN 9781107134829 (Hardback)
Subjects: LCSH: Functions of complex variables – Textbooks.
| Mathematical analysis – Textbooks.
Classification: LCC QA331.7 M365 2019 | DDC 515/.9–dc23
LC record available at https://lccn.loc.gov/2018029851

ISBN 978-1-107-13482-9 Hardback

Contents

Figures

Preface

This book provides a graduate-level introduction to complex analysis. There are four points of view for this subject due primarily to Cauchy, Weierstrass, Riemann and Runge. Cauchy thought of analytic functions in terms of a complex derivative and through his famous integral formula. Weierstrass instead stressed the importance of power series expansions. Riemann viewed analytic functions as locally rigid mappings from one region to another, a more geometric point of view. Runge showed that analytic functions are nothing more than limits of rational functions. The seminal modern text in this area was written by Ahlfors [1], which stresses Cauchy's point of view. Most subsequent texts have followed his lead. One aspect of the first-year course in complex analysis is that the material has been around so long that some very slick and elegant proofs have been discovered. The subject is quite beautiful as a result, but some theorems then may seem mysterious.

I have decided instead to start with Weierstrass's point of view for local behavior. Cartan [4] has a similar approach. Power series are elementary and give you many non-trivial functions immediately. In many cases it is a lot easier to see why certain theorems are true from this point of view. For example, it is remarkable that a function which has a complex derivative actually has derivatives of all orders. However, the derivative of a power series is just another power series and hence has derivatives of all orders.

Cauchy's theorem is a more global result concerned with integrals of analytic functions. Why integrals of the form $\int \frac{1}{z-a}dz$ are important in Cauchy's theorem is very easy to understand using partial fractions for rational functions. So we will use Runge's point of view for more global results: analytic functions are simply limits of rational functions.

As a pedagogic device we will use the term "analytic" for local power series expansion and "holomorphic" for possessing a continuous complex derivative. We will of course prove that these concepts (and several others) are equivalent eventually, but in the early chapters the reader should be alert to the different definitions.

The emphasis in Chapters 1–6 is to view analytic functions as behaving like polynomials or rational functions. Perhaps the most important elementary tool in this subject is the maximum principle, highlighted in Chapter 3. Runge's theorem is proved in Chapter 4 and is used to prove Cauchy's theorem in Chapter 5. Chapter 6 uses color to visualize complex-valued functions. Given a coloring of the complex plane, a function f can be illustrated by placing the color of $f(z)$ at the point z. See Section A.2 of the appendix for a computer program to do this.

Chapters 7 and 8 introduce harmonic and subharmonic functions and highlight their application to the study of analytic functions. Chapter 8 includes a method, called the geodesic zipper algorithm, for numerically computing conformal maps, which is fast and simple to program. Together with Harnack's principle, it is used to give a somewhat constructive proof of the Riemann mapping theorem in Chapter 8. Because it does not require the development of normal families, it is possible to give a one-quarter course that includes this proof of the Riemann mapping theorem. The standard proof based on normal families is given in Chapter 10. In Chapter 10 we also give Zalcman's remarkable characterization of non-normal families in terms of an associated convergent sequence, then use it to prove Montel's theorem and Picard's great theorem.

Complete and accessible proofs of Carathéodory's theorem and the Jordan curve theorem are included in Chapter 12. Local barriers instead of barriers are used to analyze regular points for the Dirichlet problem in Chapter 13, so that it is easier to verify that every boundary point of a simply-connected region is regular. This allows us to give another proof of the Riemann mapping theorem. The uniformization theorem and the classification of all Riemann surfaces in Chapters 14–16 tie together complex functions, topology, manifolds and groups. The proof of the uniformization theorem here uses Green's function, when available, and the dipole Green's function otherwise. This yields a very similar treatment of the two cases. The main tool is simply the maximum principle, which allows a proof that avoids the "oil speck" method of exhaustion by relatively compact surfaces, and avoids the need to prove triangulation or Green's theorem on Riemann surfaces. Another benefit of this approach is that it is then easy to construct plenty of meromorphic functions on any Riemann surface in Chapter 16. The first section in the appendix lists 15 ways developed in the text to determine whether a function is analytic.

Each of the three parts of this book can be comfortably covered in a one-quarter course. A one-semester course might include most of the material in Chapters 1–9. A list of prerequisites follows this preface. Students should be encouraged to review this material as needed, especially if they encounter difficulties in Chapter 1. Lebesgue integration is not needed in this text because, by Theorem 4.32, we can integrate an analytic function on any continuous curve using Riemann integration. The exercises at the end of each chapter are divided by difficulty, though in some cases they can be solved in more than one way. Exercises A are mostly straightforward, requiring little originality, and are designed for practice with the material. The B exercises require a good idea or non-routine use of the results in the chapter. Sometimes a creative idea or the right insight can lead to a simple solution. The C exercises are usually much more difficult. You can think of "C" as "challenge." I generally ask students to do the A exercises while reading, but focus on the B exercises for homework. Class discussions are facilitated by asking the students to read as much as they can before we discuss the material. Most of the B exercises come from the PhD qualifying exams in complex analysis at the University of Washington [18]. It is entirely possible to find solutions to problems by searching the internet. It is also possible to solve some problems using more advanced techniques or theorems than have been covered in the text. Both will defeat the purpose of developing the ability to solve problems, a goal of this book. For that reason, we also ask you not to tempt others by posting solutions.

This book is not written as a novel that can be read passively. Active involvement will increase your understanding as you read this material. You should have plenty of scratch paper at hand so that you can check all details. The ideas in a proof are at least as important as the statement of the corresponding theorem, if not more so. But the ideas are meaningful only if you can fill in all the details. View this as practice for proving your own theorems.

I am grateful to many people for their assistance in preparing this text. First are all of the students in my complex analysis classes who have pointed out errors, omissions and less than stellar explanations. But I would particularly like to thank John Garnett, Pietro Poggi-Corradini and Steffen Rohde, who have used the material in their classes and made numerous excellent suggestions for improvement. Similar thanks go to Robert Burckel, who read the first thirteen chapters with the eye of an eagle and a fine-toothed comb. I owe a great deal to all my teachers, coauthors and the books I have read for the mathematics they have taught me. Hopefully, some of the elegance, beauty and technique have been retained here. Several excellent texts on this subject are also listed in the bibliography. As with any mathematics text, errors still no doubt remain. I would appreciate receiving email at dmarshal@uw.edu about any errors you encounter. I will list corrections on the web page:
www.cambridge.org/marshall

Wov'n through carefully chosen words
Lie the strands of key support
By my truelove Marianne
Mainstay of our life journey
Fifty years with more to come.

Prerequisites

You should be on friendly terms with the following concepts. If you have only seen the corresponding proofs for real numbers and real-valued functions, check to see whether the same proofs also work when "real" is replaced by "complex," after reading the first two sections of Chapter 1. As you read the text, check all the details. If many of the concepts below are new to you, then I would recommend that you first take a senior-level analysis class.

Let $\{a_n\}_{n=0}^{\infty}$, $\{b_n\}_{n=0}^{\infty}$ be sequences of real numbers and let $\{f_n\}$ be a sequence of real-valued, continuous functions defined on some interval $I \subset \mathbb{R}$.

1. $\{a_n\}$ converges to a (notation: $a_n \to a$) "$\epsilon - \delta$" version.
2. Cauchy sequence.
3. $\sum a_n$ converges, converges absolutely (notation: $\sum |a_n| < \infty$).
4. $\sum a_n$ converges implies $a_n \to 0$, but not conversely;
5. $\limsup_{n\to\infty} a_n$, $\liminf_{n\to\infty} a_n$.
6. Comparison test for convergence.
7. Rearranging absolutely convergent series gives the same sum, but a similar statement does not hold for for conditionally convergent series.
8. If $\sum_{n=0}^{\infty} a_n = A$ and $\sum_{n=0}^{\infty} b_n = B$ then

$$A + B = \sum_{n=0}^{\infty}(a_n + b_n) \quad \text{and} \quad cA = \sum_{n=0}^{\infty} ca_n.$$

   If $\sum a_n$ converges absolutely and $c_n = \sum_{k=0}^{n} a_k b_{n-k}$ then

$$AB = \sum_{n=0}^{\infty} c_n.$$

9.

$$\sum_{n=0}^{\infty}\sum_{k=0}^{\infty} a_{n,k} = \sum_{k=0}^{\infty}\sum_{n=0}^{\infty} a_{n,k}$$

   provided at least one sum converges absolutely. Absolute convergence of either of these double sums is equivalent to the finiteness of

$$\sup_{S \text{ finite}} \sum_{n,k\in S} |a_{n,k}|.$$

10. Continuous function, uniformly continuous function.
11. $f_n(x) \to f(x)$ pointwise, $f_n(x) \to f(x)$ uniformly.
12. Uniform limit of a sequence of continuous functions is continuous.

13. $\left|\int_I f(x)dx\right| \leq \int_I |f(x)|dx$.
14. If $f_n \to f$ uniformly on a bounded interval I then

$$\lim \int_I f_n(x)dx = \int_I \lim f_n(x)dx = \int_I f(x)dx.$$

15. Corollary:

$$\sum_{n=0}^{\infty} \int_I f_n(x)dx = \int_I \sum_{n=0}^{\infty} f_n(x)dx,$$

    if the partial sums of $\sum f_n$ converge uniformly on the bounded interval I.
16. Open set, closed set, connected set, compact set, metric space.
17. f continuous on a compact set X implies f is uniformly continuous on X.
18. $X \subset \mathbb{R}^n$ is compact if and only if it is closed and bounded.
19. A metric space X is compact if and only if every infinite sequence in X has a limit (cluster) point in X. (This can fail if X is not a metric space.)
20. If f is continuous on a connected set U then $f(U)$ is connected. If f is continuous on a compact set K then $f(K)$ is compact.
21. A continuous real-valued function on a compact set has a maximum and a minimum.

All of the above can be found in the undergraduate text Rudin [22], as well as many other sources.

PART I

1 Preliminaries

1.1 Complex Numbers

The **complex numbers** $\mathbb{C}$ consist of pairs of real numbers: $\{(x, y) : x, y \in \mathbb{R}\}$. The complex number (x, y) can be represented geometrically as a point in the plane $\mathbb{R}^2$, or viewed as a vector whose tip has coordinates (x, y) and whose tail has coordinates $(0, 0)$. The complex number (x, y) can be identified with another pair of real numbers (r, θ), called the polar coordinate representation. The line from $(0, 0)$ to (x, y) has length r and forms an angle θ with the positive x axis. The angle is measured by using the distance along the corresponding arc of the circle of radius 1 (centered at $(0, 0)$). By similarity, the length of the subtended arc on the circle of radius r is $r\theta$. See Figure 1.1.

Conversion between these two representations is given by

$$x = r\cos\theta, \quad y = r\sin\theta$$

and

$$r = \sqrt{x^2 + y^2}, \quad \tan\theta = \frac{y}{x}.$$

Care must be taken to find θ from the last equality since many angles can have the same tangent. However, consideration of the quadrant containing (x, y) will give a unique $\theta \in [0, 2\pi)$, provided $r > 0$ (we do not define θ when $r = 0$).

Addition of complex numbers is defined coordinatewise:

$$(a, b) + (c, d) = (a + c, b + d),$$

and can be visualized by vector addition. See Figure 1.2.

Multiplication is given by

$$(a, b) \cdot (c, d) = (ac - bd, bc + ad)$$

and can be visualized as follows. The points $(0, 0), (1, 0), (a, b)$ form a triangle. Construct a similar triangle with corresponding points $(0, 0)$, (c, d), (x, y). Then it is an exercise in high-

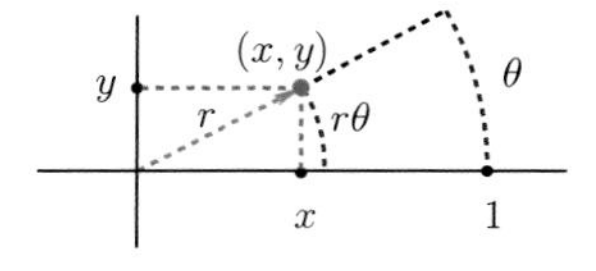

Figure 1.1 Cartesian and polar representations of complex numbers.

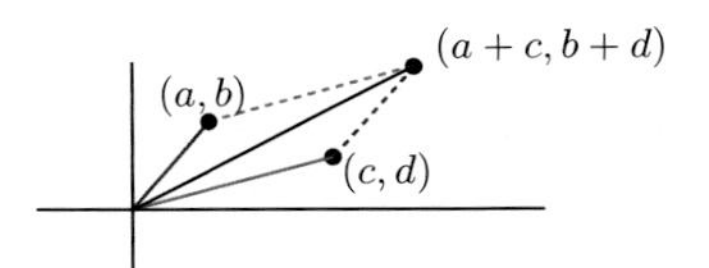

Figure 1.2 Addition.

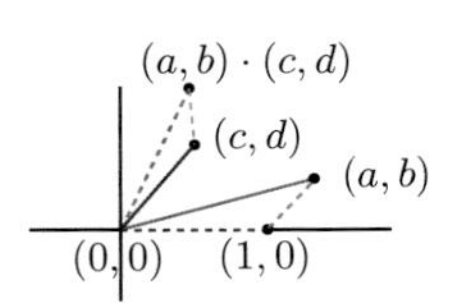

Figure 1.3 Multiplication.

school geometry to show that $(x, y) = (a, b) \cdot (c, d)$. By similarity, the length of the product is the product of the lengths and polar coordinate angles are added. See Figure 1.3.

The real number t is identified with the complex number $(t, 0)$. With this identification, complex addition and multiplication are extensions of the usual addition and multiplication of real numbers. For conciseness, when t is real, $t(x, y)$ means $(t, 0) \cdot (x, y) = (tx, ty)$. The additive identity is $0 = (0, 0)$ and $-(x, y) = (-x, -y)$. The multiplicative identity is $1 = (1, 0)$ and the multiplicative inverse of (x, y) is $(x/(x^2 + y^2), -y/(x^2 + y^2))$. It is a tedious exercise to check that the commutative and associative laws of addition and multiplication hold, as does the distributive law.

The notation for complex numbers becomes *much* easier if we use a single letter instead of a pair. It is traditional, at least among mathematicians, to use the letter i to denote the complex number $(0, 1)$. If z is the complex number given by (x, y), then, because $(x, y) = x(1, 0) + y(0, 1)$, we can write $z = x + yi$. If $z = x + iy$, then the **real part** of z is $\text{Re}z = x$ and the **imaginary part** is $\text{Im}z = y$. Note that $i \cdot i = -1$. We can now just use the usual algebraic rules for manipulating complex numbers together with the simplification $i^2 = -1$. For example, z/w means multiplication of z by the multiplicative inverse of w. To find the real and imaginary parts of the quotient, we use the analog of "rationalizing the denominator":

$$\begin{aligned}\frac{x + iy}{a + ib} &= \frac{(x + iy)(a - ib)}{(a + ib)(a - ib)} = \frac{xa - i^2 yb + iya - ixb}{a^2 + b^2} \\ &= \frac{xa + yb}{a^2 + b^2} + \frac{ya - xb}{a^2 + b^2} i.\end{aligned}$$

Here is some additional notation: if $z = x + iy$ is given in polar coordinates by the pair (r, θ) then

$$|z| = r = \sqrt{x^2 + y^2}$$

is called the **modulus** or **absolute value** of z. Note that $|z|$ is the distance from the complex number z to the origin 0. The angle θ is called the **argument** of z and is written

$$\theta = \arg z.$$

The most common convention is that $-\pi < \arg z \leq \pi$, where positive angles are measured counter-clockwise and negative angles are measured clockwise. The complex conjugate of z is given by

$$\bar{z} = x - iy.$$

The complex conjugate is the reflection of z about the **real line** $\mathbb{R}$.

It is an easy exercise to show the following:

$$\begin{aligned}
|zw| &= |z||w|, \\
|cz| &= c|z| \text{ if } c > 0, \\
z/|z| &\text{ has absolute value } 1, \\
z\bar{z} &= |z|^2, \\
\text{Re}z &= (z + \bar{z})/2, \\
\text{Im}z &= (z - \bar{z})/(2i), \\
\overline{z + w} &= \bar{z} + \overline{w}, \\
\overline{zw} &= \bar{z} \cdot \overline{w}, \\
\bar{\bar{z}} &= z, \\
|z| &= |\bar{z}|, \\
\arg zw &= \arg z + \arg w \quad \text{modulo } 2\pi, \\
\arg \bar{z} &= -\arg z = 2\pi - \arg z \quad \text{modulo } 2\pi.
\end{aligned}$$

The statement **modulo** 2π means that the difference between the left- and right-hand sides of the equality is an integer multiple of 2π.

The identity $a + (z - a) = z$ expressed in vector form shows that $z - a$ is (a translate of) the vector from a to z. Thus $|z - a|$ is the length of the complex number $z - a$ but it is also equal to the distance from a to z. The circle centered at a with radius r is given by $\{z : |z - a| = r\}$ and the disk centered at a of radius r is given by $\{z : |z - a| < r\}$. The open disks are the basic open sets generating the standard topology on $\mathbb{C}$. We will use $\mathbb{D}$ to denote the **unit disk**,

$$\mathbb{D} = \{z : |z| < 1\},$$

and use $\partial\mathbb{D}$ to denote the **unit circle**,

$$\partial\mathbb{D} = \{z : |z| = 1\}.$$

Complex numbers were around for at least 250 years before good applications were found; Cardano discussed them in his book *Ars Magna* (1545). Beginning in the 1800s, and continuing today, there has been an explosive growth in their usage. Now complex numbers are very important in the application of mathematics to engineering and physics.

It is a historical fiction that solutions to quadratic equations forced us to take complex numbers seriously. How to solve $x^2 = mx + c$ has been known for 2000 years and can be visualized as the points of intersection of the standard parabola $y = x^2$ and the line $y = mx+c$. As the line is shifted up or down by changing c, it is easy to see there are two, one or no (real) solutions. The solution to the cubic equation is where complex numbers really became important. A cubic equation can be put in the standard form

$$x^3 = 3px + 2q$$

by scaling and translating. The solutions can be visualized as the intersection of the standard cubic $y = x^3$ and the line $y = 3px + 2q$. *Every* line meets the cubic, so there will always be a solution. By formal manipulations, Cardano showed that a solution is given by

$$x = (q + \sqrt{q^2 - p^3})^{\frac{1}{3}} + (q - \sqrt{q^2 - p^3})^{\frac{1}{3}}.$$

Bombelli pointed out 30 years later that if $p = 5$ and $q = 2$, then $x = 4$ is a solution, but $q^2 - p^3 < 0$ so the above solution does not make sense. His "wild thought" was to use complex numbers to understand the solution

$$x = (2 + 11i)^{\frac{1}{3}} + (2 - 11i)^{\frac{1}{3}}.$$

He found that $(2 \pm i)^3 = 2 \pm 11i$, and so the above solution actually equals 4. In other words, complex numbers were used to find a real solution. This is not just an oddity of Cardano's formula, because, for some cubics, complex numbers must be used in any rational formula involving radicals by a theorem of O. Hölder [15]. See Exercises 1.9 and 1.10 for solutions of cubic and quartic equations.

1.2 Estimates

Here are some elementary estimates which the reader should check:

$$-|z| \leq \text{Re}z \leq |z|,$$
$$-|z| \leq \text{Im}z \leq |z|$$

and

$$|z| \leq |\text{Re}z| + |\text{Im}z|.$$

Perhaps the most useful inequality in analysis is the triangle inequality.

Theorem 1.1 (triangle inequality)

$$|z + w| \leq |z| + |w|$$

and

$$|z + w| \geq \big||z| - |w|\big|.$$

The associated picture perhaps makes this result geometrically clear. See Figure 1.4. Analysis is used to give a more rigorous proof of the triangle inequality (and it is good practice with the notation we have introduced).

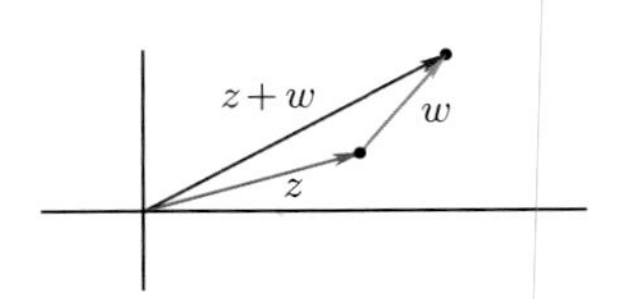

Figure 1.4 Triangle inequality.

Proof

$$\begin{aligned}|z+w|^2 &= (z+w)(\overline{z+w}) \\ &= z\bar{z} + w\bar{z} + z\bar{w} + w\bar{w} \\ &= |z|^2 + 2\mathrm{Re}(w\bar{z}) + |w|^2 \\ &\leq |z|^2 + 2|w||\bar{z}| + |w|^2 \\ &= (|z|+|w|)^2.\end{aligned}$$

To obtain the second part of the triangle inequality we use

$$|z| = |z+w+(-w)| \leq |z+w| + |-w| = |z+w| + |w|.$$

By subtracting $|w|$,

$$|z| - |w| \leq |z+w|,$$

and switching z and w,

$$|w| - |z| \leq |z+w|,$$

so that

$$||z| - |w|| \leq |z+w|.$$

□

These estimates can be used to prove that $\{z_n\}$ converges if and only if both $\{\mathrm{Re}z_n\}$ and $\{\mathrm{Im}z_n\}$ converge. The series $\sum a_n$ is said to **converge** if the sequence of partial sums

$$S_m = \sum_{n=1}^{m} a_n$$

converges, and the series **converges absolutely** if $\sum |a_n|$ converges. A series is said to **diverge** if it does not converge. Absolute convergence implies convergence because Cauchy sequences converge. We sometimes write $\sum |a_n| < \infty$ to denote absolute convergence because the partial sums are increasing. It also follows that $\sum a_n$ is absolutely convergent if and only if both $\sum \mathrm{Re}a_n$ and $\sum \mathrm{Im}a_n$ are absolutely convergent. By comparing the nth partial sum and the $(n-1)$st partial sum, if $\sum a_n$ converges then $a_n \to 0$. The converse statement is false, for example if $a_n = 1/n$.

Another useful estimate is the Cauchy–Schwarz inequality.

Theorem 1.2 (Cauchy–Schwarz inequality)

$$\left|\sum_{j=1}^{n} a_j\overline{b_j}\right| \leq \left(\sum_{j=1}^{n} |a_j|^2\right)^{\frac{1}{2}} \left(\sum_{j=1}^{n} |b_j|^2\right)^{\frac{1}{2}}.$$

If v and w are vectors in $\mathbb{C}^n$, the Cauchy–Schwarz inequality says that $|\langle v,w\rangle| \leq ||v||||w||$, where the left-hand side is the absolute value of the inner product and the right-hand side is the product of the lengths of the vectors.

Proof The square of the right-hand side minus the square of the left-hand side in the Cauchy–Schwarz inequality can be written as

$$\sum_{i=1}^{n}\sum_{j=1}^{n}\left(|a_j|^2|b_i|^2 - a_j\overline{b_j}\overline{a_i\overline{b_i}}\right).$$

We can add another copy of this quantity, switching the index i and the index j to obtain

$$= \frac{1}{2}\sum_{i=1}^{n}\sum_{j=1}^{n}\left(|a_j|^2|b_i|^2 + |a_i|^2|b_j|^2 - a_j\overline{b_j}\overline{a_i}b_i - a_i\overline{b_i}\overline{a_j}b_j\right).$$

Using the identity $|A-B|^2 = |A|^2 + |B|^2 - A\overline{B} - \overline{A}B$, with $A = a_jb_i$ and $B = a_ib_j$, we obtain

$$= \frac{1}{2}\sum_{i=1}^{n}\sum_{j=1}^{n}|a_jb_i - a_ib_j|^2. \qquad \square$$

The above proof also gives the error

$$\frac{1}{2}\sum_{j=1}^{n}\sum_{i=1}^{n}|a_jb_i - a_ib_j|^2,$$

and so equality occurs if and only if $a_j = cb_j$ for all j and some (complex) constant c, or $b_j = 0$ for all j.

The reader can use Riemann integration to deduce the following, which is also called the Cauchy–Schwarz inequality. For a complex-valued function f defined on a real interval $[a,b]$, we define $\int_a^b f dx \equiv \int_a^b \text{Re} f dx + i\int_a^b \text{Im} f dx$.

Corollary 1.3 *If f and g are continuous complex-valued functions defined on $[a,b] \subset \mathbb{R}$ then*

$$\left|\int_a^b f(t)\overline{g(t)}dt\right| \le \left(\int_a^b |f(t)|^2 dt\right)^{\frac{1}{2}}\left(\int_a^b |g(t)|^2 dt\right)^{\frac{1}{2}}.$$

This corollary can also be proved directly by expanding

$$\int_a^b\int_a^b |f(x)g(y) - f(y)g(x)|^2 dxdy \tag{1.1}$$

in a similar way, giving a proof for square integrable functions f, g. Moreover, the error term is half of the integral (1.1) and equality occurs if and only if $f = cg$, for some constant c, or g is identically zero.

1.3 Stereographic Projection

A component of Riemann's point of view of functions as mappings is that ∞ is like any other complex number. But we cannot extend the definition of complex numbers to include ∞ and still have the usual laws of arithmetic hold. However, there is another "picture" of complex

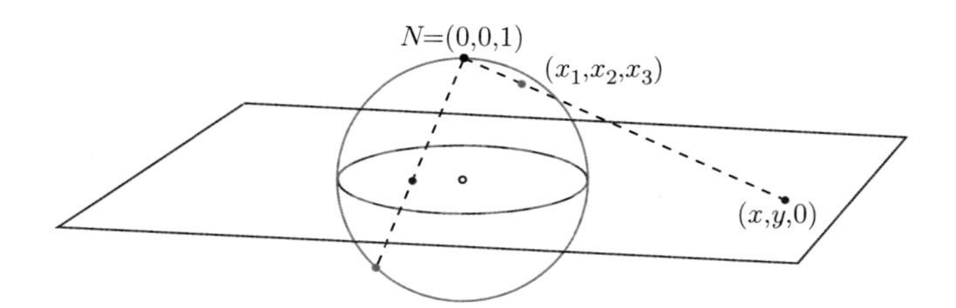

Figure 1.5 Stereographic projection.

numbers that can help us visualize this idea. The picture is called **stereographic projection**. We identify the complex numbers with the plane $\{(x,y,0) : x,y \in \mathbb{R}\}$ in $\mathbb{R}^3$. If $z = x + iy$, let z^* be the unique point on the unit sphere in $\mathbb{R}^3$ which also lies on the line from the **north pole** $(0,0,1)$ to $(x,y,0)$. Thus

$$z^* = (x_1, x_2, x_3) = (0,0,1) + t[(x,y,0) - (0,0,1)].$$

See Figure 1.5.

Then

$$|z^*| = \sqrt{(tx)^2 + (ty)^2 + (1-t)^2} = 1,$$

which gives

$$t = \frac{2}{x^2 + y^2 + 1},$$

where $0 < t \leq 2$, and

$$z^* = \left(\frac{2x}{x^2+y^2+1}, \frac{2y}{x^2+y^2+1}, \frac{x^2+y^2-1}{x^2+y^2+1} \right).$$

The reader is invited to find $z = x + iy$ from $z^* = (x_1, x_2, x_3)$. The sphere is sometimes called the **Riemann sphere** and is denoted $\mathbb{S}^2$. We can extend stereographic projection $\pi : \mathbb{C} \to \mathbb{S}^2$ to the **extended plane** $\mathbb{C}^* = \mathbb{C} \cup \{\infty\}$ by defining $\pi(\infty) = (0,0,1)$. We give $\mathbb{C}^*$ the topology inherited from $\mathbb{S}^2$. It is an explicit one-point compactification of the complex plane. The north pole corresponds to "∞."

Theorem 1.4 *Under stereographic projection, circles and straight lines in $\mathbb{C}$ correspond precisely to circles on $\mathbb{S}^2$.*

Proof Every circle on the sphere is given by the intersection of a plane with the sphere, and conversely the intersection of a plane with a sphere is a circle or a point. See Exercise 1.6. If a plane is given by

$$Ax_1 + Bx_2 + Cx_3 = D,$$

and if (x_1, x_2, x_3) corresponds to $(x, y, 0)$ under stereographic projection, then

$$A\left(\frac{2x}{x^2+y^2+1}\right) + B\left(\frac{2y}{x^2+y^2+1}\right) + C\left(\frac{x^2+y^2-1}{x^2+y^2+1}\right) = D. \tag{1.2}$$

Equivalently,

$$(C - D)(x^2 + y^2) + 2Ax + 2By = C + D. \tag{1.3}$$

If $C = D$, then this is the equation of a line, and all lines can be written this way. If $C \neq D$, then, by completing the square, we get the equation of a circle, and all circles can be put in this form. □

So we will consider a line in $\mathbb{C}$ as just a special kind of **"circle."**

The sphere $\mathbb{S}^2$ inherits a topology from the usual topology on $\mathbb{R}^3$ generated by the balls in $\mathbb{R}^3$.

Corollary 1.5 *The topology on $\mathbb{S}^2$ induces the standard topology on $\mathbb{C}$ via stereographic projection, and moreover a basic neighborhood of ∞ is of the form $\{z : |z| > r\}$.*

For later use, we note that the chordal distance between two points on the sphere induces a metric, called the **chordal metric**, on $\mathbb{C}$ which is given by

$$\chi(z, w) = |z^* - w^*| = \frac{2|z - w|}{\sqrt{1 + |z|^2}\sqrt{1 + |w|^2}}. \tag{1.4}$$

This metric is bounded (by 2). See Exercise 1.5.

1.4 Exercises

A

1.1 Check that item 9 of the prerequisites holds for complex $a_{n,k}$. Check that items 13, 14 and 15 of the prerequisites hold for complex-valued functions defined on an interval $I \subset \mathbb{R}$.

1.2 Check the details of the high-school geometry problem in the geometric version of complex multiplication.

1.3 Prove the parallelogram equality:

$$|z + w|^2 + |z - w|^2 = 2(|z|^2 + |w|^2).$$

In geometric terms, the equality says that the sum of the squares of the lengths of the diagonals of a parallelogram equals the sum of the squares of the lengths of the sides. It is perhaps a bit easier to prove it using the complex notation of this chapter than to prove it using high-school geometry.

1.4 Prove Corollary 1.3.

1.5 (a) Prove formula (1.4). An algebraic proof can be found in [1], p. 20. Alternatively, use the law of cosines for the triangles with vertices $N = (0, 0, 1)$, z, w and N, $z*$, $w*$. Compute edge lengths of these two triangles using triangles that have N and $(0, 0, 0)$ as vertices.

(b) The chordal distance is bounded by 2, by the triangle inequality. Verify analytically that the formula for this distance given in the text is bounded by 2 using the

Cauchy–Schwarz inequality, and also by directly multiplying it out using complex notation and one of the estimates at the start of Section 1.2.

1.6 Show that the intersection of a plane with the unit sphere in $\mathbb{R}^3$ is a circle or a point and conversely that every circle or point on the sphere is equal to the intersection of the sphere with a plane. Hint: Rotate the plane and sphere so that the plane is parallel to the $(x, y, 0)$ plane.

1.7 (a) Suppose w is a non-zero complex number. Choose z so that $|z| = |w|^{\frac{1}{2}}$ and $\arg z = \frac{1}{2} \arg w$ or $\arg z = \frac{1}{2} \arg w + \pi$. Show that $z^2 = w$ in both cases, and that these are the only solutions to $z^2 = w$.

(b) The quadratic formula gives two solutions to the equation $az^2 + bz + c = 0$, when a, b, c are complex numbers with $a \neq 0$ because completing the square is a purely algebraic manipulation of symbols, and there are two complex square roots of every non-zero complex number by part (a). Check the details.

(c) If w is a non-zero complex number, find n solutions to $z^n = w$ using polar coordinates.

B

1.8 Suppose that f is a continuous complex-valued function on a real interval $[a, b]$. Let

$$A = \frac{1}{b-a} \int_a^b f(x)dx$$

be the average of f over the interval $[a, b]$.

(a) Show that if $|f(x)| \leq |A|$ for all $x \in [a, b]$, then $f = A$. Hint: Rotate f so that $A > 0$. Then $\int_a^b (A - \text{Re}f)dx/(b-a) = 0$, and $A - \text{Re}f$ is continuous and non-negative.

(b) Show that if $|A| = (1/(b-a)) \int_a^b |f(x)|dx$, then $\arg f$ is constant modulo 2π on $\{z : f(z) \neq 0\}$.

1.9 Formally solve the cubic equation $ax^3+bx^2+cx+d = 0$, where $x, a, b, c, d \in \mathbb{C}$, $a \neq 0$, by the following reduction process:

(a) Set $x = u + t$ and choose the constant t so that the coefficient of u^2 is equal to zero.

(b) If the coefficient of u is also zero, then take a cube root to solve. If the coefficient of u is non-zero, set $u = kv$ and choose the constant k so that $v^3 = 3v + r$, for some constant r.

(c) Set $v = z + 1/z$ and obtain a quadratic equation for z^3. The map $z + 1/z$ is important for several reasons, including constructing what are called conformal maps. It will be examined in more detail in Section 6.4.

(d) Use the quadratic formula to find two possible values for z^3, and then take a cube root to solve for z.

(e) In Section 2.2 we will show that the cubic equation has exactly three solutions, counting multiplicity. But the process in this exercise appears to generate more solutions, if we use two solutions to the quadratic and all three cube roots. Moreover, there might be more than one valid choice for the constants used to reduce to a simpler equation. Explain.

1.10 The equation

$$a\left(z+\frac{1}{z}\right)^2+b\left(z+\frac{1}{z}\right)+c=0$$

has four solutions, which can be found by two applications of the quadratic formula. If we multiply by z^2 we obtain the quartic

$$az^4+bz^3+(2a+c)z^2+bz+a=0.$$

Which quartics $Aw^4+Bw^3+Cw^2+Dw+E$ can be put in this form after a linear change of variable $w=\alpha z+\beta$?

C

1.11 Prove that stereographic projection preserves angles between curves. In other words, if two curves γ_1, γ_2 in the plane meet at an angle θ, then their lifts γ_1^*, γ_2^* to the sphere meet at the same angle. Moreover, the direction of the angle from γ_1 to γ_2 corresponds to the direction from γ_1^* to γ_2^* when viewed from inside the sphere. Orientation is reversed when viewed from outside the sphere. Hint: This can be done without any calculations by considering intersecting planes. This exercise will be revisited in Exercise 6.16, where it is used to find the Mercator projection, a map of tremendous economic impact.

1.12 Stereographic projection combined with rigid motions of the sphere can be used to describe some transformations of the plane.

(a) Map a point $z \in \mathbb{C}$ to $\mathbb{S}^2$, apply a rotation of the unit sphere, then map the resulting point back to the plane. For a fixed rotation, find this map of the extended plane to itself as an explicit function of z. Two cases are worth working out first: rotation about the x_3 axis and rotation about the x_1 axis.

(b) Another map can be obtained by mapping a point $z \in \mathbb{C}$ to $\mathbb{S}^2$, then translating the sphere so that the origin is sent to (x_0, y_0, z_0), then projecting back to the plane. The projection to the plane is given by drawing a line through the (translated) north pole and a point on the (translated) sphere and finding the intersection with the plane $\{(x, y, 0)\}$. For a fixed translation, find this map as an explicit function of z. In this case it is worth working out a vertical translation and a translation in the plane separately. Then view an arbitrary translation as a composition of these two maps. Partial answer: the maps in parts (a) and (b) are of the form $(az+b)/(cz+d)$ with $ad-bc \neq 0$.

For an award-winning movie of these maps, see
http://www-users.math.umn.edu/~arnold/moebius/
but do the exercise before viewing this link.

2 Analytic Functions

2.1 Polynomials

This course is about complex-valued functions of a complex variable. We could think of such functions in terms of real variables as maps from $\mathbb{R}^2$ into $\mathbb{R}^2$ given by

$$f(x,y) = (u(x,y), v(x,y)),$$

and think of the graph of f as a subset of $\mathbb{R}^4$. But the subject becomes more tractable if we use a single letter z to denote the independent variable and write $f(z)$ for the value at z, where $z = x + iy$ and $f(z) = u(z) + iv(z)$. For example,

$$f(z) = z^n$$

is much simpler to write (and understand) than its real equivalent. Here z^n means the product of n copies of z.

The simplest functions are the **polynomials** in z:

$$p(z) = a_0 + a_1 z + a_2 z^2 + \ldots + a_n z^n, \tag{2.1}$$

where $a_0, \ldots, a_n$ are complex numbers. If $a_n \neq 0$, then we say that n is the **degree** of p. Note that $\bar{z}$ is not a (complex) polynomial, and neither is $\text{Re}z$ or $\text{Im}z$.

Let's take a closer look at **linear** or degree 1 polynomials. For example, if b is a (fixed) complex number, then

$$g(z) = z + b$$

translates, or shifts, the plane. If a is a (fixed) complex number then

$$h(z) = az$$

can be viewed as a dilation and rotation. To see this, recall that by Chapter 1 and Exercise 1.2, $|az| = |a||z|$ and $\arg(az) = \arg(a) + \arg(z)$ (up to a multiple of 2π). So, h dilates z by a factor of $|a|$ and rotates the point z by the angle $\arg a$. A linear function

$$f(z) = az + b$$

can then be viewed as a dilation and rotation followed by a translation. Equivalently, writing $f(z) = a(z + b/a)$ we can view f as a translation followed by a rotation and dilation.

Another instructive example is the function $p(z) = z^n$. By Chapter 1 again,

$$|p(z)| = |z|^n \quad \text{and} \quad \arg p(z) = n \arg z \mod 2\pi.$$

Each pie slice

$$S_k = \left\{ z : \left| \arg z - \frac{2\pi k}{n} \right| < \frac{\pi}{n} \right\} \cap \{z : |z| < r\},$$

$k = 0, \ldots, n-1$ is mapped to a slit disk

$$\{z : |z| < r^n\} \setminus (-r^n, 0).$$

Angles between straight-line segments issuing from the origin are multiplied by n, and, for small r, the size of the image disk is much smaller than the "radius" of the pie slice. See Figure 2.1.

The function $k(z) = b(z - z_0)^n$ can be viewed as a translation by $-z_0$, followed by the power function, and then a rotation and dilation. To put it another way, k translates a neighborhood of z_0 to the origin, then acts like the power function z^n, followed by a dilation and rotation by b.

To understand the local behavior of a polynomial (2.1) near a point z_0, write $z = (z - z_0) + z_0$ and expand (2.1) by multiplying out and collecting terms to obtain

$$p(z) = p(z_0) + b_1(z - z_0) + b_2(z - z_0)^2 + \ldots + b_n(z - z_0)^n. \tag{2.2}$$

Another way to see this is to note that $p(z) - a_n(z - z_0)^n$ is a polynomial of degree at most $n - 1$, so (2.2) follows by induction on the degree. If $b_1 \neq 0$ then $p(z)$ behaves like the linear function $p(z_0) + b_1(z - z_0)$ for z near z_0. If $b_1 = 0$ then, near z_0, $p(z)$ is closely approximated by $p(z_0) + b_k(z - z_0)^k$, where b_k is the first non-zero coefficient in the expansion (2.2). Indeed, for small $\zeta = z - z_0$,

$$|p(z_0 + \zeta) - [p(z_0) + b_k \zeta^k]| \leq C|\zeta|^{k+1},$$

for some constant C, by (2.2). Figure 2.2 is sometimes called "walking the dog," where the walking path has radius $r = |b_k||\zeta|^k$ and the leash has length $s = C|\zeta|^{k+1}$. As ζ traces a circle centered at 0 of radius ε, the function $p(z_0) + b_k\zeta^k$ winds k times around the circle centered at $p(z_0)$ with radius r. For small ε, s is much smaller than r so the function $p(z_0 + \zeta)$ also then traces a path which winds k times around $p(z_0)$.

So, for z near z_0, $p(z)$ behaves like a translation by $-z_0$, followed by a power function, a rotation and dilation, and finally a translation by $p(z_0)$.

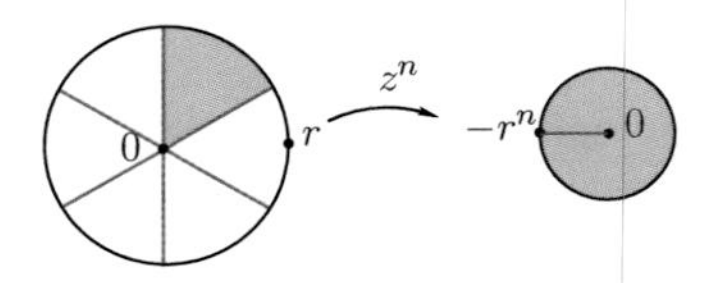

Figure 2.1 The power map.

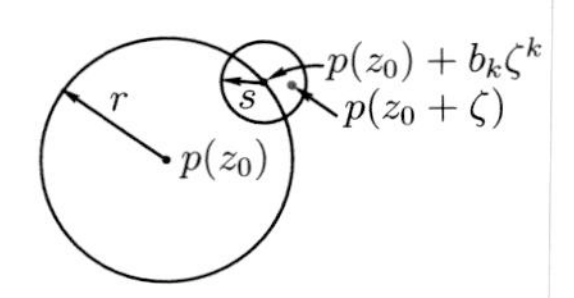

Figure 2.2 $p(z_0 + \zeta)$ lies in a small disk of radius $s = C|\zeta|^{k+1} < r = |b_k||\zeta|^k$.

2.2 Fundamental Theorem of Algebra and Partial Fractions

The local behavior of a polynomial described in Section 2.1 can be used to prove an important result you will have seen in some form or another since high school. If p is a polynomial and $p(a) = 0$ then a is called a **zero** of p.

Theorem 2.1 (fundamental theorem of algebra) *Every non-constant polynomial has a zero.*

This remarkable result says that if we extend the real numbers to the complex numbers via the solution to the equation $z^2 + 1 = 0$ then every polynomial equation has a solution.

Proof Suppose $p(z) = a_n z^n + a_{n-1} z^{n-1} + \ldots + a_1 z + a_0$, $n \geq 1$, is a polynomial which has no zeros and for which $a_n \neq 0$. We first prove that $|p(z)|$ must have a non-zero minimum value on $\mathbb{C}$. Write

$$p(z) = z^n \left(a_n + \frac{a_{n-1}}{z} + \ldots + \frac{a_0}{z^n} \right).$$

Because $a_n \neq 0$, $1/z^k \to 0$ and $|z^n| \to \infty$ as $|z| \to \infty$, we conclude that $|p(z)| \to \infty$ as $|z| \to \infty$. So, if $M = \inf_{\mathbb{C}} |p(z)|$ and $|p(w_j)| \to M$, then there is an $R < \infty$ so that $|w_j| \leq R$, for all j. Because $\{z : |z| \leq R\}$ is compact and because $|p|$ is continuous, there exists z_0 so that $|p(z_0)| = M$. Moreover, $M \neq 0$ since p has no zeros. Now, by the "walking the dog" argument in Section 2.1 (see Figure 2.2), for small $\varepsilon > 0$, as ζ traces a circle of radius ε, the function $p(z_0 + \zeta)$ traces a path about $p(z_0)$ which must intersect the open disk centered at 0 of radius M because $|p(z_0)| = M$. More explicitly, choose $\varepsilon > 0$ so that $0 < s < r < M$, where r, s are defined in Section 2.1. Because $b_k \zeta^k$ traces a circle of radius r, we can find ζ so that $|p(z_0) + b_k \zeta^k| = M - r$. Then

$$|p(z_0 + \zeta)| \leq |p(z_0 + \zeta) - (p(z_0) + b_k \zeta^k)| + M - r \leq s + M - r < M.$$

This contradiction proves that no such polynomial exists and Theorem 2.1 follows. □

Corollary 2.2 *If p is a polynomial of degree $n \geq 1$, then there are complex numbers $z_1, \ldots, z_n$ and a complex constant c so that*

$$p(z) = c \prod_{k=1}^{n} (z - z_k).$$

Corollary 2.2 does not tell us how to find the zeros, but it does say that there are exactly n zeros, counting multiplicity.

Proof The proof is by induction. First note that

$$z^k - b^k = (z - b)(b^{k-1} + z b^{k-2} + \ldots + z^{k-2} b + z^{k-1}).$$

So if $p(z) = \sum_{k=0}^{n} a_k z^k$ and $p(b) = 0$, then

$$q(z) \equiv \frac{p(z)}{z-b} = \frac{p(z)-p(b)}{z-b} = \sum_{k=1}^{n} a_k \left(\sum_{j=0}^{k-1} b^{k-1-j} z^j \right). \tag{2.3}$$

The coefficient of z^{n-1} in (2.3) is a_n so q is a polynomial of degree $n-1$. Repeating this argument n times proves the corollary. □

For example, the polynomial $z^n - 1$ has n zeros. If $z^n = 1$ then $1 = |z^n| = |z|^n$ so that $|z| = 1$. Write $z = \cos t + i \sin t = e^{it}$ (see Exercise 2.6). Then $z^n = e^{int} = 1$ so that $nt = 2\pi k$ for some integer k, and thus $t = 2\pi k/n$. The n distinct zeros of $z^n - 1$ are then $e^{i2\pi k/n}$, $k = 0, 1, \ldots, n-1$, which are equally spaced around the unit circle.

A **rational function** r is the ratio of two polynomials. By the fundamental theorem of algebra, we can write r in the form

$$r(z) = \frac{p(z)}{\prod_{j=1}^{N}(z-z_j)^{n_j}}.$$

The next corollary, also probably familiar, allows us to write a rational function in a form that is easier to analyze. The form is also of practical importance because it allows us to solve certain differential equations that arise in engineering problems using the Laplace transform and its inverse.

Corollary 2.3 (partial fraction expansion) *If p is a polynomial then there is a polynomial q and constants $c_{k,j}$ so that*

$$\frac{p(z)}{\prod_{j=1}^{N}(z-z_j)^{n_j}} = q(z) + \sum_{j=1}^{N} \sum_{k=1}^{n_j} \frac{c_{k,j}}{(z-z_j)^k}. \tag{2.4}$$

Proof There are two initial cases to consider: If p is a polynomial then

$$\frac{p(z)}{z-a} = q(z) + \frac{p(a)}{z-a}, \tag{2.5}$$

where $q(z) = (p(z) - p(a))/(z-a)$ is a polynomial, as in (2.3). Secondly, if $a \neq b$, we can write

$$\frac{1}{(z-a)(z-b)} = \frac{A}{z-a} + \frac{B}{z-b}, \tag{2.6}$$

for some constants A and B. For if this equation is true, then we can multiply each term on the right by $z-a$ and let $z \to a$ to obtain A on the right. The same process on the left yields $1/(a-b)$, and hence $A = 1/(a-b)$. Similarly $B = 1/(b-a)$. Now substitute these values for A and B into (2.6) and check that equality holds. The full corollary now follows by induction: suppose the corollary is true if the degree of the denominator is at most d. If we have an equation of the form (2.4) of degree d then we can divide each term in the equation by $z-a$. After division, the right-hand side consists of lower degree terms to which the induction hypothesis applies, with one exception: when the denominator of the left-hand side of (2.4) is $(z-b)^d$. If $a = b$, then, after division by $z-a$, each term will be of the correct form. If $a \neq b$, then we could have applied the inductive assumption to the decomposition of

$$\frac{p(z)}{(z-b)^{d-1}(z-a)}$$

and then divided this result by $z-b$ instead of $z-a$. □

The above proof also suggests an algorithm for computing the coefficients $\{c_{k,j}\}$. First apply (2.5) with $a = z_1$. Multiply each term of the result by $1/(z-b)$, where b is one of the zeros of the denominator in (2.4), and apply either (2.5) or (2.6) to each of the resulting terms on the right-hand side. Repeat this process, increasing the degree of the denominator by 1 until you have the desired expansion. At each stage, the terms to be expanded are of the form $1/[(z-a)^k(z-b)]$. These can be expanded by starting with (2.6), dividing the result by $(z-a)$, then using (2.6) again, repeating until you have reached the power k in the denominator.

The algorithm can be speeded up because we know the form of the solution. If powers in the denominator n_j are all equal to one, and if the numerator has smaller degree than the denominator, then the form is

$$\frac{p(z)}{\prod_{j=1}^{N}(z-z_j)} = \sum_{j=1}^{N} \frac{c_j}{z-z_j}. \tag{2.7}$$

If we multiply each term of the right-hand side by $z-z_1$ then let $z \to z_1$, we obtain c_1. If we multiply the left-hand side by the same factor, it cancels one of the terms in the denominator. Letting $z \to z_1$ we obtain the value of the remaining part of the left-hand side at z_1. This quickly gives c_1 and can be repeated for $c_2, \ldots, c_N$. This method is sometimes called the "cover-up method" because it can be done with less writing by observing that c_j is the value of the left-hand side at z_j when you cover $z - z_j$ with your hand. If the denominator has terms with degree bigger than one, first use a denominator with all terms of degree one as above. Then, as in the proof of Corollary 2.3, multiply everything by $1/(z-z_k)$ and simplify all terms on the right, repeating as often as needed. If the degree of the numerator of any term is not less than the degree of its denominator, polynomial division can also be used to reduce the degree instead of repeated application of (2.5).

Engineering problems typically have rational functions with real coefficients. See Exercise 2.2 for a similar technique that decomposes rational functions with real coefficients into terms whose denominators are either powers of linear terms with real zeros or powers of irreducible quadratics with real coefficients.

2.3 Power Series

More complicated functions are found by taking limits of polynomials. Here is the primary example:

$$\sum_{n=0}^{\infty} z^n.$$

This series is important to understand because its behavior is typical of all power series (defined shortly) and because it is one of the few series we can actually add up explicitly.

If $|z| \geq 1$ then $|z^n| = |z|^n \geq 1$, so the terms of the series do not tend to 0 as $n \to \infty$. The series then diverges. If $|z| < 1$, then the partial sums

$$S_m = \sum_{n=0}^{m} z^n = 1 + z + z^2 + \ldots + z^m$$

satisfy

$$(1 - z)S_m = 1 - z^{m+1},$$

as can be seen by multiplying out the left-hand side and cancelling. If $|z| < 1$, then $|z^{m+1}| = |z|^{m+1} \to 0$ as $m \to \infty$ and so $S_m(z) \to 1/(1 - z)$. We conclude that if $|z| < 1$ then

$$\sum_{n=0}^{\infty} z^n = \frac{1}{1 - z}. \tag{2.8}$$

It is important to note that the left- and right-hand sides of (2.8) are different objects. They agree in $|z| < 1$, the right-hand side is defined for all $z \neq 1$, but the left-hand side is defined only for $|z| < 1$.

The formal power series

$$f(z) = \sum_{n=0}^{\infty} a_n(z - z_0)^n = a_0 + a_1(z - z_0) + a_2(z - z_0)^2 + \ldots$$

is called a **convergent power series centered (or based) at** z_0 if there is an $r > 0$ so that the series converges for all z such that $|z - z_0| < r$. Note: If we plug $z = z_0$ into the formal power series, then we always get $a_0 = f(z_0)$. More formally, the definition of the summation notation includes the convention that the $n = 0$ term equals a_0, so that we are not raising 0 to the power 0. The requirement for a power series to converge is stronger than convergence at just the one point z_0.

A variant of the primary example is:

$$\frac{1}{z - a} = \frac{1}{z - z_0 - (a - z_0)} = \frac{1}{-(a - z_0)(1 - (\frac{z - z_0}{a - z_0}))}.$$

Substituting

$$w = \frac{z - z_0}{a - z_0}$$

into (2.8) we obtain, when $|w| = |(z - z_0)/(a - z_0)| < 1$,

$$\frac{1}{z - a} = \sum_{n=0}^{\infty} \frac{-1}{(a - z_0)^{n+1}}(z - z_0)^n. \tag{2.9}$$

By (2.8), this series converges if $|z - z_0| < |a - z_0|$ and diverges if $|z - z_0| \geq |a - z_0|$. The domain of convergence is an open disk, and it is the largest disk centered at z_0 which is contained in the domain of definition of $1/(z - a)$. In particular, this function has a power series expansion based at every $z_0 \neq a$, but different series for different base points z_0.

Theorem 2.4 (Weierstrass M-test) *If* $|a_n(z - z_0)^n| \leq M_n$ *for* $|z - z_0| \leq r$ *and if* $\sum M_n < \infty$ *then* $\sum_{n=0}^{\infty} a_n(z - z_0)^n$ *converges uniformly and absolutely in* $\{z : |z - z_0| \leq r\}$.

Proof If $M > N$ then the partial sums $S_n(z)$ satisfy

$$|S_M(z) - S_N(z)| = \left| \sum_{n=N+1}^{M} a_n(z-z_0)^n \right| \le \sum_{n=N+1}^{M} M_n.$$

Since $\sum M_n < \infty$, we deduce $\sum_{n=N+1}^{M} M_n \to 0$ as $N, M \to \infty$, and so $\{S_n\}$ is a Cauchy sequence converging uniformly. The same proof also shows absolute convergence. □

Note that the convergence depends only on the "tail" of the series so that we need only satisfy the hypotheses in the Weierstrass M-test for $n \ge n_0$ to obtain the conclusion.

The primary example (2.8) converges on a disk and diverges outside the disk. The next result says that disks are the only kind of domain in which a power series can converge.

Theorem 2.5 (root test) *Suppose $\sum a_n(z-z_0)^n$ is a formal power series. Let*

$$R = \liminf_{n\to\infty} |a_n|^{-\frac{1}{n}} = \frac{1}{\limsup\limits_{n\to\infty} |a_n|^{\frac{1}{n}}} \in [0, +\infty].$$

Then $\sum_{n=0}^{\infty} a_n(z-z_0)^n$

(a) converges absolutely in $\{z : |z-z_0| < R\}$,
(b) converges uniformly in $\{z : |z-z_0| \le r\}$, for all $r < R$, and
(c) diverges in $\{z : |z-z_0| > R\}$.

See Figure 2.3; R is called the **radius of convergence** of the series $\sum a_n(z-z_0)^n$. It has an interpretation in terms of the decay rate for the coefficients. If $S < R$ then $|a_n| \le S^{-n}$, for large n, and this statement fails for any $S > R$.

Proof The idea is to compare the given series with the example (2.8), $\sum z^n$. If $|z-z_0| \le r < R$, then choose r_1 with $r < r_1 < R$. Thus $r_1 < \liminf |a_n|^{-\frac{1}{n}}$, and there is an $n_0 < \infty$ so that $r_1 < |a_n|^{-\frac{1}{n}}$ for all $n \ge n_0$. This implies that $|a_n(z-z_0)^n| \le (\frac{r}{r_1})^n$. But, by (2.8),

$$\sum_{n=0}^{\infty} \left(\frac{r}{r_1}\right)^n = \frac{1}{1 - r/r_1} < \infty$$

since $r/r_1 < 1$. Applying Weierstrass's M-test to the tail of the series ($n \ge n_0$) proves (b). This same proof also shows absolute convergence (a) for each z with $|z-z_0| < R$. If $|z-z_0| > R$, fix z and choose r so that $R < r < |z-z_0|$. Then $|a_n|^{-\frac{1}{n}} < r$ for infinitely many n, and hence

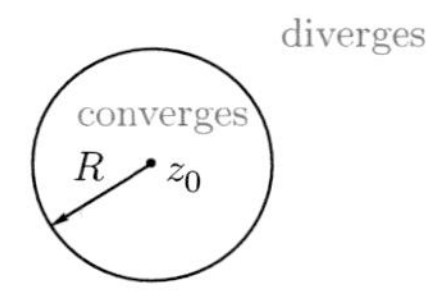

Figure 2.3 Convergence of a power series.

$$|a_n(z-z_0)^n| > \left(\frac{|z-z_0|}{r}\right)^n$$

for infinitely many n. Since $(|z-z_0|/r)^n \to \infty$ as $n \to \infty$, (c) holds. □

The proof of the root test also shows that if the terms $a_n(z-z_0)^n$ of the formal power series are bounded when $z = z_1$ then the series converges on $\{z : |z-z_0| < |z_1 - z_0|\}$.

The root test does not give any information about convergence on the circle of radius R. The series can converge at none, some or all points of $\{z : |z-z_0| = R\}$, as the following examples illustrate.

Examples

$$\text{(i)} \sum_{n=1}^{\infty} \frac{z^n}{n}, \quad \text{(ii)} \sum_{n=1}^{\infty} \frac{z^n}{n^2}, \quad \text{(iii)} \sum_{n=1}^{\infty} nz^n, \quad \text{(iv)} \sum_{n=1}^{\infty} 2^{n^2} z^n, \quad \text{(v)} \sum_{n=1}^{\infty} 2^{-n^2} z^n.$$

The reader should verify the following facts about these examples. The radius of convergence of each of the first three series is $R = 1$. When $z = 1$, the first series is the harmonic series which diverges, and when $z = -1$ the first series is an alternating series whose terms decrease in absolute value and hence converges. The second series converges uniformly and absolutely on $\{|z| = 1\}$. The third series diverges at all points of $\{|z| = 1\}$. The fourth series has radius of convergence $R = 0$ and hence is not a convergent power series. The fifth example has radius of convergence $R = \infty$ and hence converges for all $z \in \mathbb{C}$.

The radius of convergence of the series $\sum a_n z^n$, where

$$a_n = \begin{cases} 3^{-n}, & \text{if } n \text{ is even} \\ 4^n, & \text{if } n \text{ is odd} \end{cases}$$

is $R = 1/4$ by the root test. This is an example where ratios of successive terms in the series do not provide sufficient information to determine convergence.

2.4 Analytic Functions

Definition 2.6 *A function f is* **analytic at** z_0 *if f has a power series expansion valid in a neighborhood of z_0. This means that there is an $r > 0$ and a power series $\sum a_n(z-z_0)^n$ which converges in $B = \{z : |z-z_0| < r\}$ and satisfies*

$$f(z) = \sum_{n=0}^{\infty} a_n(z-z_0)^n,$$

for all $z \in B$. A function f is **analytic on an open set** Ω *if f is analytic at each $z_0 \in \Omega$.*

Note that we do not require one series for f to converge in all of Ω. The function $(z-a)^{-1}$ is analytic on $\mathbb{C} \setminus \{a\}$, as shown in (2.9), and is not given by one series. Note that if f is analytic on Ω then f is continuous in Ω. Indeed, continuity is a local property. To check continuity near z_0, use the series based at z_0. Since the partial sums are continuous and converge uniformly on a closed disk centered at z_0, the limit function f is continuous on that disk. Occasionally it is convenient to say that a function f is analytic on a set E which is not open. This means that there is an open set $\Omega \supset E$ and an analytic function g defined on Ω with $g = f$ on E.

A natural question at this point is: where is a power series analytic?

Theorem 2.7 *If $f(z) = \sum a_n(z-z_0)^n$ converges on $\{z : |z-z_0| < r\}$ then f is analytic on $\{z : |z-z_0| < r\}$.*

Proof Fix z_1 with $|z_1 - z_0| < r$. We need to prove that f has a power series expansion based at z_1. We saw in (2.2) how to rearrange a polynomial. The idea is the same for power series, but we need to prove convergence. By the binomial theorem,

$$(z-z_0)^n = (z-z_1+z_1-z_0)^n = \sum_{k=0}^{n} \binom{n}{k} (z_1-z_0)^{n-k}(z-z_1)^k.$$

Hence

$$f(z) = \sum_{n=0}^{\infty} \left[\sum_{k=0}^{n} a_n \binom{n}{k} (z_1-z_0)^{n-k}(z-z_1)^k \right]. \tag{2.10}$$

Suppose, for the moment, that we can interchange the order of summation, then

$$\sum_{k=0}^{\infty} \left[\sum_{n=k}^{\infty} a_n \binom{n}{k} (z_1-z_0)^{n-k} \right] (z-z_1)^k$$

will be the power series expansion for f based at z_1. To justify this interchange of summation, it suffices to prove absolute convergence of (2.10). By the root test,

$$\sum_{n=0}^{\infty} |a_n||w-z_0|^n$$

converges if $|w-z_0| < r$. Set

$$w = |z-z_1| + |z_1-z_0| + z_0.$$

Then $|w-z_0| = |z-z_1| + |z_1-z_0| < r$ provided $|z-z_1| < r - |z_1-z_0|$. See Figure 2.4.

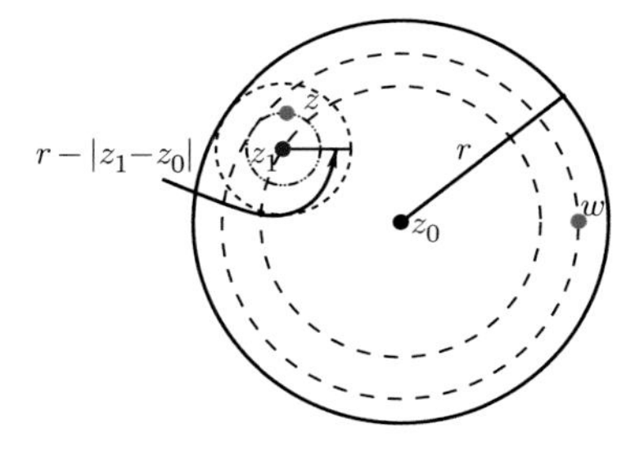

Figure 2.4 Proof of Theorem 2.7.

Thus if $|z - z_1| < r - |z_1 - z_0|$ then

$$\begin{aligned}\infty &> \sum_{n=0}^{\infty} |a_n||w - z_0|^n \\ &= \sum_{n=0}^{\infty} |a_n| \left(|z - z_1| + |z_1 - z_0|\right)^n \\ &= \sum_{n=0}^{\infty} \left[\sum_{k=0}^{n} |a_n| \binom{n}{k} |z_1 - z_0|^{n-k} |z - z_1|^k\right]\end{aligned}$$

as desired. □

Another natural question is: can an analytic function have more than one power series expansion based at z_0?

Theorem 2.8 (uniqueness of series) *Suppose*

$$\sum_{n=0}^{\infty} a_n(z - z_0)^n = \sum_{n=0}^{\infty} b_n(z - z_0)^n,$$

for all z such that $|z - z_0| < r$, where $r > 0$. Then $a_n = b_n$ for all n.

Proof Set $c_n = a_n - b_n$. The hypothesis implies that $\sum_{n=0}^{\infty} c_n(z - z_0)^n = 0$, for all z with $|z-z_0| < r$. We need to show that $c_n = 0$ for all n. Suppose c_m is the first non-zero coefficient. If $0 < |z - z_0| < r$ then

$$(z - z_0)^{-m} \sum_{n=m}^{\infty} c_n(z - z_0)^n = \sum_{k=0}^{\infty} c_{m+k}(z - z_0)^k \equiv F(z).$$

The series for F converges in $0 < |z - z_0| < r$ because we can multiply the terms of the series on the left-hand side by the non-zero number $(z - z_0)^{-m}$ and not affect convergence. By the root test, the series for F converges in a disk and hence in $\{|z - z_0| < r\}$. Since F is continuous and $c_m \neq 0$, there is a $\delta > 0$ so that, if $|z - z_0| < \delta$ then

$$|F(z) - F(z_0)| = |F(z) - c_m| < |c_m|/2.$$

If $F(z) = 0$, then we obtain the contradiction $|-c_m| < |c_m|/2$. Thus $F(z) \neq 0$ when $|z - z_0| < \delta$. But $(z - z_0)^m = 0$ only when $z = z_0$, and thus

$$\sum_{n=0}^{\infty} c_n(z - z_0)^n = (z - z_0)^m F(z) \neq 0$$

when $0 < |z - z_0| < \delta$, contradicting our assumption on $\sum c_n(z - z_0)^n$. □

Note that the proof of Theorem 2.8 shows that if f is analytic at z_0 then, for some $\delta > 0$, either $f(z) \neq 0$ when $0 < |z - z_0| < \delta$ or $f(z) = 0$ for all z such that $|z - z_0| < \delta$. If $f(a) = 0$, then a is called a **zero** of f. A **region** is a connected open set.

Corollary 2.9 *If f is analytic on a region Ω then either $f \equiv 0$ or the zeros of f are isolated in Ω.*

Corollary 2.9 is sometimes called the **uniqueness theorem** or **identity theorem** because of the consequence that if two analytic functions on a region Ω agree on a set with an accumulation point in Ω then they must be identical.

Proof Let E denote the set of non-isolated zeros of f. Since f is continuous, the set of zeros of f is closed in Ω. Each isolated zero is contained in an open disk containing no other zero and hence E is closed in Ω. In the proof of Theorem 2.8, we showed that if z_0 is a non-isolated zero of f then f is identically zero in a neighborhood of z_0. Thus E is open. By connectedness, either $E = \Omega$ or $E = \emptyset$. □

There are plenty of continuous functions for which the corollary is false, for example $x \sin(1/x)$. The corollary is true because, near z_0, the function $f(z)$ behaves like the first non-zero term in its power series expansion about z_0. See Exercise 2.5.

2.5 Elementary Operations

Theorem 2.10 *If f and g are analytic at z_0 then so are*

$$f + g, \quad f - g, \quad cf \text{ (where } c \text{ is a constant) and } fg.$$

If h is analytic at $f(z_0)$ then $(h \circ f)(z) \equiv h(f(z))$ is analytic at z_0.

Proof The first three follow from the associative, commutative and distributive laws applied to the partial sums. To prove that the product of two analytic functions is analytic, multiply $f(z) = \sum a_n(z - z_0)^n$ and $g = \sum b_n(z - z_0)^n$ as if they were polynomials to obtain

$$\sum_{n=0}^{\infty} a_n(z - z_0)^n \sum_{k=0}^{\infty} b_k(z - z_0)^k = \sum_{n=0}^{\infty}\left(\sum_{k=0}^{n} a_k b_{n-k}\right)(z - z_0)^n, \tag{2.11}$$

which is called the **Cauchy product** of the two series. Why is this formal computation valid? The series for f and the series for g converge absolutely for $|z - z_0|$ sufficiently small. Because we can rearrange non-negative convergent series,

$$\infty > \sum_{n=0}^{\infty} |a_n||z - z_0|^n \sum_{k=0}^{\infty} |b_k||z - z_0|^k = \sum_{n=0}^{\infty}\left(\sum_{k=0}^{n} |a_k||b_{n-k}|\right)|z - z_0|^n.$$

This says that the series on the right-hand side of (2.11) is absolutely convergent and therefore can be arranged to give the left-hand side of (2.11). To put it another way, the doubly indexed sequence $a_n b_k(z - z_0)^{n+k}$ can be added up two ways: if we add along diagonals: $n + k = m$, for $m = 0, 1, 2, \ldots$, we obtain the partial sums of the right-hand side of (2.11). If we add along partial rows and columns $n = m$, $k = 0, \ldots, m$, and $k = m$, $n = 0, \ldots, m - 1$, for $m = 1, 2, \ldots$, we obtain the product of the partial sums for the series on the left-hand side of

(2.11). Since the series is absolutely convergent (as can be seen by using the latter method of summing the doubly indexed sequence of absolute values), the limits are the same.

To prove that we can compose analytic functions where it makes sense, suppose $f(z) = \sum a_n(z-z_0)^n$ is analytic at z_0 and suppose $h(z) = \sum b_n(z-a_0)^n$ is analytic at $a_0 = f(z_0)$. The sum

$$\sum_{m=1}^{\infty} |a_m||z-z_0|^{m-1} \tag{2.12}$$

converges in $\{z : 0 < |z-z_0| < r\}$ for some $r > 0$ since the series for f is absolutely convergent, and $|z-z_0|$ is non-zero. By the root test (set $k = m-1$), this implies that the series (2.12) converges uniformly in $\{|z-z_0| \le r_1\}$, for $r_1 < r$, and hence is bounded in $\{|z-z_0| \le r_1\}$. Thus there is a constant $M < \infty$ so that

$$\sum_{m=1}^{\infty} |a_m||z-z_0|^m \le M|z-z_0|,$$

if $|z-z_0| < r_1$. We conclude that

$$\sum_{m=0}^{\infty} |b_m| \left(\sum_{n=1}^{\infty} |a_n||z-z_0|^n \right)^m \le \sum_{m=0}^{\infty} |b_m|(M|z-z_0|)^m < \infty,$$

for $|z-z_0|$ sufficiently small, by the absolute convergence of the series for h. This proves absolute convergence for the composed series, and thus we can rearrange the doubly indexed series for the composition so that it is a (convergent) power series. □

As a consequence, if f is analytic at z_0 and $f(z_0) \neq 0$ then $1/f$ is analytic at z_0. Indeed, the function $1/z$ is analytic on $\mathbb{C} \setminus \{0\}$ by (2.9) with $a = 0$, and $1/f$ is the composition of $1/z$ with f. A **rational function** r is the ratio

$$r(z) = \frac{p(z)}{q(z)},$$

where p and q are polynomials. The rational function r is then analytic on $\{z : q(z) \neq 0\}$ by Theorem 2.10. Rational functions and their limits are really what this whole book is about.

Definition 2.11 *If f is defined in a neighborhood of z then*

$$f'(z) = \lim_{w \to z} \frac{f(w) - f(z)}{w - z}$$

is called the (complex) derivative of f, provided the limit exists.

The function $\bar{z}$ does not have a (complex) derivative. If n is a non-negative integer,

$$(z^n)' = nz^{n-1}.$$

The next theorem says that you can differentiate power series term-by-term.

Theorem 2.12 *If* $f(z) = \sum_{n=0}^{\infty} a_n(z - z_0)^n$ *converges in* $B = \{z : |z - z_0| < r\}$ *then* $f'(z)$ *exists for all* $z \in B$ *and*

$$f'(z) = \sum_{n=1}^{\infty} na_n(z - z_0)^{n-1} = \sum_{n=0}^{\infty}(n+1)a_{n+1}(z - z_0)^n,$$

for $z \in B$. *Moreover, the series for* f' *based at* z_0 *has the same radius of convergence as the series for* f.

Proof If $0 < |h| < r$ then

$$\frac{f(z_0 + h) - f(z_0)}{h} - a_1 = \frac{\sum_{n=0}^{\infty} a_n h^n - a_0}{h} - a_1 = \sum_{n=2}^{\infty} a_n h^{n-1} = \sum_{n=1}^{\infty} a_{n+1} h^n.$$

By the root test, the region of convergence for the series $\sum a_{n+1}h^n$ is a disk centered at 0 and hence it converges uniformly in $\{h : |h| \le r_1\}$, if $r_1 < r$. In particular, $\sum a_{n+1}h^n$ is continuous at 0 and hence

$$\lim_{h \to 0} \sum_{n=1}^{\infty} a_{n+1} h^n = 0.$$

This proves that $f'(z_0)$ exists and equals a_1.

By Theorem 2.7, f has a power series expansion about each z_1, with $|z_1 - z_0| < r$ given by

$$\sum_{k=0}^{\infty} \left[\sum_{n=k}^{\infty} a_n \binom{n}{k} (z_1 - z_0)^{n-k} \right] (z - z_1)^k.$$

Therefore $f'(z_1)$ exists and equals the coefficient of $z - z_1$:

$$f'(z_1) = \sum_{n=1}^{\infty} a_n \binom{n}{1} (z_1 - z_0)^{n-1} = \sum_{n=1}^{\infty} a_n n (z_1 - z_0)^{n-1}.$$

By the root test and the fact that $n^{\frac{1}{n}} \to 1$, the series for f' has exactly the same radius of convergence as the series for f. □

Since the series for f' has the same radius of convergence as the series for f, we obtain the following corollary.

Corollary 2.13 *An analytic funtion* f *has derivatives of all orders. Moreover, if* f *is equal to a convergent power series on* $B = \{z : |z - z_0| < r\}$ *then the power series is given by*

$$f(z) = \sum_{n=0}^{\infty} \frac{f^{(n)}(z_0)}{n!}(z - z_0)^n,$$

for $z \in B$.

By definition of the symbols, the $n = 0$ term in the series is $f(z_0)$.

Proof If $f(z) = \sum_{n=0}^{\infty} a_n(z-z_0)^n$, then we proved in Theorem 2.12 that $a_1 = f'(z_0)$ and

$$f'(z) = \sum_{n=1}^{\infty} n a_n (z-z_0)^{n-1}.$$

Applying Theorem 2.12 to $f'(z)$, we obtain $2a_2 = (f')'(z_0) \equiv f''(z_0)$, and, by induction,

$$n!\, a_n = f^{(n)}(z_0).$$ □

If f is analytic in a region Ω with $f'(z) = 0$ for all z in a neighborhood of $z_0 \in \Omega$, then, by Corollary 2.13 and Corollary 2.9 applied to $f(z) - f(z_0)$, f is constant in Ω. A useful consequence is that if f and g are analytic with $f' = g'$, then $f - g$ is constant.

A closer examination of the idea of the proof of Theorem 2.12 shows that power series satisfy a stronger notion of differentiability at a point. Corollary 2.14 will be used in Chapter 3 for understanding the local behavior of power series.

Corollary 2.14 *If $f(z) = \sum a_n(z-z_0)^n$ converges in $B = \{z : |z-z_0| < r\}$ then*

$$f'(z_0) = \lim_{z,w \to z_0} \frac{f(z)-f(w)}{z-w}.$$

Proof Set $z = z_0 + h$ and $w = z_0 + k$. Then, for $h - k \neq 0$ and $\varepsilon = \max(|h|,|k|) < r$,

$$\frac{f(z_0+h)-f(z_0+k)}{h-k} - a_1 = \sum_{n=2}^{\infty} a_n \frac{h^n - k^n}{h-k} = \sum_{n=2}^{\infty} a_n \sum_{j=0}^{n-1} h^j k^{n-j-1}. \tag{2.13}$$

But

$$\lim_{N,M\to\infty} \sum_{n=N}^{M} |a_n| \sum_{j=0}^{n-1} |h|^j |k|^{n-j-1} \leq \lim_{N,M\to\infty} \sum_{n=N}^{M} |a_n| n \varepsilon^{n-1} = 0,$$

by the root test. Because of uniform convergence, the right-hand side of (2.13) is a continuous function of (h,k) when $\varepsilon < r$, vanishing at $(0,0)$, and hence

$$\lim_{z,w\to z_0} \frac{f(z)-f(w)}{z-w} = \lim_{h,k\to 0} \frac{f(z_0+h)-f(z_0+k)}{h-k} = a_1 = f'(z_0).$$ □

Corollary 2.14 fails for the real-valued differentiable function $x^2 \sin(1/x)$ but holds for continuously differentiable real-valued functions by the mean-value theorem of calculus. Exercise 2.17 shows, however, that the mean-value theorem does not hold for all analytic functions.

Corollary 2.15 *If $f(z) = \sum a_n(z-z_0)^n$ converges in $B = \{z : |z-z_0| < r\}$ then the power series*

$$F(z) = \sum_{n=0}^{\infty} \frac{a_n}{n+1}(z-z_0)^{n+1}$$

converges in B and satisfies

$$F'(z) = f(z),$$

for $z \in B$.

The series for F has the same radius of convergence as the series for f, by Theorem 2.12 or by direct calculation.

2.6 Exercises

A

2.1 Check that Examples (i)–(v) in Section 2.3 are correct.

2.2 (a) If p is a polynomial with real coefficients, prove that p can be factored into a product of linear and quadratic factors, each of which has real coefficients, such that the quadratic factors are non-zero on $\mathbb{R}$. Most engineering problems involving polynomials only need polynomials with real coefficients.

(b) For rational functions with real coefficients, such as those that typically occur in applications, it is sometimes preferable to use a partial fraction expansion without complex numbers in the expression. The cover-up method can also be used in this case. Here is an example to illustrate the idea. If a, b, c, d and e are real, show that

$$\frac{z^2 + dz + e}{(z-a)((z-b)^2 + c^2)} = \frac{A}{z-a} + \frac{B(z-b) + D}{(z-b)^2 + c^2},$$

where A, B and D are real. Here we have completed the square for the irreducible quadratic factor. Note also that we have written the numerator of the last term as $B(z-b)+D$, not $Bz+D$. We can find A by the usual cover-up method. Then, to find B and D, we multiply by $(z-b)^2 + c^2$ and let it tend to 0. Thus $z \to b \pm ic$. Cover up the quadratic factor in the denominator on the left and let $z \to b + ic$. On the right-hand side, when we multiply by the quadratic factor, the first term will tend to 0, the denominator of the second term will be cancelled and $B(z-b)+D$ tends to $Bic + D$. Thus, the real part of the result on the left equals D and the imaginary part equals Bc, and then we can immediately write down the coefficients B and D. Try this process with two different irreducible quadratic factors in the denominator, and you will see how much faster and accurate it is than solving many equations with many unknowns. The choice of the form of the numerator as $B(z-b)+D$ instead of $Bz+D$ made this computation a bit easier. It also turns out that it makes it a bit easier to compute inverse Laplace transforms of these rational functions, because the resulting term is a shift in the domain of a simpler function.

2.3 For what values of z is

$$\sum_{n=0}^{\infty} \left(\frac{z}{1+z}\right)^n$$

convergent? Draw a picture of the region.

2.4 Prove the sum, product, quotient and chain rules for differentiation of analytic functions and find the derivative of $(z-a)^{-n}$, where n is a positive integer and $a \in \mathbb{C}$.

2.5 (a) Prove that f has a power series expansion about z_0 with radius of convergence $r > 0$ if and only if $g(z) = \frac{f(z)-f(z_0)}{z-z_0}$ has a power series expansion about z_0, with the same radius of convergence. (How must you define $g(z_0)$, in terms of the coefficients of the series for f, to make this a true statement?)

(b) It follows from (a) that if f has a power series expansion at z_0 with radius of convergence R, and if $r < R$, then there is a constant C so that $|f(z) - f(z_0)| \le C|z - z_0|$, provided $|z - z_0| \le r$. Use the same idea to show that if $f(z) = \sum a_n(z - z_0)^n$ then

$$\left| f(z) - \sum_{n=0}^{k} a_n(z - z_0)^n \right| \le D_k |z - z_0|^{k+1},$$

where D_k is a constant and $|z - z_0| \le r < R$.

(c) Use the proof of the root test to give an explicit estimate of D_k (for large k) and therefore an estimate of the rate of convergence of the series for f if $|z - z_0| < r < R$.

2.6 Define $e^z = \exp(z) = \sum_{n=0}^{\infty} \frac{z^n}{n!}$.

(a) Show that this series converges for all $z \in \mathbb{C}$.
(b) Show that $e^z e^w = e^{z+w}$.
(c) Define $\cos\theta = \frac{1}{2}(e^{i\theta} + e^{-i\theta})$ and $\sin\theta = \frac{1}{2i}(e^{i\theta} - e^{-i\theta})$, so that $e^{i\theta} = \cos\theta + i\sin\theta$. Using the series for e^z show that you obtain the same series expansions for sin and cos that you learned in calculus. Check that $\cos^2\theta + \sin^2\theta = 1$, by multiplying out the definitions, so that $e^{i\theta}$ is a point on the unit circle corresponding to the cartesian coordinate $(\cos\theta, \sin\theta)$.
(d) Show that $|e^z| = e^{\text{Re}z}$ and $\arg e^z = \text{Im}z$. If z is a non-zero complex number then show that $z = re^{it}$, where $r = |z|$ and $t = \arg z$. Moreover, show that $z^n = r^n e^{int}$.
(e) Show that $e^z = 1$ only when $z = 2\pi ki$, for some integer k.

2.7 (a) Using the definitions in Exercise 2.6, prove $\frac{d}{dz}e^z = e^z$.
(b) Use (a) and the chain rule to compute the indefinite integral

$$\int e^{nt} \cos mt \, dt.$$

Hint: Use $\text{Re} \int e^{(n+im)t} dt$, which results in a lot less work than the standard calculus trick of integrating by parts twice.

(c) Use (a), the chain rule and the fundamental theorem of calculus to prove $\int_0^{2\pi} e^{int} dt = 0$, if n is a non-zero integer.
(d) Suppose $a = e^b$. If $f'(z) = 1/z$ and $f(a) = b$, find the series expansion for f about a valid in $|z - a| < |a|$. Use (a) to prove $f(e^z) = z$. The function f is called the complex logarithm with $f(a) = b$. See Corollary 5.8 and Definition 5.9.

2.8 Prove Theorems 2.7 and 2.10 without explicitly exhibiting the rearrangements, by using the first sentence of Exercise 1.1. While this may yield slightly easier proofs, the proofs in the text were chosen because the explicit rearrangements are useful, as in Exercise 2.6(b).

B

2.9 (a) Suppose p and q are polynomials with no common zero, and suppose $q(z_0) \neq 0$. Let d denote the distance from z_0 to the nearest zero of q. Then the rational function $r = p/q$ has a power series expansion which converges in $\{z : |z - z_0| < d\}$ and no larger disk. Hint: Use the partial fraction expansion, Exercise 2.4, Theorem 2.12, and (2.9).

(b) Find the series expansion and radius of convergence of

$$\frac{z+2i}{(z-6)^2(z^2+6z+10)}$$

about the point 1. Hint: Set $z = 1 + w$, then expand in powers of w.

2.10 Let n be a positive integer. Prove that $z^{\frac{1}{n}}$ is analytic in $B = \{z : |z-1| < 1\}$ in the following sense: there is a convergent power series f in B with the property that $f(z)^n = z$ and $f(1) = 1$. Hint: Write $z = 1 + w$, $|w| < 1$, and let $g(w) = \sum a_k w^k$ be the (formal) Taylor series for $(1+w)^{\frac{1}{n}}$. Then prove $|a_k| \leq 1/k$ so that g is analytic in $|w| < 1$. Use Taylor's theorem to show that $g(x) = (1+x)^{1/n}$ for $-1 < x < 1$, and then use the uniqueness theorem, Corollary 2.9, to show that $g(w)^n = 1 + w$. Alternatively, prove that $z^{1/n}$ can be defined so that it has derivatives of all orders and prove that Taylor's theorem is true for complex differentiable functions using complex integration.

2.11 Suppose $\sum_{j=0}^{\infty} |a_j|^2 < \infty$. Show that $f(z) = \sum_{j=0}^{\infty} a_j z^j$ is analytic in $\{z : |z| < 1\}$. Compute (and prove your answer):

$$\lim_{r \nearrow 1} \int_0^{2\pi} |f(re^{i\theta})|^2 \frac{d\theta}{2\pi}.$$

2.12 Suppose f has a power series expansion at 0 which converges in all of $\mathbb{C}$. Suppose also that $\int_{\mathbb{C}} |f(x+iy)| dxdy < \infty$. Prove $f \equiv 0$. Hint: Use polar coordinates to prove $f(0) = 0$.

2.13 Suppose f is analytic in a connected open set U such that, for each $z \in U$, there exists an n (depending upon z) such that $f^{(n)}(z) = 0$. Prove f is a polynomial.

2.14 Let f be analytic in a region U containing the point $z = 0$. Suppose $|f(1/n)| < e^{-n}$ for $n \geq n_0$. Prove $f(z) \equiv 0$.

2.15 (**Newton's method** for solving $f(z) = 0$.) Suppose f is analytic in a neighborhood of a and suppose $f(z) = (z-a)g(z)$, where $g(a) \neq 0$. Set $N_f(z) = z - f(z)/f'(z)$.

(a) Prove that, for δ sufficiently small,

$$\sup_{|z-a|\leq\delta} \left| \frac{g'(z)}{g(z)} \right| \leq \frac{1}{2\delta}.$$

(b) Prove that if $|z-a| < \delta$ then

$$\left| \frac{N_f(z)-a}{\delta} \right| \leq \left| \frac{z-a}{\delta} \right|^2.$$

Thus if $|z_0 - a| < \delta$ and $z_n = N_f(z_{n-1})$ then

$$\left| \frac{z_n - a}{\delta} \right| \leq \left| \frac{z_0 - a}{\delta} \right|^{2^n}.$$

The function N_f is called the **Newton iterate** of f. We say that z_n **converges to** a **quadratically** because the error is squared at each step. Roughly, if z_n has the same first k digits as a then z_{n+1} will have the same first $2k$ digits. Numerically, just a few steps of Newton's method gives virtually a formula for a, provided z_0 is sufficiently close to a. If z_0 is not sufficiently close to a, then Newton's method may not converge.

Studying the behavior of the iteration $z_n = r(z_{n-1})$ for analytic functions r is the basis of the field called complex dynamics.

2.16 If f is analytic in a neighborhood of a then we can approximate $f'(a)$ by the difference quotient $D_\delta = \frac{f(a+\delta)-f(a)}{\delta}$ with error on the order of $|\delta|$. However, on a computer, where functions can only be evaluated approximately, there is a loss in precision called **round-off error**. If functions can be evaluated with 16 decimal digits of accuracy and if $\delta = 10^{-k}$ then the evaluation of D_δ will only have roughly $16 - k$ digits of accuracy. The approximation to $f'(a)$ by D_δ will have roughly $\min(16 - k, k)$ digits of accuracy. Choosing $k = 8$ gives the best estimate of 8 digits of accuracy. Consider instead

$$\widetilde{D}_\delta = \frac{1}{4\delta} \sum_{j=0}^{3} f(a + \delta i^j) i^{-j}.$$

The numerical approximation of $f'(a)$ by $\widetilde{D}_\delta$ has $\min(16 - k, 4k)$ digits of accuracy, roughly. When $k = 3$ this gives roughly 12 digits of accuracy. Explain this reasoning using the power series expansion of f at a with estimates in terms of the derivatives of f. An **approximate Newton's method** involves combining Exercises 2.15 and 2.16 when an explicit formula for f' is not available.

C

2.17 Suppose f is analytic in a convex open set U. Suppose that for each $z, w \in U$ there exists a point ζ on the line segment between z and w with

$$\frac{f(z) - f(w)}{z - w} = f'(\zeta).$$

Prove f is a polynomial of degree at most 2. (The point is that you have to be careful: not all calculus theorems extend to similar complex versions.) Hint: First prove it for a degree 3 polynomial on a very small disk.

2.18 Let $f(z) = \sum_{n=0}^{\infty} a_n z^n$ have radius of convergence 1 and suppose $a_n \geq 0$ for all n. Prove that $z = 1$ is a singular point of f. That is, there is no function g analytic in a neighborhood U of $z = 1$ such that $f = g$ on $U \cap \mathbb{D}$.

To illustrate that not everything is known about polynomials, here is an unsolved problem posed by the Fields Medalist S. Smale [25] in 1981, which arose in connection with the complex version of Newton's method. Find the smallest constant K so that, for each polynomial p and each $z \in \mathbb{C}$, there exists $c \in \mathbb{C}$ so that $p'(c) = 0$ and

$$\left| \frac{p(z) - p(c)}{z - c} \right| \leq K|p'(z)|.$$

This is known to hold for all polynomials if $K = 4$ and fails for some polynomials if $K < K_d = 1 - 1/d$, where d is the degree of p. The conjectured best constant is K_d, but even $K = 1$ would be an interesting result.

3 The Maximum Principle

3.1 The Maximum Principle

We can apply the same local ("walking the dog") analysis to analytic functions that we applied to polynomials in Section 2.1. If

$$f(z) - f(z_0) = \sum_{n=k}^{\infty} a_n(z - z_0)^n, \tag{3.1}$$

where $a_k \neq 0$, then, setting $z = z_0 + h$,

$$|f(z_0 + h) - [f(z_0) + a_k h^k]| \leq C|h|^{k+1}, \tag{3.2}$$

for $|h|$ sufficiently small by Exercise 2.5(b). This inequality says that the value $f(z)$ lies inside a disk centered at $f(z_0) + a_k h^k$ of radius at most $C|h|^{k+1}$, which is much smaller than $|a_k h^k|$. See Figure 2.2. In other words,

$$f(z) \approx f(z_0) + a_k(z - z_0)^k. \tag{3.3}$$

The next result is perhaps the most important elementary result in complex analysis. It follows from the "walking the dog" analysis much like the proof of the fundamental theorem of algebra. See Exercise 3.1. Because of its importance, we give another proof. The proof is less geometric, but we will apply it to a more general class of functions in Chapter 7.

Theorem 3.1 (maximum principle) *Suppose f is analytic in a region Ω. If there exists a $z_0 \in \Omega$ such that*

$$|f(z_0)| = \sup_{z \in \Omega} |f(z)|$$

then f is constant in Ω.

In particular, a non-constant analytic function has no local maximum absolute value, by Theorem 3.1 and the identity theorem, Corollary 2.9.

Proof If f has a series expansion given by (3.1) which converges in $\{z : |z - z_0| < r_0\}$, then for $r < r_0$

$$f(z_0) = \int_0^{2\pi} f(z_0 + re^{it}) \frac{dt}{2\pi}, \tag{3.4}$$

because we can interchange the order of summation and integration by uniform convergence. Equation (3.4) is called the **mean-value property** for analytic functions. It says that $f(z_0)$ is the average of f over a circle centered at z_0. Intuitively, if $f(z_0+re^{it})$ lies in a disk of radius M, then the only way that an average of its values can have absolute value M is for the function to be a constant of absolute value M. A more rigorous argument follows.

Suppose $|f(z)| \leq |f(z_0)|$ for all $z \in \Omega$. If $|f(z_0)| = 0$, then f is the constant function 0. If $|f(z_0)| \neq 0$, set $\lambda = |f(z_0)|/f(z_0)$, so that $|\lambda| = 1$ and $|f(z_0)| = \lambda f(z_0)$. By the mean-value property,

$$0 = \mathrm{Re} \int_0^{2\pi} \left(\lambda f(z_0) - \lambda f(z_0 + re^{it})\right) \frac{dt}{2\pi}$$
$$= \int_0^{2\pi} \left(|f(z_0)| - \mathrm{Re}\lambda f(z_0 + re^{it})\right) \frac{dt}{2\pi}, \tag{3.5}$$

for $r < r_0$. But the right-hand integrand in (3.5) is continuous and non-negative and hence

$$|f(z_0)| = \mathrm{Re}\lambda f(z),$$

for all $|z - z_0| < r_0$. Moreover, for $|z - z_0| < r_0$,

$$|f(z_0)|^2 = (\mathrm{Re}\lambda f(z))^2 \leq (\mathrm{Re}\lambda f(z))^2 + (\mathrm{Im}\lambda f(z))^2 = |\lambda f(z)|^2 \leq |f(z_0)|^2$$

so that $\mathrm{Im}\lambda f(z) = 0$ when $|z - z_0| < r_0$. Thus $\lambda f(z) = |f(z_0)|$ in a neighborhood of z_0. By the identity theorem, Corollary 2.9, $\lambda f(z) - |f(z_0)| = 0$ for all $z \in \Omega$. □

A colleague calls this the "nobody is above average" proof.

If $f(z_0) \neq 0$ then the same argument shows that $|f(z_0)|$ is not a local minimum if f is non-constant. This fact can also be derived from the statement of the maximum principle by considering the function $1/f$, which is analytic off the zeros of f.

Another form of the maximum principle is Corollary 3.2.

Corollary 3.2 *If f is a non-constant analytic function in a bounded region Ω, and if f is continuous on $\overline{\Omega}$, then*

$$\max_{z \in \overline{\Omega}} |f(z)|$$

occurs on $\partial\Omega$ but not in Ω.

The reader should verify the alternative form: if f is analytic on Ω then

$$\limsup_{z \to \partial\Omega} |f(z)| = \sup_{\Omega} |f(z)|. \tag{3.6}$$

We say that a sequence tends to $\partial\Omega$ if it is eventually outside each compact subset of Ω. The lim sup is then the largest subsequential limit of the values of $|f|$ over all sequences that are eventually outside of each compact subset of Ω. If Ω is unbounded, we view our sets as lying on the Riemann sphere, so that the boundary includes the north pole (the point "at ∞"). Equivalently, we can measure the distance to the boundary using the chordal metric given at the end of Section 1.3.

The function $f(z) = e^{-iz}$ is analytic in the upper half-plane $\mathbb{H} = \{z : \mathrm{Im} z > 0\}$, continuous on $\{z : \mathrm{Im} z \geq 0\}$ and has absolute value 1 on the real line $\mathbb{R}$, but is not bounded by 1 in $\mathbb{H}$. See Exercise 3.2.

3.2 Local Behavior

The maximum principle allows us to give an improved description of the local mapping property of analytic functions.

Corollary 3.3 *A non-constant analytic function defined on a region is an open map.*

In other words, if f is analytic and non-constant on a region Ω, and if $U \subset \Omega$ is open, then $f(U)$ is an open set.

Proof Suppose f has a power series expansion which converges on $\{z : |z - z_0| < R\}$. Pick $r < R$ and set

$$\delta = \inf_{|z-z_0|=r} |f(z) - f(z_0)|.$$

Since the zeros of $f - f(z_0)$ are isolated by Corollary 2.9, we may suppose that $\delta > 0$ by decreasing r if necessary. If $|w - f(z_0)| < \delta/2$ and if $f(z) \neq w$ for all z such that $|z - z_0| \leq r$, then $1/(f - w)$ is analytic in $|z - z_0| \leq r$ and

$$\left|\frac{1}{f(z) - w}\right| \leq \frac{1}{|f(z) - f(z_0)| - |w - f(z_0)|} < \frac{1}{\delta - \delta/2} = \frac{2}{\delta}$$

on $|z - z_0| = r$. By the maximum principle, the inequality $|1/(f(z) - w)| < 2/\delta$ persists in $|z - z_0| < r$. But evaluating this expression at z_0 we obtain the contradiction $2/\delta < 2/\delta$. Thus the image of the disk of radius r about z_0 contains a disk of radius $\delta/2$ about $f(z_0)$. This implies that the image of an open set contains a neighborhood of each of its points. □

The main ingredient in the proof of Corollary 3.3 is the maximum principle. An open continuous function on a region always satisfies the maximum principle.

Definition 3.4 *A function f is* **one-to-one** *if $f(z) = f(w)$ only when $z = w$.*

A more enlightened terminology proposed by one of my teachers, Richard Arens, is **two-to-two** since two points go to two points. "One-to-one" should really be the definition of a function, but the inertia of common usage is too large to overcome.

Proposition 3.5 *If f is analytic at z_0 with $f'(z_0) \neq 0$, then f is one-to-one in a sufficiently small neighborhood of z_0.*

Proof By Corollary 2.14, if $z_n, w_n \to z_0$ with $f(z_n) = f(w_n)$ then $f'(z_0) = 0$. □

Corollary 3.3 and Proposition 3.5 show that if f is analytic at z_0 with $f'(z_0) \neq 0$, then f is a homeomorphism of a neighborhood of z_0 onto a neighborhood of $f(z_0)$. That is, f has a continuous inverse function in a neighborhood of $f(z_0)$. We will prove in Corollary 4.17 that the inverse is analytic.

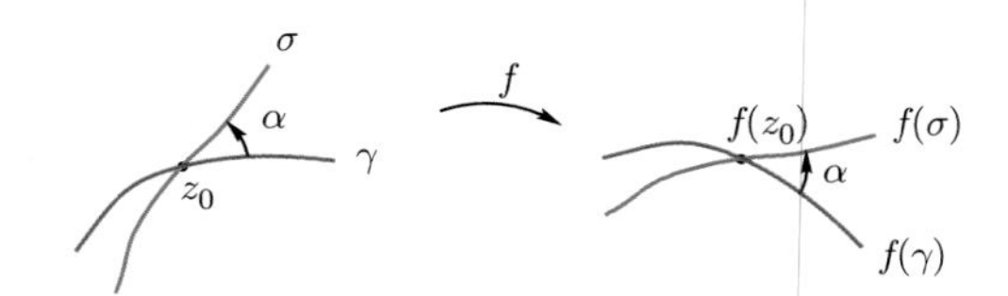

Figure 3.1 Conformality.

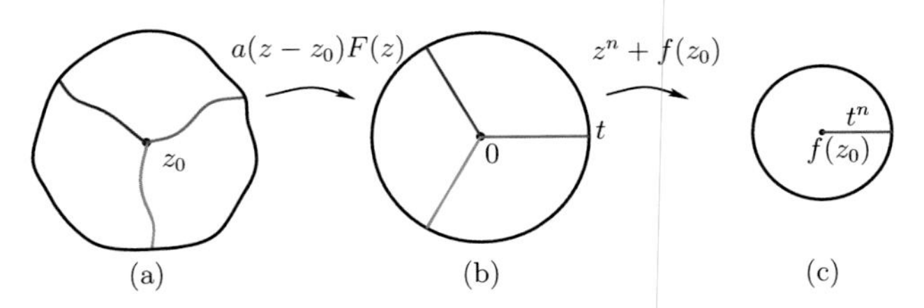

Figure 3.2 Local behavior of an analytic function.

Returning to our local analysis, if $f'(z_0) \neq 0$ and if $|z - z_0|$ is small then by (3.3) the map f approximately translates by $-z_0$, dilates by the factor $|f'(z_0)|$ and rotates by $\arg f'(z_0)$, then translates by $f(z_0)$, and this approximation becomes more and more accurate as $|z - z_0| \to 0$.

To put it another way, if γ and σ are two curves passing through z_0 with angle α from γ to σ, then $f(\gamma)$ and $f(\sigma)$ will be curves passing through $f(z_0)$ and the angle from $f(\gamma)$ to $f(\sigma)$ will also be equal to α. See Figure 3.1.

Definition 3.6 *We say that f is* **locally conformal** *if it preserves angles (including direction) between curves.*

An analytic function f with $f' \neq 0$ on a region Ω is locally conformal on Ω, though not necessarily one-to-one. Some books use the term "conformal" instead of locally conformal. But modern usage of the phrase "conformal map" usually means a one-to-one and analytic function. For that reason we add the qualifier "locally."

We can now picture the local correspondence given by an analytic function. By repeated division as in Exercise 2.5, if f is analytic at z_0 we can write

$$f(z) - f(z_0) = a_n(z - z_0)^n g(z),$$

where a_n is the first non-zero power series coefficient after a_0 and g is analytic at z_0 with $g(z_0) = 1$. Choose a so that $a^n = a_n$. By Exercise 2.10, we can define $z^{\frac{1}{n}}$ to be analytic in a neighborhood of 1. Set $F(z) = g(z)^{\frac{1}{n}}$. Then

$$f(z) = f(z_0) + [a(z - z_0)F(z)]^n.$$

By Proposition 3.5, $a(z - z_0)F(z)$ is one-to-one in a neighborhood of z_0.

The case when $n = 3$ is illustrated in Figure 3.2. The composed function is equal to $f(z)$ near z_0. Each of the three regions in Figure 3.2(a) is mapped one-to-one onto the slit disk in Figure 3.2(c). Asymptotically (as the radius tends to 0) the map "looks" like z^3, translated, rotated and dilated. Note that Figure 3.2 was constructed right-to-left by taking preimages.

By the analysis above, if γ and σ are two curves passing through z_0 with angle $\alpha < 2\pi/n$ from γ to σ, then $f(\gamma)$ and $f(\sigma)$ will be two curves passing through $f(z_0)$ and the angle from $f(\gamma)$ to $f(\sigma)$ will be $n\alpha$.

Corollary 3.7 *Suppose f is analytic at z_0. Then*

$$f(z) - f(z_0) = \sum_{m=n}^{\infty} a_m (z - z_0)^m$$

with $a_n \neq 0$ if and only if, for ε sufficiently small, there exists $\delta > 0$ so that $f(z) - w$ has exactly n distinct roots in $\{z : 0 < |z - z_0| < \varepsilon\}$, provided $0 < |w - f(z_0)| < \delta$.

The condition in Corollary 3.7 states that $f(B_\varepsilon(z_0))$ covers $B_\delta(w_0) \setminus \{w_0\}$ exactly n times, where $B_r(\zeta)$ is the ball centered at ζ with radius r, and $w_0 = f(z_0)$. In particular, f is one-to-one in a neighborhood of z_0 if and only if $f'(z_0) \neq 0$.

3.3 Growth on $\mathbb{C}$ and $\mathbb{D}$

In this section we will use the maximum principle to draw some conclusions about the growth of analytic functions defined on the plane $\mathbb{C}$ or on the open unit disk $\mathbb{D} = \{z : |z| < 1\}$.

Corollary 3.8 (Liouville's theorem) *If f is analytic in $\mathbb{C}$ and bounded, then f is constant.*

Proof Suppose $|f| \leq M < \infty$. Set $g(z) = (f(z) - f(0))/z$. Then g is analytic and $|g| \to 0$ as $|z| \to \infty$. By the maximum principle, $g \equiv 0$ and hence $f \equiv f(0)$. □

Here is a typical use of Liouville's theorem. If p is a polynomial of degree n, then $|p(z)| \leq C|z|^n$ for sufficiently large $|z|$. If f is analytic in $\mathbb{C}$, and if $|f(z)| \leq C|z|^n$ for $|z| > M$, then let $p(z) = \sum_{k=0}^{n} a_k z^k$ be the terms of the power series expansion of f at 0 up to degree n. Then $g(z) = (f(z) - p(z))/z^n$ is analytic in $\mathbb{C}$ and bounded. By Liouville's theorem, g is constant and hence f must be a polynomial. In fact, $f = p$ since $g(0) = 0$.

Corollary 3.9 (Schwarz's lemma) *Suppose f is analytic in $\mathbb{D}$ and suppose $|f(z)| \leq 1$ and $f(0) = 0$. Then*

$$|f(z)| \leq |z|, \tag{3.7}$$

for all $z \in \mathbb{D}$, and

$$|f'(0)| \leq 1. \tag{3.8}$$

Moreover, if equality holds in (3.7) *for some $z \neq 0$ or if equality holds in* (3.8), *then $f(z) = cz$, where c is a constant with $|c| = 1$.*

In some sense, Schwarz's lemma says that a bounded analytic function cannot grow too fast in the disk.

Proof By Exercise 2.5, the function g given by

$$g(z) = \begin{cases} \dfrac{f(z)}{z}, & \text{if } z \in \mathbb{D} \setminus \{0\} \\ f'(0), & \text{if } z = 0 \end{cases}$$

is analytic in $\mathbb{D}$ and, for $0 < r < 1$,

$$\sup_{|z|=r} |g(z)| \leq \frac{1}{r}.$$

Fix $z_0 \in \mathbb{D}$, then for $r > |z_0|$ the maximum principle implies $|g(z_0)| \leq \frac{1}{r}$, so that, letting $r \to 1$, we obtain (3.7) and (3.8). If equality holds in (3.7) at z_0 or holds in (3.8) then $g(z)$ has a maximum at z_0 or 0 and hence is constant. □

Corollary 3.10 (invariant form of Schwarz's lemma) *Suppose f is analytic in $\mathbb{D} = \{z : |z| < 1\}$ and suppose $|f(z)| < 1$. If $z, a \in \mathbb{D}$ then*

$$\left| \frac{f(z) - f(a)}{1 - \overline{f(a)}f(z)} \right| \leq \left| \frac{z - a}{1 - \bar{a}z} \right| \tag{3.9}$$

and

$$\frac{|f'(z)|}{1 - |f(z)|^2} \leq \frac{1}{1 - |z|^2}. \tag{3.10}$$

Proof If $|c| < 1$, the function

$$T_c(z) = \frac{z - c}{1 - \bar{c}z}$$

is analytic except at $z = 1/\bar{c}$. Moreover, for $z = e^{it} \in \partial\mathbb{D}$,

$$|T_c(e^{it})| = \left| \frac{e^{it} - c}{1 - \bar{c}e^{it}} \right| = \frac{|e^{it} - c|}{|e^{-it} - \bar{c}|} = 1.$$

By the maximum principle (or direct computation), $|T_c| \leq 1$ on $\mathbb{D}$. Setting $c = f(a)$, the composition $T_c \circ f$ is analytic on $\mathbb{D}$ and bounded by 1. Furthermore,

$$\frac{T_c \circ f(z)}{T_a(z)} = \left(\frac{f(z) - f(a)}{1 - \overline{f(a)}f(z)} \right) \left(\frac{1 - \bar{a}z}{z - a} \right)$$

is analytic on $\mathbb{D}$, and

$$\limsup_{|z| \to 1} \left| \frac{T_c \circ f(z)}{T_a(z)} \right| = \limsup_{|z| \to 1} |T_c \circ f(z)| \leq 1.$$

By the maximum principle, (3.9) holds. Inequality (3.10) follows by dividing both sides of (3.9) by $|z - a|$ and letting $z \to a$. □

An alternative proof is to apply Schwarz's lemma to $g(w) = T_c \circ f \circ T_{-a}(w)$ then set $w = T_a(z)$. The details are almost the same.

Corollary 3.11 *If f is analytic on $\mathbb{D}$, $|f| \leq 1$ and $f(z_j) = 0$, for $j = 1, \ldots, n$, then*

$$f(z) = \prod_{j=1}^{n} \left(\frac{z - z_j}{1 - \overline{z_j} z} \right) g(z),$$

where g is analytic in $\mathbb{D}$ and $|g(z)| \leq 1$ on $\mathbb{D}$.

Proof If $f(a) = 0$, then, by the proof of Corollary 3.10, $g(z) = f(z)/T_a(z)$ is analytic in the disk and bounded by 1. Repeating this argument n times proves the corollary. □

By the uniqueness theorem, if f is analytic on $\mathbb{D}$, then the zeros of f do not cluster in $\mathbb{D}$, unless f is identically zero. If we have more restrictions on f, such as boundedness, then the zeros cannot approach the unit circle too slowly by the next corollary.

Corollary 3.12 *If f is non-constant, bounded and analytic in $\mathbb{D}$, and if $\{z_j\}$ are the zeros of f, then*

$$\sum_j (1 - |z_j|) < \infty.$$

The convention we adopt here is that if z_j is a zero of order k, then $(1 - |z_j|)$ occurs k times in the sum in the statement of Corollary 3.12.

Proof We may suppose $|f| \leq 1$, by dividing f by a constant if necessary. If $f(0) \neq 0$, then, using the notation of the proof of Corollary 3.11,

$$|f(0)| = \left(\prod_{j=1}^{n} |z_j| \right) |g(0)| \leq \prod_{j=1}^{n} |z_j|,$$

so that, by taking logarithms (base e),

$$\ln \frac{1}{|f(0)|} \geq \sum_{j=1}^{n} \ln \frac{1}{|z_j|} \geq \sum_{j=1}^{n} (1 - |z_j|).$$

If $f(0) = 0$, then write $f(z) = z^k h(z)$, where $h(0) \neq 0$. Applying the preceding argument to h, we obtain

$$\sum_{j=1}^{n} (1 - |z_j|) \leq \ln \frac{1}{|h(0)|} + k.$$

The corollary follows by letting $n \to \infty$. □

Much of what appears in this book takes place on $\mathbb{D}$ or on $\mathbb{C}$, which look like rather special domains, but we know that a power series converges on a disk, and, by translating and scaling the domain, we can assume it is $\mathbb{D}$ or $\mathbb{C}$. Also, in Chapter 15 we shall prove the uniformization theorem which says that in some sense the only analytic functions we need to understand can be defined on $\mathbb{D}$ or $\mathbb{C}$.

3.4 Boundary Behavior

We conclude this chapter with some examples and a theorem about boundary behavior of analytic functions on the unit disk.

The first example is

$$I(z) = e^{\frac{z+1}{z-1}}.$$

(See Exercise 2.6 for the definition of e^z.) K. Hoffman called this the "world's greatest function." Since I is the composition of an analytic function on $\mathbb{C}\setminus\{1\}$ and the exponential function, which is analytic in $\mathbb{C}$, I is analytic on $\mathbb{C}\setminus\{1\}$ by Theorem 2.10. Moreover, $|e^z| = e^{\mathrm{Re}z}$ so that by a computation

$$|I(z)| = e^{\frac{|z|^2-1}{|z-1|^2}}.$$

Thus $|I(z)| \le 1$ on $\mathbb{D}$. On the unit circle $I(e^{it}) = e^{-i\cot(t/2)}$, for $0 < t < 2\pi$. In particular, if $\zeta \in \partial\mathbb{D}\setminus\{1\}$ then

$$\lim_{z\to\zeta} I(z)$$

exists and has absolute value 1. However, for $0 < r < 1$, $I(r) = e^{\frac{r+1}{r-1}} \to 0$ as $r \to 1$. On the unit circle, $I(e^{it})$ is spinning rapidly as $t \to 0$. Hence $I(z)$ does not have a limit as $z \in \mathbb{D} \to 1$. The proper way to take limits in the disk is through cones. For $\zeta \in \partial\mathbb{D}$ and $\alpha > 1$, define

$$\Gamma_\alpha(\zeta) = \{z \in \mathbb{D} : |z-\zeta| \le \alpha(1-|z|)\}$$

to be a **Stolz angle** or **non-tangential cone** at ζ. See Figure 3.3.

The precise shape of Γ_α is not important except that it is symmetric about the line segment $[0, \zeta]$ and forms an angle less than π at ζ. For $z \in \Gamma_\alpha(1)$ we have

$$|I(z)| = e^{\frac{-(1-|z|)}{|z-1|}\frac{(1+|z|)}{|z-1|}} \le e^{-\frac{1}{\alpha|z-1|}} \to 0$$

as $z \in \Gamma_\alpha(1) \to 1$. Thus $I(z) \to 0$ in *every* cone, and $\cup_\alpha \Gamma_\alpha(1) = \mathbb{D}$, but there is still no limit as $z \in \mathbb{D} \to 1$. The image of the disk by the function I can be better explained once we understand linear fractional transformations. See Sections 6.1 and 6.6.

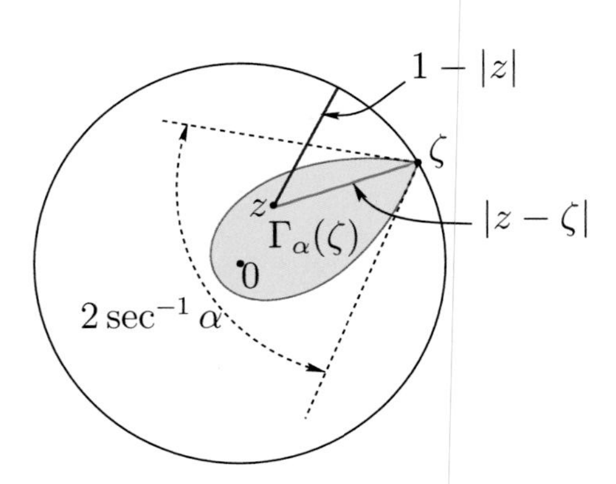

Figure 3.3 Stolz angle, $\Gamma_\alpha(\zeta)$.

Another example is the function given by

$$L(z) = \sum_{n=0}^{\infty} z^{2^n}.$$

This series converges uniformly and absolutely on compact subsets of $\mathbb{D}$. It is sometimes called a lacunary series because the spacing between non-zero coefficients increases as n increases. The function L does not extend continuously to any point of $\partial\mathbb{D}$. For example, if $\zeta^{2^k} = 1$ then

$$L(r\zeta) = \sum_{n=0}^{k-1} (r\zeta)^{2^n} + \sum_{n=k}^{\infty} r^{2^n}.$$

The second sum is clearly positive and increasing to ∞ as $r \to 1$. Since the 2^k roots of 1 are evenly spaced around $\partial\mathbb{D}$, if $e^{it} \in \partial\mathbb{D}$, then we can find a ζ as close to e^{it} as we like with $\zeta^{2^k} = 1$, for some k. Thus, in any neighborhood in $\mathbb{D}$ of e^{it}, f is unbounded.

The next theorem gives a connection between Fourier series and analytic functions in $\mathbb{D}$.

Theorem 3.13 (Abel's limit theorem) *If $\sum_{n=0}^{\infty} a_n e^{int}$ converges for some $\zeta = e^{it} \in \partial\mathbb{D}$, then $f(z) = \sum_{n=0}^{\infty} a_n z^n$ converges for all $z \in \mathbb{D}$. Moreover, if $\Gamma = \Gamma_\alpha(\zeta)$ is any Stolz angle at ζ then*

$$\lim_{z \in \Gamma \to \zeta} f(z) = \sum_{n=0}^{\infty} a_n e^{int}. \tag{3.11}$$

The convergence in Theorem 3.13 is called **non-tangential convergence**. Abel's limit theorem says that you get what you would expect for a limit, but only non-tangentially. Note that the limit does not depend on which Stolz angle we choose. When written as a power series, the function

$$f(z) = \sum_{n=1}^{\infty} \frac{1}{n} \left(z^{3^n} - (z^2)^{3^n} \right)$$

converges at $z = 1$ and hence converges in $|z| < 1$ by the root test. By Abel's limit theorem, f has non-tangential limit 0 at 1. Set $\zeta_k = e^{i\pi/3^k}$. Then $|f(r\zeta_k)| \to \infty$ as $r \uparrow 1$, so we may choose $z_k = r_k\zeta_k \to 1$ so that $|f(z_k)| \to \infty$. This says that we cannot conclude unrestricted convergence in Abel's limit theorem.

Proof By the root test, the series for f converges in $\mathbb{D}$ since it converges at a point on the unit circle. Replacing $f(z)$ by $f(\zeta z)$, we may suppose $\zeta = 1$, and by subtracting a constant from a_0 we may suppose $\sum_{n=0}^{\infty} a_n = 0$. Suppose $z \in \Gamma_\alpha(1)$. Then

$$\frac{f(z)}{1-z} = \sum_{n=0}^{\infty} z^n \sum_{k=0}^{\infty} a_k z^k = \sum_{n=0}^{\infty} \left(\sum_{k=0}^{n} a_k \right) z^n.$$

Set $s_n = \sum_{k=0}^{n} a_k$. Then

$$|f(z)| \le |1-z| \left| \sum_{n=0}^{N-1} s_n z^n \right| + |1-z| \left| \sum_{n=N}^{\infty} s_n z^n \right|.$$

Given $\varepsilon > 0$, there exists $N < \infty$ so that $|s_n| < \varepsilon$ for $n \ge N$. Thus

$$|f(z)| \le |1-z| \sum_{n=0}^{N-1} |s_n| + |1-z|\varepsilon \sum_{n=N}^{\infty} |z|^n = |1-z| \sum_{n=0}^{N-1} |s_n| + \frac{|1-z||z|^N \varepsilon}{1-|z|}.$$

We can choose $\delta > 0$ so that, if $z \in \Gamma_\alpha(1)$ and $|z-1| < \delta$,

$$|f(z)| \le \varepsilon + \alpha\varepsilon.$$

Since $\varepsilon > 0$ was arbitrary,

$$\lim_{z \in \Gamma_\alpha \to 1} |f(z)| = 0. \qquad \square$$

For example, the series $\sum_{n=1}^{\infty} z^n/n$ converges at $z = -1$ by the alternating series test. By Taylor's theorem (or by integrating the derivative on $[0, x]$), this series converges to $-\ln(1-x)$ for $-1 < x < 1$. By Abel's limit theorem, $\sum_{n=1}^{\infty}(-1)^n/n = -\ln 2$.

A Fourier series is a function of the form

$$F(t) = \sum_{n=-\infty}^{\infty} a_n e^{int}.$$

We say F converges at t if $\sum_{n=-M}^{N} a_n e^{int}$ converges as $M, N \to +\infty$ independently. If $F(t)$ converges, then $|a_n| \to 0$ as $|n| \to \infty$. Thus

$$f(z) = \sum_{n=0}^{\infty} a_n z^n \text{ and } g(z) = \sum_{n=1}^{\infty} \overline{a_{-n}} z^n$$

converge and are analytic on $\mathbb{D}$ by the root test. By Abel's limit theorem, $f(re^{it}) + \overline{g(re^{it})}$ converges to $F(t)$ at each t, where the series for F converges. Thus the function $f + \overline{g}$ "extends" F to $\mathbb{D}$, and the infinitely differentiable functions defined on $[0, 2\pi]$ by

$$f_r(t) = f(re^{it}) \text{ and } g_r(t) = g(re^{it})$$

satisfy

$$f_r + \overline{g_r} \to F$$

as $r \to 1$, provided the series for F converges at t.

Fourier series arose from attempting to solve certain differential equations. Each square integrable function on $[0, 2\pi]$ has a Fourier series, and it was a famous problem for many years to prove that the Fourier series converges almost everywhere on $[0, 2\pi]$. The proof that was eventually found remains perhaps one of the hardest proofs in analysis, but it used this connection between Fourier series and analytic functions on the disk, and the relation between convergence on the circle and nontangential convergence.

3.5 Exercises

A

3.1 (a) Show geometrically why the maximum principle holds using a "walking the dog" argument. Make it rigorous by imitating the last half of the proof of the fundamental theorem of algebra.

(b) Use the maximum principle to prove the fundamental theorem of algebra by applying it to $1/p$.

3.2 (a) Prove Corollary 3.2.

(b) Prove the alternative form (3.6) of Corollary 3.2.

3.3 Suppose f is analytic in a connected open set U. If $|f(z)|$ is constant on U, prove that f is constant on U. Likewise, prove that f is constant if $\text{Re}f$ is constant.

3.4 Suppose f and g are analytic in $\mathbb{C}$ and $|f(z)| \le |g(z)|$ for all z. Prove that there exists a constant c so that $f(z) = cg(z)$ for all z.

3.5 Prove that if f is non-constant and analytic on all of $\mathbb{C}$ then $f(\mathbb{C})$ is dense in $\mathbb{C}$.

3.6 Let f be analytic in $\mathbb{D}$ and suppose $|f(z)| < 1$ on $\mathbb{D}$. Let $a = f(0)$. Show that f does not vanish in $\{z : |z| < |a|\}$

B

3.7 Prove that if f is a one-to-one (two-to-two!) analytic map of an open set Ω onto $f(\Omega)$, and if $z_n \in \Omega \to \partial\Omega$, then $f(z_n) \to \partial f(\Omega)$, in the sense that $f(z_n)$ eventually lies outside each compact subset of $f(\Omega)$. A function with this property is called **proper**.

3.8 (a) Prove that φ is a one-to-one analytic map of $\mathbb{D}$ onto $\mathbb{D}$ if and only if

$$\varphi(z) = c\left(\frac{z-a}{1-\bar{a}z}\right),$$

for some constants c and a, with $|c| = 1$ and $|a| < 1$. What is the inverse map?

(b) Let f be analytic in $\mathbb{D}$ and satisfy $|f(z)| \to 1$ as $|z| \to 1$. Prove that f is rational.

3.9 The **pseudohyperbolic metric** on $\mathbb{D}$ is defined by

$$\rho(z,w) = \left|\frac{z-w}{1-\bar{w}z}\right|.$$

The **hyperbolic metric** on $\mathbb{D}$ is given by

$$\delta(z,w) = \frac{1}{2}\ln\left(\frac{1+\rho(z,w)}{1-\rho(z,w)}\right).$$

(a) Prove that if τ is a map of the form in Exercise 3.8(a) then

$$\rho(z,w) = \rho(\tau(z),\tau(w)).$$

(b) Prove the identity

$$1-\rho^2(z,w) = \frac{(1-|z|^2)(1-|w|^2)}{|1-\overline{w}z|^2}.$$

Parts (a) and (b) should be used as much as possible to reduce the work in subsequent parts of this exercise.

(c) Prove $\rho(|z|, |w|) \leq \rho(z, w) \leq \rho(|z|, -|w|)$.

(d) Prove ρ and δ are metrics on $\mathbb{D}$. Hint: To prove the triangle inequality for ρ, assume one point is 0 and use (c). Then apply (a). The identity $1 - r = (1 - r^2)/(1 + r)$ is useful in proving that δ is a metric.

(e) Suppose f is analytic on $\mathbb{D}$ and $|f(z)| \leq 1$. Schwarz's lemma says that f is a contraction in the pseudohyperbolic metric:

$$\rho(f(z), f(w)) \leq \rho(z, w).$$

Deduce the same fact for the hyperbolic metric.

The shortest curves or geodesics in the pseudohyperbolic geometry will be identified in Exercise 6.8, where it will be shown that the hyperbolic geometry does not satisfy Euclid's parallel postulate.

3.10 (a) Suppose p is a non-constant polynomial with all its zeros in the upper half-plane $\mathbb{H} = \{z : \mathrm{Im} z > 0\}$. Prove that all of the zeros of p' are contained in $\mathbb{H}$. Hint: Look at the partial fraction expansion of p'/p.

(b) Use (a) to prove that if p is a polynomial then the zeros of p' are contained in the (closed) convex hull of the zeros of p. (The closed convex hull is the intersection of all half-planes containing the zeros.)

3.11 Suppose f is analytic in $\mathbb{D}$ and $|f(z)| \leq 1$ in $\mathbb{D}$ and $f(0) = 1/2$. Prove that $|f(1/3)| \geq 1/5$.

3.12 Suppose f is analytic and non-constant in $\mathbb{D}$ and $|f(z)| \leq M$ on $\mathbb{D}$. Prove that the number of zeros of f in the disk of radius $1/4$, centered at 0, does not exceed

$$\frac{1}{\ln 4} \ln \left| \frac{M}{f(0)} \right|.$$

C

3.13 Suppose f is bounded and analytic in the right half-plane $\{z : \mathrm{Re} z > 0\}$, and $\limsup_{z \to iy} |f(z)| \leq M$ for all iy on the imaginary axis. Prove that $|f(z)| \leq M$ on the right half-plane. Check that $f(z) = e^z$ satisfies all the hypotheses above, except for boundedness, and fails to be bounded in the right half-plane.

4 Integration and Approximation

In this chapter we prove several important properties of analytic functions including equivalent ways of determining analyticity in terms of complex derivatives, integrals around rectangles and uniform approximation by rational functions.

4.1 Integration on Curves

In this section we give the basic definitions for integration along curves in $\mathbb{C}$.

Definition 4.1 *A **curve** is a continuous mapping of an interval $I \subset \mathbb{R}$ into $\mathbb{C}$.*

Different curves can have the same image. For example, if $\gamma(t) : [0, 1] \to \mathbb{C}$ then $\gamma(t^2) : [0, 1] \to \mathbb{C}$ and both curves have the same image. We will also use the symbol γ to denote the image or range $\gamma(I)$ of a curve $\gamma : I \to \mathbb{C}$ when it is clear from the context that we mean a set in the plane, not a function. Arrows, as in Figure 4.1, show how a parameterization $\gamma(t)$ traces the image as $t \in I$ increases.

Definition 4.2 (i) *A curve γ is called an* **arc** *if it is* one-to-one.
(ii) *A curve $\gamma : [a, b] \to \mathbb{C}$ is called* **closed** *if $\gamma(a) = \gamma(b)$.*
(iii) *A closed curve $\gamma : [a, b] \to \mathbb{C}$ is called* **simple** *if γ restricted to $[a, b)$ is one-to-one.*

A simple closed curve $\gamma : [0, 2\pi] \to \mathbb{C}$ can also be viewed as a one-to-one (two-to-two!) continuous mapping of the unit circle given by $\psi(e^{it}) = \gamma(t)$.

Definition 4.3 *A curve $\gamma(t) = x(t) + iy(t)$ is called* **piecewise continuously differentiable** *if $\gamma'(t) = x'(t) + iy'(t)$ exists and is continuous except for finitely many t, and x' and y' have one-sided limits at the exceptional points.*

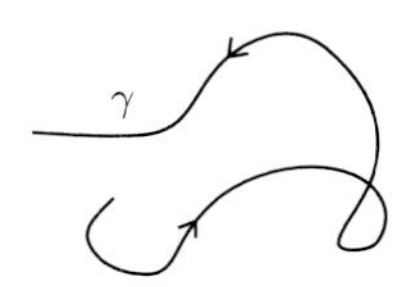

Figure 4.1 A curve, γ.

If γ is piecewise continuously differentiable then

$$\gamma(t_2) - \gamma(t_1) = (x(t_2) - x(t_1)) + i(y(t_2) - y(t_1)) = \int_{t_1}^{t_2} x'(t)dt + i\int_{t_1}^{t_2} y'(t)dt.$$

Note that $\gamma(t_2) - \gamma(t_1)$ corresponds to the vector from $\gamma(t_1)$ to $\gamma(t_2)$, so that

$$\gamma'(t_1) = \lim_{t_2 \to t_1} \frac{\gamma(t_2) - \gamma(t_1)}{t_2 - t_1}$$

is tangent to the curve γ at t_1, provided $\gamma'(t_1)$ exists.

Definition 4.4 *A curve $\psi : [c, d] \to \mathbb{C}$ is called a* **reparameterization** *of a curve $\gamma : [a, b] \to \mathbb{C}$ if there exists a one-to-one, onto, increasing function $\alpha : [a, b] \to [c, d]$ such that $\psi(\alpha(t)) = \gamma(t)$.*

For example, $\psi(t) = t^2 + it^4$, for $0 \le t \le 1$, is a reparameterization of $\gamma(t) = t + it^2$, for $0 \le t \le 1$, with $\alpha(t) = t^{\frac{1}{2}}$. Any curve can be reparameterized to be defined on $[0, 1]$. If $\sigma : [0, 1] \to \mathbb{C}$ is a curve then the curve β, defined by $\beta(t) = \sigma(1-t)$, is not a reparameterization of σ because $1 - t$ is decreasing. If ψ is a reparameterization of a piecewise continuously differentiable curve γ with $\psi(\alpha(t)) = \gamma(t)$, where α is also piecewise continuously differentiable, then $\psi'(\alpha(t))\alpha'(t) = \gamma'(t)$, by the chain rule applied to the real and imaginary parts, or by taking limits of difference quotients.

Definition 4.5 *If $\gamma : [a, b] \to \mathbb{C}$ is a piecewise continuously differentiable curve, and if f is a continuous complex-valued function defined on (the image of) γ, then*

$$\int_\gamma f(z)dz \equiv \int_a^b f(\gamma(t))\gamma'(t)dt.$$

The reader can check that a piecewise continuously differentiable reparameterization of γ will not change the integral by the chain rule, and so the integral really depends on the image of γ, not the choice of parameterization. For that reason we use the notation $\int_\gamma f dz$. Note, however, that the direction of the image curve is important.

Definition 4.6 *If $\gamma : [a, b] \to \mathbb{C}$ is a curve, then $-\gamma : [-b, -a] \to \mathbb{C}$ is the curve defined by*

$$-\gamma(t) = \gamma(-t).$$

The curve $-\gamma$ has the same geometric image as γ, but it is traced in the opposite direction. If f is continuous on a piecewise continuously differentiable curve γ, then

$$\int_{-\gamma} f(z)dz = -\int_\gamma f(z)dz.$$

If a closed piecewise continuously differentiable curve γ is split into a curve γ_1 followed by another curve γ_2, as in Figure 4.2, then

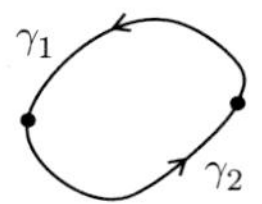

Figure 4.2 A closed curve, $\gamma = \gamma_1 + \gamma_2$.

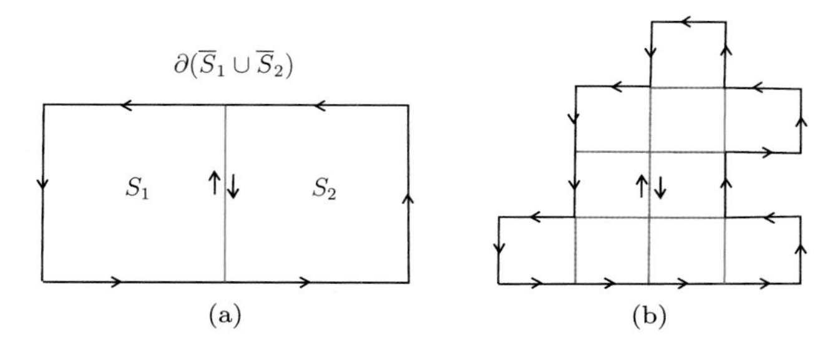

Figure 4.3 Integrals around squares.

$$\int_{\gamma} f(z)dz = \int_{\gamma_1} f(z)dz + \int_{\gamma_2} f(z)dz.$$

The integral is the same if we use the curve formed by γ_2 followed by γ_1. It follows that the integral around a closed curve of a continuous function does not depend on the choice of the "starting point."

For the purposes of computing integrals, it is useful to extend the notion of a curve to allow finite unions of curves. If $\gamma_1, \ldots, \gamma_n$ are curves defined on $[0, 1]$, then we can define $\gamma : [0, n) \to \mathbb{C}$ by $\gamma(t) = \gamma_j(t - j + 1)$ for $j - 1 \leq t < j, j = 1, \ldots, n$. If f is continuous on (the image of) each γ_j, and if each γ_j is piecewise continuously differentiable, then

$$\int_{\gamma} f(z)dz = \sum_{j=1}^{n} \int_{\gamma_j} f(z)dz.$$

For this reason we define $\sum_j \gamma_j \equiv \gamma$. The associative and commutative laws hold for sums (unions) of curves in this sense. We do not require the union to be connected.

In particular, $\int_{\gamma+(-\gamma)} f(z)dz = \int_{\gamma} f(z)dz - \int_{\gamma} f(z)dz = 0$, if γ is piecewise continuously differentiable. This idea can be used to simplify some integrals. For example, the integral around two adjacent squares, each in the counter-clockwise direction, is equal to the integral around the boundary of the union of the squares.

In Figure 4.3(a), the boundaries of the squares, ∂S_1 and ∂S_2, are parameterized in the counter-clockwise direction and

$$\int_{\partial S_1} f(z)dz + \int_{\partial S_2} f(z)dz = \int_{\partial(\overline{S_1} \cup \overline{S_2})} f(z)dz,$$

for every continuous function f defined on $\partial S_1 \cup \partial S_2$. This can be seen by writing the integrals around each square as the sum of integrals on the bounding line segments. The common boundary edge is traced in opposite directions, so the corresponding integrals will cancel. A similar argument applies to a finite union of squares, so that, after cancellation, the sum of the integrals around the boundaries of all the squares in Figure 4.3(b) is equal to the integral around the boundary of the union of the squares.

For the most part, we will deal with finite unions of *closed* curves.

Definition 4.7 *A **cycle** $\gamma = \sum_{j=1}^{n} \gamma_j$ is a finite union of closed curves $\gamma_1, \ldots, \gamma_n$.*

Another reason for using the notation $\int_\gamma f dz$ is the following. Suppose $\gamma : [a,b] \to \mathbb{C}$ is a piecewise continuously differentiable curve, and suppose $a = t_0 < t_1 < t_2 < \cdots < t_n = b$. Set $\gamma(t_j) = z_j$. Then

$$\begin{aligned} \sum_{j=0}^{n-1} f(z_j)(z_{j+1} - z_j) &= \sum_{j=0}^{n-1} f(\gamma(t_j))[\gamma(t_{j+1}) - \gamma(t_j)] \\ &\approx \sum_{j=0}^{n-1} f(\gamma(t_j))\gamma'(t_j)[t_{j+1} - t_j]. \end{aligned} \tag{4.1}$$

The left-hand side looks like a Riemann sum for $\int_\gamma f(z)dz$ with independent variable z, and the last sum is a Riemann sum, using the independent variable t, for

$$\int_a^b f(\gamma(t))\gamma'(t)dt.$$

The left-hand side of (4.1) also converges to $\int_\gamma f(z)dz$ as the **mesh** $\mu(\{t_j\}) = \max_j(t_{j+1} - t_j)$ of this partition tends to 0.

A closely related notion is integration with respect to arc-length.

Definition 4.8 *If $\gamma : [a,b] \to \mathbb{C}$ is a piecewise continuously differentiable curve, and if f is a continuous complex-valued function defined on (the image of) γ, then we define*

$$\int_\gamma f(z)|dz| = \int_a^b f(\gamma(t))|\gamma'(t)|dt.$$

Thus

$$\begin{aligned} \left|\int_\gamma f(z)dz\right| &= \left|\int_a^b f(\gamma(t))\gamma'(t)dt\right| \\ &\le \int_a^b |f(\gamma(t))||\gamma'(t)|dt = \int_\gamma |f(z)||dz|. \end{aligned}$$

Note that if $z_j = \gamma(t_j)$ then

$$\sum f(z_j)|z_{j+1} - z_j|$$

is approximately a Riemann sum for $\int_\gamma f(z)|dz| = \int_a^b f(\gamma(t))|\gamma'(t)|dt$.

Definition 4.9 *If $\gamma : [a,b] \to \mathbb{C}$ is a piecewise continuously differentiable curve then the **length** of γ is defined to be*

$$\ell(\gamma) = |\gamma| = \int_\gamma |dz| = \int_a^b |\gamma'(t)|dt.$$

The following important estimate follows immediately from the definitions because all piecewise continuously differentiable curves have finite length. If γ is piecewise continuously differentiable, and if f is continuous on γ, then

$$\left|\int_\gamma f(z)dz\right| \le \left(\sup_\gamma |f(z)|\right) \ell(\gamma). \tag{4.2}$$

Consequently, if f_n converges uniformly to f on γ, then

$$\lim_n \int_\gamma f_n(z)dz = \int_\gamma f(z)dz.$$

To prove this, just use the following estimate:

$$\left|\int_\gamma (f_n - f)dz\right| \le \left(\sup_\gamma |f_n - f|\right) \ell(\gamma).$$

Finally, we note that integration is linear on piecewise continuously differentiable curves: $\int_\gamma (f(z) + g(z))dz = \int_\gamma f(z)dz + \int_\gamma g(z)dz$, and if C is constant then $\int_\gamma Cf(z)dz = C\int_\gamma f(z)dz$, for continuous functions f and g.

4.2 Equivalence of Analytic and Holomorphic

Definition 4.10 *A complex-valued function f is said to be* **holomorphic** *on an open set U if*

$$f'(z) = \lim_{w\to z} \frac{f(w) - f(z)}{w - z}$$

exists for all $z \in U$ and is continuous on U. A complex-valued function f is said to be holomorphic on a set S if it is holomorphic on an open set $U \supset S$.

There are various a priori weaker conditions for analyticity. For example, many books do not require continuity of the derivative in the definition of a holomorphic function. In almost every situation encountered in practice, however, verifying that the derivative is continuous, once you have proved it exists, is not hard. Exercise 4.12 removes the requirement that the derivative is continuous. An important advance in partial differential equations was to consider "weak" derivatives in the sense of distributions. Indeed, it led to the development of functional analysis. See Exercise 7.13 for the corresponding definition of "weakly-analytic."

As we saw in Section 2.5, analytic functions are holomorphic. In particular, polynomials are holomorphic. A rational function is holomorphic except where the denominator is zero. Linear combinations of holomorphic functions are holomorphic. The reader is invited to verify that the chain rule for complex differentiation holds for the composition of two holomorphic functions, and so the composition of two holomorphic functions is holomorphic, wherever the composition is defined.

It also follows from the usual chain rule applied to real and imaginary parts that if γ : $[a, b] \to \mathbb{C}$ is a piecewise continuously differentiable curve, and if f is holomorphic on a neighborhood of γ, then $f \circ \gamma$ is a piecewise continuously differentiable curve, and

$$\frac{d}{dt}f(\gamma(t)) = f'(\gamma(t))\gamma'(t),$$

except at finitely many points $t_1, \dots, t_n$. Then, by the fundamental theorem of calculus,

$$\int_\gamma f'(z)dz = \int_a^b f'(\gamma(t))\gamma'(t)dt = \int_a^b \frac{d}{dt}f(\gamma(t))dt = f(\gamma(b)) - f(\gamma(a)).$$

For example, the line segment from z to ζ can be parameterized by $\gamma(t) = z + t(\zeta - z)$, for $t \in [0, 1]$, so that, if f is holomorphic in a neighborhood of γ, then

$$f(\zeta) - f(z) = \int_0^1 f'(z + t(\zeta - z))(\zeta - z)dt. \tag{4.3}$$

Corollary 4.11 *If $\gamma : [a, b] \to \mathbb{C}$ is a closed, piecewise continuously differentiable curve, and if f is holomorphic in a neighborhood of γ, then*

$$\int_\gamma f'(z)dz = 0.$$

Proof Because γ is closed, $f(\gamma(b)) - f(\gamma(a)) = 0$. □

For example, if γ is a closed curve then $\int_\gamma p(z)dz = 0$ for every polynomial p.

Corollary 4.12 *If $f(z) = \sum_{n=0}^\infty a_n(z - z_0)^n$ converges in $B = \{z : |z - z_0| < r\}$, and if $\gamma \subset B$ is a closed, piecewise continuously differentiable curve, then*

$$\int_\gamma f(z)dz = 0.$$

Proof Corollary 4.12 follows immediately from Corollaries 2.15 and 4.11. □

Much of this chapter and the next center around extending Corollary 4.12 to larger sets than disks B and more general curves.

If γ is a piecewise continuously differentiable closed curve and $a \notin \gamma$ then, by Corollary 4.11, for $n \neq 1$,

$$\int_\gamma \frac{1}{(z - a)^n}dz = 0.$$

By the partial fraction expansion, Corollary 2.3, in order to integrate a rational function along γ we need only to be able to compute $\int_\gamma (z - a)^{-1}dz$ for various values of a. The next example will be key to understanding integrals of analytic functions. If $r > 0$ set

$$C_r = \{z_0 + re^{it} : 0 \le t \le 2\pi\}.$$

Then we have the following proposition.

Proposition 4.13

$$\frac{1}{2\pi i}\int_{C_r} \frac{1}{z - a}dz = \begin{cases} 1, & \textit{if } |a - z_0| < r \\ 0, & \textit{if } |a - z_0| > r. \end{cases}$$

Proof Suppose $|a - z_0| < r$. Then $C_r'(t) = ire^{it}$ and

$$\begin{aligned}
\frac{1}{2\pi i}\int_{C_r}\frac{1}{z-a}dz &= \frac{1}{2\pi i}\int_0^{2\pi}\frac{1}{re^{it}-(a-z_0)}ire^{it}dt\\
&= \frac{1}{2\pi}\int_0^{2\pi}\frac{1}{1-(\frac{a-z_0}{re^{it}})}dt\\
&= \frac{1}{2\pi}\int_0^{2\pi}\sum_{n=0}^{\infty}\left(\frac{a-z_0}{re^{it}}\right)^n dt\\
&= \sum_{n=0}^{\infty}\frac{(a-z_0)^n}{r^n}\frac{1}{2\pi}\int_0^{2\pi}e^{-int}dt = 1.
\end{aligned}$$

Interchanging the order of summation and integration is justified because $|(a-z_0)/(re^{it})| < 1$ implies uniform convergence of the series.

If $|a-z_0| > r$, then write

$$\frac{re^{it}}{re^{it}-(a-z_0)} = \left(\frac{re^{it}}{z_0-a}\right)\frac{1}{1-\frac{re^{it}}{a-z_0}} = -\sum_{n=1}^{\infty}\frac{r^n e^{int}}{(a-z_0)^n},$$

so that

$$\begin{aligned}
\frac{1}{2\pi i}\int_{C_r}\frac{1}{z-a}dz &= \frac{1}{2\pi}\int_0^{2\pi}\frac{re^{it}}{re^{it}-(a-z_0)}dt\\
&= -\sum_{n=1}^{\infty}\frac{r^n}{(a-z_0)^n}\frac{1}{2\pi}\int_0^{2\pi}e^{int}dt = 0.
\end{aligned}$$

□

An immediate consequence of Corollary 4.11 and Proposition 4.13 is that there is no function f defined in a neighborhood of $\partial\mathbb{D}$ satisfying $f'(z) = 1/z$.

Theorem 4.14 *If f is holomorphic on $\{z : |z-z_0| \le r\}$ then, for $|z-z_0| < r$,*

$$f(z) = \frac{1}{2\pi i}\int_{C_r}\frac{f(\zeta)}{\zeta - z}d\zeta,$$

where C_r is the circle of radius r centered at z_0, parameterized in the counter-clockwise direction.

Theorem 4.14 shows that it is possible to find the values of a holomorphic function inside a disk from the values on the bounding circle.

Proof By (4.3) and Corollary 4.11, for $|z-z_0| < r$,

$$\begin{aligned}
\int_{C_r}\frac{f(\zeta)-f(z)}{\zeta - z}d\zeta &= \int_{C_r}\int_0^1 f'(z+t(\zeta - z))dtd\zeta\\
&= \int_0^1\int_{C_r} f'(z+t(\zeta - z))d\zeta dt\\
&= \lim_{\varepsilon\to 0}\int_\varepsilon^1\int_{C_r}\frac{d}{d\zeta}f(z+t(\zeta - z))d\zeta\frac{dt}{t} = 0.
\end{aligned}$$

Thus

$$\frac{1}{2\pi i}\int_{C_r}\frac{f(\zeta)}{\zeta-z}d\zeta = f(z)\cdot\frac{1}{2\pi i}\int_{C_r}\frac{d\zeta}{\zeta-z} = f(z),$$

by Proposition 4.13. □

A consequence of Theorem 4.14 is the converse to Theorem 2.12.

Corollary 4.15 *A complex-valued function f is holomorphic on a region Ω if and only if f is analytic on Ω. Moreover, the series expansion for f based at $z_0 \in \Omega$ converges on the largest open disk centered at z_0 and contained in Ω.*

See Exercise 2.9 for a proof of the second statement in Corollary 4.15 for rational functions.

Proof If f is analytic in Ω then f is holomorphic in Ω, by Theorem 2.12. To prove the converse, suppose f is holomorphic on $\{z : |z - z_0| \le r\}$. If $|z - z_0| < r$, then, by (2.9),

$$\begin{aligned} f(z) = \frac{1}{2\pi i}\int_{C_r}\frac{f(\zeta)}{\zeta-z}d\zeta &= \frac{1}{2\pi i}\int_{C_r}\left(\sum_{n=0}^{\infty}\frac{1}{(\zeta-z_0)^{n+1}}(z-z_0)^n\right)f(\zeta)d\zeta \\ &= \sum_{n=0}^{\infty}\left(\frac{1}{2\pi i}\int_{C_r}\frac{f(\zeta)}{(\zeta-z_0)^{n+1}}d\zeta\right)(z-z_0)^n. \end{aligned}$$

Interchanging the order of the summation and integral is justified by the uniform convergence in $\zeta \in C_r$ of the series for z fixed. Thus f has a power series expansion convergent in $\{z : |z - z_0| < r\}$, provided the closed disk is contained in Ω. By Theorem 2.7, f is analytic in Ω. □

In particular, if f is analytic in $\mathbb{C}$ then f has a power series expansion which converges in all of $\mathbb{C}$. Such functions are called **entire**. From now on, we will use the words analytic and holomorphic interchangably. Note that, when interchanging the order of summation and integration it is useful to think of C_r as a set rather than a function.

The Bernoulli numbers B_n are given by

$$\frac{z}{e^z-1} = \sum_{n=0}^{\infty}\frac{B_n}{n!}z^n.$$

By Corollary 4.15 this series converges in $\{|z| < 2\pi\}$ and no larger disk, so that, by the root test,

$$\limsup_{n\to\infty}\left(\frac{|B_n|}{n!}\right)^{\frac{1}{n}} = \frac{1}{2\pi}.$$

The rate of convergence of the series on $|z| = r < 2\pi$ can also be deduced from this estimate.

The proof of Corollary 4.15 yields a bit more information. Not only can we find the values of an analytic function inside a disk from its values on the boundary, but also we have a formula for each of its derivatives in the disk.

Corollary 4.16 *If f is analytic in $\{z : |z - z_0| \leq r\}$ and $C_r(z_0) = \{z_0 + re^{it} : 0 \leq t \leq 2\pi\}$, then*

$$\frac{f^{(n)}(z_0)}{n!} = \frac{1}{2\pi i}\int_{C_r(z_0)} \frac{f(\zeta)}{(\zeta - z_0)^{n+1}} d\zeta \tag{4.4}$$

and

$$\left|\frac{f^{(n)}(z_0)}{n!}\right| \leq \frac{\sup_{C_r(z_0)} |f|}{r^n}. \tag{4.5}$$

Inequality (4.5) is called **Cauchy's estimate**. We will show in Lemma 4.30 that (4.4) holds for all $z \in C_r(z_0)$.

Proof Equation (4.4) follows from Corollary 2.13, the proof of Corollary 4.15, and the uniqueness of series, Theorem 2.8. Inequality (4.5) follows from (4.4) by using inequality (4.2). □

The equivalence of analytic and holomorphic makes it easy to prove that the inverse of a one-to-one analytic function is analytic.

Corollary 4.17 *If f is analytic and one-to-one in a region Ω then the inverse of f, defined on $f(\Omega)$, is analytic.*

Proof Since analytic functions are open by Corollary 3.3, f has a continuous inverse. Take $z_0 \in \Omega$ and set $w_0 = f(z_0)$. Then $f(\Omega)$ contains a disk centered at w_0. If $w \in f(\Omega)$ tends to w_0, then $z = f^{-1}(w)$ tends to z_0. By Corollary 3.7, $f'(z_0) \neq 0$, so that

$$\frac{f^{-1}(w) - f^{-1}(w_0)}{w - w_0} = \frac{z - z_0}{f(z) - f(z_0)} \to \frac{1}{f'(z_0)}.$$

This proves, f^{-1} has a complex derivative at w_0 equal to $1/f'(f^{-1}(w_0))$. This derivative is continuous, so f^{-1} is holomorphic and hence analytic. □

As an application of the second sentence of Corollary 4.15, we obtain a local version of a theorem we will encounter in the next chapter, called Cauchy's theorem.

Corollary 4.18 *If f is analytic in an open disk B, and if $\gamma \subset B$ is a closed, piecewise continuously differentiable curve, then*

$$\int_\gamma f(z)dz = 0.$$

Proof By Corollary 4.15, f has a power series expansion which converges on all of B. Now apply Corollary 4.12. □

Morera's theorem is a useful converse to Corollary 4.18.

Theorem 4.19 (Morera) *If f is continuous in an open disk B, and if*

$$\int_{\partial R} f(\zeta)d\zeta = 0$$

for all closed rectangles $R \subset B$ with sides parallel to the axes, then f is analytic on B.

Proof We may suppose $B = \mathbb{D}$. Define

$$F(z) = \int_{\gamma_z} f(\zeta)d\zeta,$$

where γ_z is a curve from 0 to z consisting of a horizontal line segment followed by a vertical line segment.

If $|h| < 1 - |z|$ then $\gamma_{z+h} = \gamma_z + \sigma + \partial R$, where σ is a curve from z to $z+h$ consisting of a horizontal line segment followed by a vertical line segment and $R \subset B$ is a closed rectangle. See Figure 4.4. By assumption, $\int_{\partial R} f(\zeta)d\zeta = 0$, so that

$$F(z+h) - F(z) = \int_{\gamma_{z+h}} f(\zeta)d\zeta - \int_{\gamma_z} f(\zeta)d\zeta = \int_\sigma f(\zeta)d\zeta.$$

By the fundamental theorem of calculus, since the identity function has derivative equal to 1, $\int_\sigma d\zeta = z + h - z = h$, and so

$$\frac{F(z+h) - F(z)}{h} - f(z) = \frac{1}{h}\int_\sigma (f(\zeta) - f(z))d\zeta.$$

By (4.2),

$$\left|\frac{1}{h}\int_\sigma (f(\zeta) - f(z))d\zeta\right| \leq \sqrt{2}\sup_{\zeta\in\sigma}|f(\zeta) - f(z)|,$$

because $|\sigma| \leq \sqrt{2}|h|$. Since f is continuous, letting $h \to 0$ proves that F is holomorphic on B with $F' = f$. By Corollary 4.15, F is analytic on B, and, by Theorem 2.12, $f = F'$ is analytic on B. Finally, apply Corollary 4.11. □

One consequence of Morera's theorem is that the definition of holomorphic does not need to include the continuity of the derivative. See Exercise 4.12.

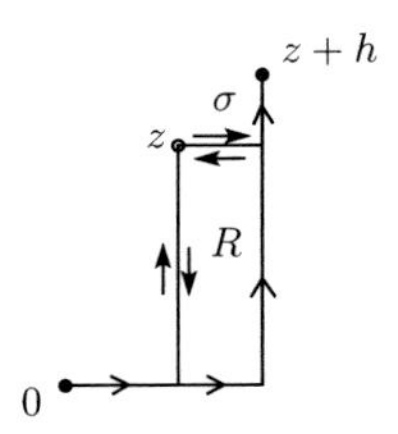

Figure 4.4 Proof of Morera's theorem.

4.3 Approximation by Rational Functions

In this section we will show that Theorem 4.14 also holds if the circle C_r is replaced by the boundary of a square, and then use it to prove Runge's theorem that analytic functions can be uniformly approximated by rational functions.

Proposition 4.20 *If S is an open square with boundary ∂S parameterized in the counter-clockwise direction then*

$$\frac{1}{2\pi i}\int_{\partial S}\frac{1}{z-a}dz=\begin{cases}1, & \text{if } a\in S\\ 0, & \text{if } a\in\mathbb{C}\setminus\overline{S}.\end{cases}$$

Proof If $a\in\mathbb{C}\setminus\overline{S}$, then we can find a disk B which contains $\overline{S}$ and does not contain a. See Figure 4.5.

By (2.9) and Corollary 4.18,

$$\int_{\partial S}\frac{1}{z-a}dz=0.$$

If $a\in S$, then let C be the circumscribed circle to ∂S parameterized in the *clockwise* direction. See Figure 4.6. Then we can write

$$\partial S=s_1+s_2+s_3+s_4,$$

where $s_j, j=1,\ldots,4$, are the sides of ∂S and

$$C=c_1+c_4+c_3+c_2,$$

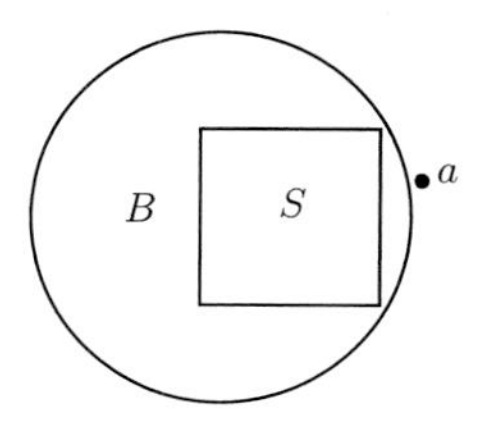

Figure 4.5 If a is outside $\overline{S}$.

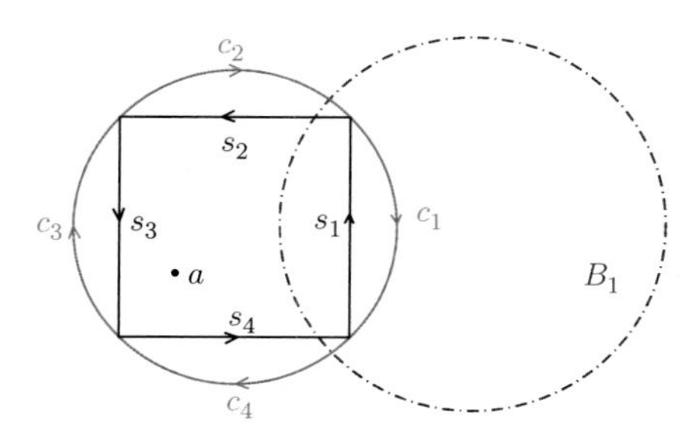

Figure 4.6 The square S and its circumscribed circle C.

where $c_j, j = 1, \ldots, 4$, are the arcs of C subtended by the corresponding sides of ∂S. Then $s_j + c_j$ is a closed curve contained in a disk B_j with $a \notin \overline{B_j}$, for $j = 1, \ldots, 4$. By Corollary 4.18,

$$\int_{s_j+c_j} \frac{1}{z-a} dz = 0, \tag{4.6}$$

for $j = 1, \ldots, 4$. Adding the four integrals (4.6) we obtain

$$\int_{\partial S} \frac{1}{z-a} dz + \int_C \frac{1}{z-a} dz = \int_{\partial S + C} \frac{1}{z-a} dz = 0.$$

But we also have, by Proposition 4.13,

$$\int_C \frac{1}{z-a} dz = -2\pi i,$$

because $-C$ is the circle parameterized in the counter-clockwise direction. This proves Proposition 4.20. □

Proposition 4.20 can also be proved by explicit computation, but we chose this proof because the idea will be used later to compute the integral of $1/(z-a)$ for other curves.

Theorem 4.21 *If f is analytic in a neighborhood of the closure $\overline{S}$ of an open square S, then, for $z \in S$,*

$$f(z) = \frac{1}{2\pi i} \int_{\partial S} \frac{f(\zeta)}{\zeta - z} d\zeta,$$

where ∂S is parameterized in the counter-clockwise direction.

Proof The proof of Theorem 4.21 is exactly like the proof of Theorem 4.14 except that Proposition 4.20 is used instead of Proposition 4.13. □

Corollary 4.22 *If f is analytic in a neighborhood of the closure $\overline{S}$ of an open square S, then*

$$\frac{1}{2\pi i} \int_{\partial S} f(\zeta) d\zeta = 0.$$

Proof Fix $z \in S$ and apply Theorem 4.21 to $g(\zeta) = f(\zeta)(\zeta - z)$. □

Theorem 4.23 (Runge) *If f is analytic on a compact set K, and if $\varepsilon > 0$, then there is a rational function r so that*

$$\sup_{z \in K} |f(z) - r(z)| < \varepsilon.$$

Proof Suppose f is analytic on U open, with $U \supset K$. Let

$$d = \text{dist}(\partial U, K) = \inf\{|z - w| : z \in \partial U, w \in K\}.$$

Construct a grid of closed squares with side length $d/2$.

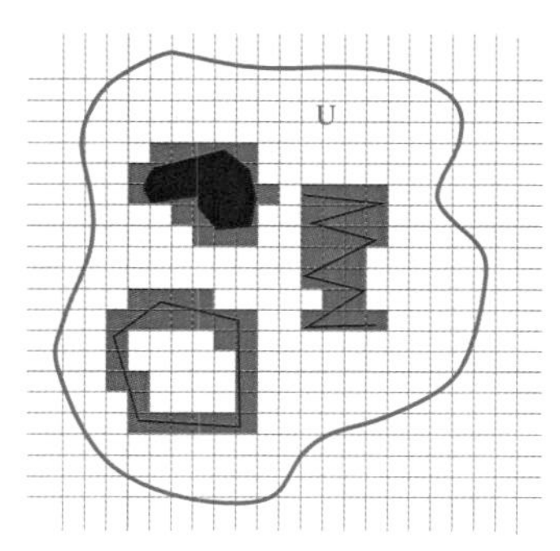

Figure 4.7 A union of closed squares covering K and contained in U.

Shade each square in the grid which intersects K. See Figure 4.7. Note that each (closed) shaded square is contained in U because each has diameter $d/\sqrt{2}$. Let $\{S_k\}$ be the collection of shaded squares and let Γ denote the boundary of the union of the shaded squares,

$$\Gamma = \partial\left(\cup_j S_j\right),$$

formed from $\cup \partial S_j$, each parameterized in the counter-clockwise direction, by cancelling edges which are traced in opposite directions. Then $\Gamma \subset U$ is a cycle, a finite union of closed curves, with $\Gamma \cap K = \emptyset$. See Exercise 4.1. If z is in the interior of one of the squares, S_{j_0}, then $f(\zeta)/(\zeta - z)$ is analytic as a function of ζ on S_j, for $j \neq j_0$. Apply Theorem 4.21 to S_{j_0} and Corollary 4.22 to all of the other squares, then cancel edges traced in opposite directions to obtain

$$f(z) = \frac{1}{2\pi i}\int_\Gamma \frac{f(\zeta)}{\zeta - z}d\zeta. \tag{4.7}$$

Both sides of (4.7) are continuous functions of z, for $z \notin \Gamma$, so that equality holds for all $z \in \cup S_k \setminus \Gamma$. Fix $z_0 \in K$ and write $\Gamma = \cup_1^n \Gamma_j$, where each Γ_j is a closed curve. Choose a Riemann sum on each Γ_j for the integral on the right-hand side of (4.7), as in (4.1), to obtain

$$\left| f(z_0) - \sum_{j=1}^{n}\sum_{k=1}^{m_j} \frac{f(\zeta_{j,k})(\zeta_{j,k+1} - \zeta_{j,k})}{2\pi i(\zeta_{j,k} - z_0)} \right| < \varepsilon.$$

We can in fact choose the partition so that the inequality remains true for all refinements of the partition. By uniform continuity, this inequality remains true for all z in a small disk containing z_0 and all refinements of the partition, if we replace ε with 4ε. See Exercise 4.5. Cover K by finitely many such disks, and take a common refinement. □

Definition 4.24 *If r is a rational function, by the fundamental theorem of algebra we can write $r(z) = p(z)/q(z)$, where p and q are polynomials with no common zeros. The zeros of q are called the* **poles** *of the rational function r.*

If b is a pole of $r(z)$ then $|r(z)| \to \infty$ as $z \to b$. A rational function is analytic everywhere in the plane, except at its poles.

It is possible to improve the statement of Runge's theorem by restricting where the poles of the rational function need to be placed using the next lemma.

Lemma 4.25 *Suppose U is open and connected, and suppose $b \in U$. Then a rational function with poles only in U can be uniformly approximated on $\mathbb{C} \setminus U$ by a rational function with poles only at b.*

Proof Suppose $a, c \in U$ and suppose $|c - a| < \text{dist}(a, \partial U)$. If $z \in \mathbb{C} \setminus U$ then $|z - a| \geq \text{dist}(a, \partial U) > |c - a|$, so that

$$\frac{1}{z-c} = \frac{1}{z-a-(c-a)} = \frac{1}{(z-a)(1-(\frac{c-a}{z-a}))} = \sum_{n=0}^{\infty} \frac{(c-a)^n}{(z-a)^{n+1}}, \tag{4.8}$$

where the sequence of partial sums approximates $1/(z-c)$ uniformly on $\mathbb{C} \setminus U$. By taking products we can also approximate $(z-c)^{-n}$ for $n \geq 1$ on $\mathbb{C} \setminus U$, and, by taking finite linear combinations, we can uniformly approximate on $\mathbb{C} \setminus U$ any rational function with poles only at c by rational functions with poles only at a.

Write $c \in R_d$ if every rational function with poles only at c can be uniformly approximated on $\mathbb{C} \setminus U$ by rational functions with poles only at d. This relation is transitive: if $c \in R_d$ and $d \in R_e$ then $c \in R_e$. Set $E = \{a \in U : a \in R_b\}$. By transitivity and the argument above, if $a \in E$ then E contains a disk centered at a with radius $\text{dist}(a, \partial U)$. Thus E is open. Moreover, if $a_n \in E$ converges to $a_\infty \in U$, then we can choose n so large that, for all $z \in \mathbb{C} \setminus U$,

$$|z - a_n| \geq |z - a_\infty| - |a_n - a_\infty| \geq \text{dist}(a_\infty, \partial U) - |a_n - a_\infty| > |a_n - a_\infty|.$$

By (4.8), $a_\infty \in R_{a_n}$ and, by transitivity, $a_\infty \in E$. This proves E is closed in U and, by connectedness, $E = U$.

Now suppose that r is rational with poles only in U and fix $b \in U$. Each term $1/(z-c)^k$ in the partial fraction expansion of r can be approximated by a rational function with poles only at b. Adding the approximations gives an approximation of r by a rational function with poles only at b. □

Corollary 4.26 *Suppose U is connected and open and suppose $\{z : |z| > R\} \subset U$ for some $R < \infty$. Then a rational function with poles only in U can be uniformly approximated on $\mathbb{C} \setminus U$ by a polynomial.*

Proof By Lemma 4.25, we need only prove that, if $|b| > R$, then a rational function with poles at b can be uniformly approximated by a polynomial on $\mathbb{C} \setminus U$. But

$$\frac{1}{z-b} = \frac{1}{-b(1-\frac{z}{b})} = -\frac{1}{b} \sum_{n=0}^{\infty} \left(\frac{z}{b}\right)^n,$$

where the sum converges uniformly on $|z| \leq R$. As in the proof of Lemma 4.25, we can approximate $(z-b)^{-n}$ for $n \geq 1$, and, by taking finite linear combinations, we can approximate any rational function with poles only at b by a polynomial, uniformly on $\{z : |z| \leq R\} \supset \mathbb{C} \setminus U$. □

Theorem 4.23, Lemma 4.25 and Corollary 4.26 combine to give the following improvement of Runge's theorem ("one pole in each hole").

Theorem 4.27 (Runge) *Suppose K is a compact set. Choose one point a_n in each bounded component U_n of $\mathbb{C} \setminus K$. If f is analytic on K and $\varepsilon > 0$, then we can find a rational function r with poles only in the set $\{a_n\}$ such that*

$$\sup_{z \in K} |f(z) - r(z)| < \varepsilon.$$

If $\mathbb{C} \setminus K$ has no bounded components, then we may take r to be a polynomial.

See Exercise 4.2(a) for the definition of components of an open set. For example, if K_1 and K_2 are disjoint compact sets such that $\mathbb{C} \setminus (K_1 \cup K_2)$ is connected and $\varepsilon > 0$, then we can find a polynomial p so that $|p| < \varepsilon$ on K_1 and $|p - 1| < \varepsilon$ on K_2 because the function which is equal to 0 on K_1 and equal to 1 on K_2 is analytic on $K_1 \cup K_2$.

Corollary 4.28 (Runge) *If f is analytic on an open set $\Omega \neq \mathbb{C}$ then there is a sequence of rational functions r_n with poles in $\partial\Omega$ so that r_n converges to f uniformly on compact subsets of Ω.*

Proof Set

$$K_n = \left\{ z \in \Omega : \text{dist}(z, \partial\Omega) \geq \frac{1}{n} \text{ and } |z| \leq n \right\}.$$

Then K_n is compact, $\cup K_n = \Omega$ and each bounded component U, of $\mathbb{C} \setminus K_n$ contains a point of $\partial\Omega$. Indeed $\partial U \subset K_n \subset \Omega$ so that $U \cap \Omega \neq \phi$. If $z \in U \cap \Omega$, then $|z| < n$ and $|z - \zeta| < 1/n$ for some $\zeta \in \partial\Omega$. Let L be the line segment from z to ζ. If $\alpha \in L$ then $|\alpha - \zeta| < 1/n$, so that $\alpha \notin K_n$. Thus L is a connected subset of $\mathbb{C} \setminus K_n$, so L must be contained in one component of $\mathbb{C} \setminus K_n$. Because $z \in L \cap U$, we must have $L \subset U$. But then $\zeta \in L \cap \partial\Omega \subset U$.

By Theorem 4.27 we can choose the rational functions approximating f to have poles only in $\partial\Omega$. □

The improvement of Corollary 4.28 over Theorem 4.23 is that the poles of r_n are outside of Ω, not just outside the compact subset of Ω on which r_n is close to f.

Corollary 4.28 says that every analytic function is a limit of rational functions, uniformly on compact subsets. Weierstrass proved that the set of analytic functions on a region is closed under uniform convergence on compact sets.

Theorem 4.29 (Weierstrass) *Suppose $\{f_n\}$ is a collection of analytic functions on a region Ω such that $f_n \to f$ uniformly on compact subsets of Ω. Then f is analytic on Ω. Moreover, $f_n' \to f'$ uniformly on compact subsets of Ω.*

Lemma 4.30 *If G is integrable on a piecewise continuously differentiable curve γ, then*

$$g(z) \equiv \int_\gamma \frac{G(\zeta)}{\zeta - z} d\zeta$$

is analytic in $\mathbb{C} \setminus \gamma$ and

$$g'(z) = \int_\gamma \frac{G(\zeta)}{(\zeta - z)^2} d\zeta.$$

Proof There are at least two ways to prove this lemma. One way is to write out a power series expansion for $1/(\zeta - z)$ based at z_0, where $z_0 \notin \gamma$, then interchange the order of summation and integration to obtain a power series expansion for g based at z_0. The derivative of g at z_0 is the coefficient of $z - z_0$ by Corollary 2.13. The second proof is to write

$$\frac{g(z+h)-g(z)}{h} - \int_\gamma \frac{G(\zeta)}{(\zeta - z)^2}d\zeta = \int_\gamma G(\zeta)\frac{h}{(\zeta - z)^2(\zeta - (z+h))}d\zeta,$$

which $\to 0$ as $h \to 0$. Thus

$$g'(z) = \int_\gamma \frac{G(\zeta)}{(\zeta - z)^2}d\zeta$$

exists and is continuous on $\mathbb{C} \setminus \gamma$. By Corollary 4.15, g is analytic on $\mathbb{C} \setminus \gamma$. □

A similar proof shows that

$$\frac{g^{(n)}(z)}{n!} = \int_\gamma \frac{G(\zeta)}{(\zeta - z)^{n+1}}d\zeta.$$

So, by Theorem 4.14, formula (4.4) in Corollary 4.16 holds with z_0 replaced by any z such that $|z - z_0| < r$.

Proof of Theorem 4.29 Analyticity is a local property, so to prove the first statement we may suppose B is a disk with $\overline{B} \subset \Omega$. Then, by Theorem 4.14, if $z \in B$,

$$f_n(z) = \frac{1}{2\pi i}\int_{\partial B} \frac{f_n(\zeta)}{\zeta - z}d\zeta.$$

The limit function f is continuous. Set

$$F(z) = \frac{1}{2\pi i}\int_{\partial B} \frac{f(\zeta)}{\zeta - z}d\zeta.$$

Then

$$|f_n(z) - F(z)| \to 0$$

for each $z \in B$, because $f_n \to f$ uniformly on ∂B. Thus $F = f$ on B and, by Lemma 4.30, F is analytic on B.

By Theorem 4.14 and Lemma 4.30,

$$f_n'(z) = \frac{1}{2\pi i}\int_{\partial B} \frac{f_n(\zeta)}{(\zeta - z)^2}d\zeta$$

and

$$f'(z) = F'(z) = \frac{1}{2\pi i}\int_{\partial B} \frac{f(\zeta)}{(\zeta - z)^2}d\zeta.$$

Again, since $f_n \to f$ uniformly on ∂B, we have that f_n' converges uniformly to f' on compact subsets of B. Thus f_n' converges uniformly to f' on closed disks contained in Ω. Given a compact subset K of Ω, we can cover K by finitely many closed disks contained in Ω and hence f_n' converges uniformly on K to f'. □

This proof of Theorem 4.29, which implicitly uses the equivalence of analytic and holomorphic, is easier than the original proof of Weierstrass, which used doubly indexed sums.

The reason for using piecewise continuously differentiable curves γ is that we want to integrate continuous functions. If f is continuous, then the left-hand side of (4.1) converges to $\int_\gamma f(z)dz$ for piecewise continuously differentiable γ, but the limit can fail to exist if γ is assumed to be continuous only. However, Corollary 4.18 actually allows us to define the integral of an analytic function over any continuous curve. See Section 12.1 for some badly behaved curves.

We first need an elementary but useful covering lemma.

Lemma 4.31 *Suppose Ω is a region and suppose $\gamma : [0,1] \to \Omega$ is continuous. Given ε, $0 < \varepsilon < \text{dist}(\gamma, \partial\Omega)$, we can find a finite partition $0 = t_0 < t_1 < \cdots < t_n = 1$ so that $\gamma([t_{j-1}, t_j]) \subset B_j \equiv \{z : |z - \gamma(t_j)| < \varepsilon\} \subset \Omega$.*

Proof If $0 < \varepsilon < \text{dist}(\gamma, \partial\Omega)$, then by the uniform continuity of γ, we can choose $\delta > 0$ so that $|\gamma(t) - \gamma(s)| < \varepsilon$ whenever $|t - s| < \delta$. Then any finite partition $\{t_j\}$ with mesh size $\mu(\{t_j\}) = \sup_j |t_{j-1} - t_j| < \delta$ will satisfy the conclusion of Lemma 4.31. □

Using the notation of Lemma 4.31, let σ be the polygonal curve $\sum \sigma_j$, where σ_j is straight-line segment from $\gamma(t_{j-1})$ to $\gamma(t_j)$. Let γ_j be the subarc $\gamma([t_{j-1}, t_j])$ of γ. Then $\beta_j = \gamma_j - \sigma_j$ is a closed curve contained in B_j. If γ is piecewise continuously differentiable, and if f is analytic on Ω, then, by Corollary 4.18, $\int_{\gamma_j} f(z)dz = \int_{\sigma_j} f(z)dz$. If γ is not piecewise continuously differentiable, we use this as the definition of the integral.

Theorem 4.32 *Suppose Ω is a region and $\gamma : [0,1] \to \mathbb{C}$ is continuous with $\gamma \subset \Omega$. Let σ be the polygonal curve defined above. If f is analytic on Ω, define*

$$\int_\gamma f(z)dz = \int_\sigma f(z)dz.$$

Then this definition of $\int_\gamma f(z)dz$ does not depend on the choice of the polygonal curve σ and it agrees with our prior definition if γ is piecewise continuously differentiable.

Proof As seen above, this definition of $\int_\gamma f(z)dz$ agrees with our prior definition if γ is piecewise continuously differentiable. For an arbitrary continuous γ, suppose σ is the polygonal curve associated with a partition $\{t_j\}$ chosen as above. Let α be the polygonal curve associated with a finite refinement $\{s_k\} \supset \{t_j\}$ of the partition $\{t_j\}$. We can write $\sigma - \alpha = \sum \beta_j$, where each $\beta_j : [0,1] \to \mathbb{C}$ is a closed polygonal curve contained in B_j. By Corollary 4.18 again, $\int_\alpha f(z)dz = \int_\sigma f(z)dz$. Given any two partitions satisfying the conditions of the definition of the integral, we can find a common refinement. Thus the definition does not depend on the choice of the partition. □

Theorem 4.32 says that if we want to prove something about $\int_\gamma f(z)dz$, where f is analytic on a region Ω and $\gamma \subset \Omega$, then it is enough to prove it for all polygonal curves γ. Note that the choice of the polygonal curve σ depends on γ and the region Ω but not on the analytic function f. For example, if f_n is a sequence of analytic functions converging uniformly on compact subsets of Ω to f, then f is analytic by Weierstrass's Theorem 4.29 and

$$\lim_{n\to\infty}\int_\gamma f_n(z)dz = \int_\gamma f(z)dz$$

for every continuous curve $\gamma \subset \Omega$, even if γ has infinite length. We simply apply (4.2) to the analytic function $f_n - f$ on the polygonal curve σ which has finite length.

4.4 Exercises

A

4.1 (a) A finite union of boundaries of squares, oriented in the usual counter-clockwise direction, is a cycle, by definition. Prove that if an edge in the union is traced twice, in opposite directions, then, after removal of the common edge, the union is still a cycle.

(b) The boundary of a finite union of squares is a cycle, oriented so that the region lies on the left. Hint: Use (a) and induction.

4.2 (a) Let U be an open set in $\mathbb{C}$. A **polygonal curve** is a curve consisting of a finite union of line segments. Define an equivalence relation on the points of U as follows: $a \sim b$ if and only if there is a polygonal curve contained in U with edges parallel to the axes and with endpoints a and b. Show that each equivalence class is open and closed in U and connected. Show that there are at most countably many equivalence classes. The equivalence classes are called the **components** of U. For open sets in $\mathbb{C}^*$, allow polygonal curves to contain a half-line and obtain a similar result.

(b) Let K be a compact set. Define an equivalence relation on the points of K as follows: $a \sim b$ if and only if there is a connected subset of K containing both a and b. Prove that the equivalence classes are connected and closed. The equivalence classes are called the **(closed) components** of K. There can be uncountably many (closed) components. In both parts (a) and (b) the components are the maximal connected subsets.

4.3 Prove the following version of Weierstrass's theorem. Suppose $\{U_n\}$ is an increasing sequence of regions and suppose f_n is defined and analytic on U_n. If the sequence f_n converges uniformly on compact subsets of $U \equiv \cup_n U_n$ to a function f, then f is analytic on U and the sequence f_n' converges to f' uniformly on compact subsets of U (even though perhaps none of the f_n is defined on all of U).

4.4 (a) Use Cauchy's estimate (4.5) to prove Liouville's theorem.

(b) Use Cauchy's estimate (4.5) to compute a lower bound on the radius of convergence of the power series representation of a bounded holomorphic function.

4.5 At the end of the proof of Runge's Theorem 4.23, we stated that an inequality remains true for all z in a disk containing z_0 and for all refinements of the partition. Supply the details. Hint: If $I = [a,b] \subset \Gamma$ is a line segment, show that we can choose a partition

$I = \cup I_j = \cup_1^n [\zeta_{j-1}, \zeta_j]$ with $\zeta_0 = a$ and $\zeta_n = b$ such that

$$\left| \frac{f(\zeta)}{\zeta - z_0} - \frac{f(\zeta_j)}{\zeta_j - z_0} \right| < \varepsilon, \tag{4.9}$$

for all $\zeta \in [\zeta_{j-1}, \zeta_j]$, $j = 1, \ldots, n$. Then (4.9) holds for any refinement of $\cup I_j$ by the triangle inequality, if ε is replaced by 2ε. Moreover, if δ is sufficiently small then (4.9) holds for all z with $|z - z_0| < \delta$, replacing 2ε with 4ε, because $z_0 \notin \Gamma$. Now use

$$\left| \int_I \frac{f(\zeta)}{\zeta - z} d\zeta - \sum_j \frac{f(\zeta_j)}{\zeta_j - z}(\zeta_j - \zeta_{j-1}) \right| = \left| \sum \int_{I_j} \left(\frac{f(\zeta)}{\zeta - z} - \frac{f(\zeta_j)}{\zeta_j - z} \right) d\zeta \right|.$$

B

4.6 Show that Theorem 4.21 and Corollary 4.22 hold for bounded convex sets S with piecewise continuously differentiable boundary, ∂S. A set S is convex if, for all $z, w \in S$ and $0 < t < 1$, we have $tz + (1 - t)w \in S$. Hint: To prove the analog of Proposition 4.20, use Corollary 4.18 if there is an appropriate disk B or show that you can replace the integral around ∂S with an integral around the boundary of a small square S_0 centered at a by writing $\partial S - \partial S_0$ as the sum of four curves chosen so that Corollary 4.18 applies to each.

4.7 Suppose Ω is a region which is symmetric about $\mathbb{R}$. Set $\Omega^+ = \Omega \cap \mathbb{H}$ and $\Omega^- = \Omega \cap (\mathbb{C} \setminus \overline{\mathbb{H}})$. If f is analytic on Ω^+, continuous on $\Omega^+ \cup (\Omega \cap \mathbb{R})$ and $\text{Im} f = 0$ on $\Omega \cap \mathbb{R}$, then the function defined by

$$F(z) = \begin{cases} f(z), & \text{for } z \in \Omega \setminus \Omega^- \\ \overline{f(\overline{z})}, & \text{for } z \in \Omega^- \end{cases}$$

is analytic on Ω. Hint: Divide a small rectangle with sides parallel to the axes and intersecting $\mathbb{R}$ into three rectangles. This is the analytic version of the Schwarz reflection principle, which we will encounter in Section 8.3.

4.8 (a) For $x > 0$, define $x^{-z} = e^{-z\ln(x)}$. Prove that the **Riemann zeta function**

$$\zeta(z) = \sum_{n=1}^{\infty} n^{-z}$$

converges and is analytic in $\{z : \text{Re}z > 1\}$.

(b) Show that, for $\text{Re}z > 1$,

$$\zeta(z) - \int_1^{\infty} x^{-z} dx = \sum_{n=1}^{\infty} \int_n^{n+1} \int_n^x z t^{-z-1} dt dx.$$

(c) Use (b) to prove that $(z - 1)\zeta(z)$ has a unique analytic extension to $\{z : \text{Re}z > 0\}$.

(d) Use the fundamental theorem of calculus to give a series for $\zeta(z) - 1/(z - 1)$, valid in $\text{Re}z > 0$, which does not involve an integral.

Probably the most famous problem in all of mathematics is to prove that if $\zeta(z) = 0$ and $\text{Re}z > 0$, then $\text{Re}z = 1/2$.

4.9 Show that there is a constant $C < \infty$ so that if f is analytic on $\mathbb{D}$ then

$$|f'(z)| \leq C \int_{\mathbb{D}} |f(x+iy)| dxdy$$

for all $|z| \leq 1/2$.

4.10 Prove that there exists a sequence of polynomials p_k such that

$$\lim_{k \to \infty} p_k(z) = \begin{cases} 1, & \text{if } \mathrm{Re}(z) > 0 \\ 0, & \text{if } \mathrm{Re}(z) = 0 \\ -1, & \text{if } \mathrm{Re}(z) < 0. \end{cases}$$

4.11 (Universal entire function) Prove that there is an entire function f with the property that, if g is entire, and if K is compact and $\varepsilon > 0$, then there is a number $r > 0$, depending on g, K and ε, so that $\sup_K |g(z) - f(r+z)| < \varepsilon$.

C

4.12 Suppose f has a complex derivative at each point of a region Ω. Prove f is analytic in Ω. Outline (Goursat): Suppose $R \subset \Omega$ is a closed rectangle. Prove $\int_{\partial R} f(z)dz = 0$ in the following way. First show that the conclusion holds for linear functions $Az + B$. Set $C = |\int_{\partial R} f(z)dz|$. Divide R into four equal rectangles. Then at least one of them, call it R_1, satisfies $|\int_{\partial R_1} f(z)dz| \geq C/4$. Repeat this idea to obtain a nested sequence of rectangles R_n with $|\int_{\partial R_n} f(z)dz| \geq C4^{-n}$ and perimeter $2^{-n}|\partial R|$. Set $b = \cap R_n$. Since f has a derivative at b, we have for $z \in R_n$, with n sufficiently large,

$$|f(z) - [f(b) + f'(b)(z-b)]| < \varepsilon|z-b| \leq \varepsilon 2^{-n}|\partial R|. \tag{4.10}$$

Estimate the integral of the left-hand side of (4.10) on ∂R_n to obtain $C4^{-n} \leq 4^{-n}\varepsilon|\partial R|^2$. Finally, deduce that f is analytic. Exercise 4.13 gives an application.

4.13 Suppose f is a function defined on $\mathbb{D}$ with the property that, given any three points $a, b, c \in \mathbb{D}$, there is an analytic function g (possibly depending on a, b, c) so that $|g| \leq 1$ on $\mathbb{D}$ and $g(a) = f(a)$, $g(b) = f(b)$ and $g(c) = f(c)$. Then f is analytic on $\mathbb{D}$ and bounded by 1. Hint: Suppose $f(0) = 0$. Prove $f(z)/z$ is Cauchy as $z \to 0$. Use the maps T_c of the proof of Corollary 3.10 to deduce that f has a complex derivative at each point of $\mathbb{D}$, then apply Exercise 4.12.

5 Cauchy's Theorem

5.1 Cauchy's Theorem

Recall that if γ is a piecewise continuously differentiable closed curve then

$$\int_\gamma q(z)dz = 0,$$

for all polynomials q, and

$$\int_\gamma \frac{1}{(z-a)^k} dz = 0,$$

when $k \geq 2$ and $a \notin \gamma$, because the integrand in each case is the derivative of an analytic function. See Corollary 4.11. These also hold for all continuous curves γ by Theorem 4.32. If r is a rational function, then it has a partial fraction expansion

$$r(z) = \sum_{k=1}^{N} \sum_{j=1}^{n_k} \frac{c_{k,j}}{(z-p_k)^j} + q(z),$$

where q is a polynomial. So, if γ is a closed curve which does not intersect any of the poles of r, then

$$\int_\gamma r(\zeta)d\zeta = \sum_{k=1}^{N} c_{k,1} \int_\gamma \frac{1}{(\zeta - p_k)} d\zeta. \tag{5.1}$$

Theorem 5.1 (Cauchy) *Suppose γ is a cycle contained in a region Ω, and suppose*

$$\int_\gamma \frac{d\zeta}{\zeta - a} = 0 \tag{5.2}$$

for all $a \notin \Omega$. If f is analytic on Ω then

$$\int_\gamma f(\zeta)d\zeta = 0.$$

Corollary 4.18 is a local version of Cauchy's theorem for piecewise continuously differentiable curves.

Proof By Runge's Corollary 4.28, we can find a sequence of rational functions r_n with poles in $\mathbb{C} \setminus \Omega$ so that r_n converges to f uniformly on the compact set $\gamma \subset \Omega$. By Theorem 4.32,

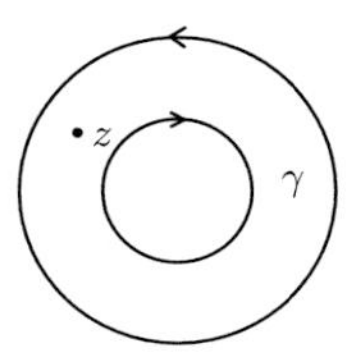

Figure 5.1 Integration on a cycle.

we may suppose γ is piecewise continuously differentiable and has finite length. By (5.1) and (5.2), $\int_\gamma r_n(z)dz = 0$. But then

$$\left|\int_\gamma f(z)dz\right| = \left|\int_\gamma (f(z) - r_n(z))dz\right| \leq \sup_\gamma |f - r_n|\ell(\gamma) \to 0. \qquad \square$$

We can use Cauchy's theorem to extend Theorems 4.14 and 4.21.

Theorem 5.2 (Cauchy's integral formula) *Suppose γ is a cycle contained in a region Ω, and suppose*

$$\int_\gamma \frac{d\zeta}{\zeta - a} = 0$$

for all $a \notin \Omega$. If f is analytic on Ω and $z \in \mathbb{C} \setminus \gamma$ then

$$\frac{1}{2\pi i}\int_\gamma \frac{f(\zeta)}{\zeta - z}d\zeta = f(z) \cdot \frac{1}{2\pi i}\int_\gamma \frac{1}{\zeta - z}d\zeta.$$

Proof For each $z \in \Omega$, the function $g(\zeta) = (f(\zeta) - f(z))/(\zeta - z)$ extends to be analytic on Ω, by Exercise 2.5(a). By Cauchy's theorem, it has integral over γ equal to 0. Theorem 5.2 follows by splitting the integral of g along γ into two pieces. $\square$

The cycle γ in Figure 5.1 consists of two circles, parameterized in opposite directions as indicated. If f is analytic on the closed region bounded by the two circles, then, by Proposition 4.13, the hypotheses in Cauchy's integral formula are satisfied. Moreover,

$$f(z) = \frac{1}{2\pi i}\int_\gamma \frac{f(\zeta)}{\zeta - z}d\zeta,$$

when z is between the two circles, again using Proposition 4.13 in addition to Theorem 5.2. When z is outside the larger circle or inside the inner circle, the integral is equal to 0 by Cauchy's theorem, because $f(\zeta)/(\zeta - z)$ is then an analytic function of ζ on the closure of the region between the two circles.

5.2 Winding Number

The important integrals (5.2) have a geometric interpretation which we will next explore.

Lemma 5.3 *If γ is a cycle and $a \notin \gamma$, then*

$$\frac{1}{2\pi i}\int_\gamma \frac{1}{\zeta - a}d\zeta$$

is an integer.

Proof Without loss of generality, we may suppose γ is a closed curve parameterized by a continuous, piecewise differentiable function $\gamma : [0,1] \to \mathbb{C}$. See the comments after the proof of Theorem 4.32. Define

$$h(x) = \int_0^x \frac{\gamma'(t)}{\gamma(t) - a}dt.$$

Then $h'(x)$ exists and equals $\gamma'(x)/(\gamma(x) - a)$, except at finitely many points x. Then

$$\begin{aligned}\frac{d}{dx}e^{-h(x)}(\gamma(x) - a) &= -h'(x)e^{-h(x)}(\gamma(x) - a) + e^{-h(x)}\gamma'(x)\\ &= -\gamma'(x)e^{-h(x)} + \gamma'(x)e^{-h(x)} = 0,\end{aligned}$$

except at finitely many points. Since $e^{-h(x)}(\gamma(x) - a)$ is continuous, it must be constant. Thus

$$e^{-h(1)}(\gamma(1) - a) = e^{-h(0)}(\gamma(0) - a) = 1 \cdot (\gamma(1) - a).$$

Since $\gamma(1) - a \neq 0$, $e^{-h(1)} = 1$ and $h(1) = 2\pi ki$, where k is an integer. Thus

$$\frac{1}{2\pi i}\int_\gamma \frac{d\zeta}{\zeta - a} = \frac{h(1)}{2\pi i} = k,$$

an integer. □

Definition 5.4 *If γ is a cycle, then the* **index** *or* **winding number** *of γ about a (or with respect to a) is*

$$n(\gamma, a) = \frac{1}{2\pi i}\int_\gamma \frac{d\zeta}{\zeta - a},$$

for $a \notin \gamma$.

Note the following properties of $n(\gamma, a)$:

(a) $n(\gamma, a)$ is an analytic function of a, for $a \notin \gamma$, by Lemma 4.30 and Theorem 4.32. In particular, it is continuous and integer valued by Lemma 5.3. Thus $n(\gamma, a)$ is constant in each component of $\mathbb{C} \setminus \gamma$.
(b) $n(\gamma, a) \to 0$ as $a \to \infty$. Thus $n(\gamma, a) = 0$ in the unbounded component of $\mathbb{C} \setminus \gamma$.
(c) $n(-\gamma, a) = -n(\gamma, a)$.
(d) $n(\gamma_1 + \gamma_2, a) = n(\gamma_1, a) + n(\gamma_2, a)$.
(e) If $\gamma(t) = e^{ikt}$, for $0 \leq t \leq 2\pi$, where k is an integer, then

$$n(\gamma, 0) = \frac{1}{2\pi i}\int_\gamma \frac{dz}{z} = \frac{1}{2\pi i}\int_0^{2\pi} \frac{ike^{ikt}}{e^{ikt}}dt = k.$$

We say that the curve γ in (e) **winds** k times around 0.

For example, in the proof of Runge's Theorem 4.23 we constructed a finite union of shaded squares $\{S_j\}$. Let $\gamma = \partial(\cup S_j)$. If a is in the interior of a square S_k then $n(\partial S_k, a) = 1$ and $n(\partial S_j, a) = 0$ for $j \neq k$, by Proposition 4.20. By (d), $n(\gamma, a) = 1$. Similarly, if $a \notin \overline{\cup S_j}$ then $n(\gamma, a) = 0$. If $a \in \partial S_k$, but $a \notin \gamma$ then, by (a), $n(\gamma, a) = 1$.

There are a couple of ways to compute winding numbers that may provide more intuition. Suppose $\gamma(t) = r(t)e^{i\theta(t)}$, $0 \leq t \leq 1$, where $r > 0$ and θ are continuous and piecewise continuously differentiable, with $\gamma(0) = \gamma(1)$. Then

$$\begin{aligned} n(\gamma, 0) = \text{Re}\left[\frac{1}{2\pi i}\int_\gamma \frac{dz}{z}\right] &= \frac{1}{2\pi}\text{Im}\int_0^1 \left(\frac{r'}{r} + i\theta'\right) dt \\ &= \frac{1}{2\pi}\left[\theta(1) - \theta(0)\right]. \end{aligned} \tag{5.3}$$

In this case the winding number of γ about 0 is equal to the total change in $\arg \gamma(t)$ divided by 2π. More generally, if $\gamma : [0, 1] \to \mathbb{C}$ is continuous, with $\gamma(0) = \gamma(1)$ and $0 \notin \gamma$, choose $\delta > 0$ so that if $|t-s| < \delta$ then $|\gamma(t)-\gamma(s)| < \text{dist}(\gamma, 0)$. Take a partition $0 = t_0 < t_1 < \cdots < t_n = 1$ with $|t_{j-1} - t_j| < \delta$. Then $\gamma([t_{j-1}, t_j])$ lies in a disk B_j which does not contain 0. Thus the line segment σ_j from $\gamma(t_{j-1})$ to $\gamma(t_j)$ also lies in B_j. As in Theorem 4.32 with $f(z) = 1/z$,

$$n(\gamma, 0) = n(\sigma, 0),$$

where $\sigma : [0, 1] \to \mathbb{C}$ is the polygonal curve $\sigma = \sigma_1 + \ldots + \sigma_n$. By (5.3)

$$n(\sigma, 0) = \frac{1}{2\pi}\left[\arg \sigma(1) - \arg \sigma(0)\right],$$

where $\arg \sigma(t)$ is defined so as to be continuous on $[0, 1]$. In fact, because $\gamma([t_{j-1}, t_j])$ is contained in a disk B_j which omits 0, and hence omits a ray from 0 to ∞, we can define $\arg \gamma(t)$ to be continuous on $[t_{j-1}, t_j]$, and hence $\arg \gamma(t)$ can be defined as a continuous function on $[0, 1]$. Then

$$n(\gamma, 0) = \frac{1}{2\pi}\left[\arg \gamma(1) - \arg \gamma(0)\right],$$

the total change in $\arg \gamma(t)$.

If $z_0 \in \mathbb{C}$ then $n(\gamma, z_0) = n(\gamma - z_0, 0)$. If we imagine standing at z_0, watching a point trace the closed curve γ then $n(\gamma, z_0)$ is the net number of revolutions we make, counting positive if counter-clockwise and negative if clockwise.

There is an even easier way to compute winding numbers in most cases. Let R_a be any ray, or half-line, from a to ∞. Suppose that γ is a cycle such that $a \notin \gamma$, and suppose that $\gamma \cap R_a$ consists of finitely many points z_j. Each connected component of $\gamma \setminus R_a$ contributes 1, 0 or

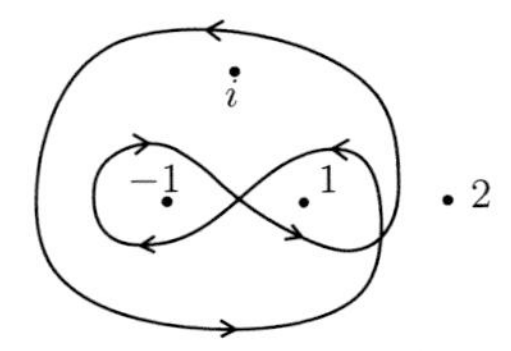

Figure 5.2 Calculating the winding number.

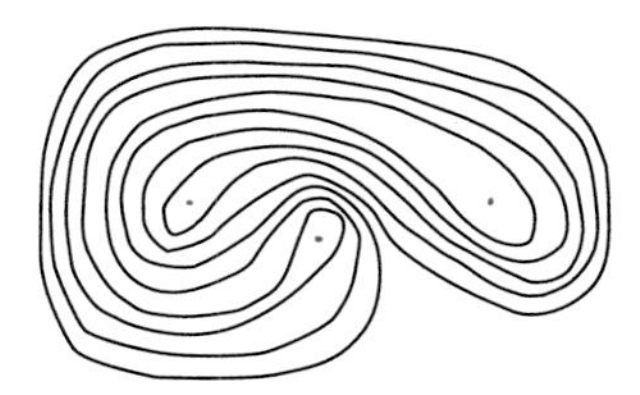

Figure 5.3 Parity: which red point is not in the same component as the other two?

-1 to the winding number of γ about a. Because γ is oriented, we can count the winding of γ about a by adding $+1$ if the crossing at z_j is counter-clockwise and -1 if the crossing is clockwise. See Exercise 5.1. If γ is the curve in Figure 5.2 then $n(\gamma, -1) = 0$, $n(\gamma, 1) = 2$, $n(\gamma, 2) = 0$ and $n(\gamma, i) = 1$.

A two-meter by four-meter version of Figure 5.3 was painted by W. Thurston and D. Sullivan on the wall of the Berkeley Mathematics department, where it remained for decades. By counting the number of crossings of three rays, it can be shown that one red point is not in the same component of the complement as either of the remaining two. Are the remaining two points in the same component? Be careful, how many components are there and why? See Section 12.1 for some badly behaved curves.

Definition 5.5 *Closed curves γ_1 and γ_2 are* **homologous** *in a region Ω if $n(\gamma_1 - \gamma_2, a) = 0$ for all $a \notin \Omega$, and in this case we write*

$$\gamma_1 \sim \gamma_2.$$

Homology is an equivalence relation on the curves in Ω (Exercise 5.2). A closed curve $\gamma \subset \Omega$ is said to be **homologous to 0 in** Ω if $n(\gamma, a) = 0$ for all $a \notin \Omega$. In this case we write $\gamma \sim 0$. Cauchy's theorem says that if $\gamma \sim 0$ in Ω, and if f is analytic in Ω, then

$$\int_\gamma f(z)dz = 0.$$

Thus, if γ_1 is homologous to γ_2 in Ω, then

$$\int_{\gamma_1} f(z)dz = \int_{\gamma_2} f(z)dz.$$

The most common application of Cauchy's integral formula is when $\gamma \subset \Omega$ with $\gamma \sim 0$ and $n(\gamma, z) = 1$. Then, for f analytic on Ω,

$$f(z) = \frac{1}{2\pi i}\int_\gamma \frac{f(\zeta)}{\zeta - z}d\zeta.$$

Figure 5.4 shows a closed curve σ contained in $\Omega = \mathbb{C} \setminus \{0, 1\}$ with the property that $n(\sigma, a) = 0$ for all $a \notin \Omega$. If you are familiar with homotopy (we will treat this subject in Section 14.1), this curve shows that homotopy and homology are different because σ cannot be shrunk to a point while remaining in Ω. See Exercise 14.5. By Cauchy's theorem, $\int_\sigma f(z)dz = 0$ for every analytic function f on Ω.

If Ω is a bounded region in $\mathbb{C}$ bounded by finitely many piecewise differentiable curves, then we can parameterize $\partial\Omega$ so that as we trace each boundary component, the region Ω

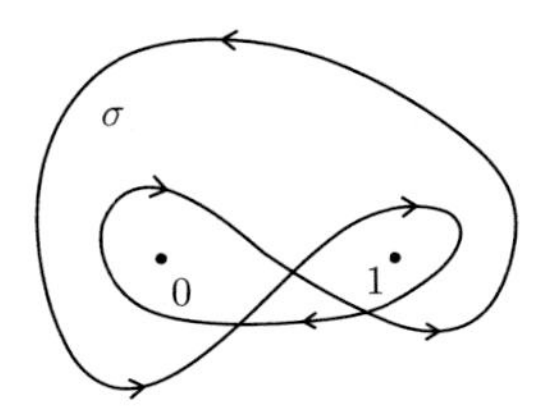

Figure 5.4 $n(\sigma, 0) = n(\sigma, 1) = 0$.

lies on the left. In other words, $i\gamma'(t)$, where it exists, is an *inner* normal, rotated counterclockwise by $\pi/2$ from the tangential direction $\gamma'(t)$. In Exercise 5.4 you are asked to show that $n(\partial\Omega, a) = 1$ for all $a \in \Omega$ and $n(\partial\Omega, a) = 0$ for all $a \notin \overline{\Omega}$. We call this the **positive orientation** of $\partial\Omega$. Thus, for all such regions, $\partial\Omega \sim 0$ in any region containing $\overline{\Omega}$. So, by Cauchy's theorem, if Ω is a bounded region, bounded by finitely many piecewise differentiable curves, and if f is analytic on $\overline{\Omega}$, then

$$f(z) = \int_{\partial\Omega} \frac{f(\zeta)}{\zeta - z} \frac{d\zeta}{2\pi i}, \tag{5.4}$$

for all $z \in \Omega$, where $\partial\Omega$ has positive orientation. A rigorous proof of the corresponding fact for regions bounded by finitely many simple closed curves is given in Corollary 12.16 and Exercise 12.10. If Ω is the unbounded region $\mathbb{C} \setminus \overline{\mathbb{D}}$, then $\partial\Omega$ is not homologous to zero in a region containing $\overline{\Omega}$ and (5.4) does not hold.

Definition 5.6 *A region $\Omega \subset \mathbb{C}^*$ is called* **simply-connected** *if $\mathbb{C}^* \setminus \Omega$ is connected in $\mathbb{C}^*$.*

Equivalently, a region Ω is simply-connected if $\mathbb{S}^2 \setminus \pi(\Omega)$ is connected, where π is stereographic projection and $\pi(\infty)$ is defined to be the "north pole" $(0, 0, 1)$ in $\mathbb{S}^2$. Simply-connected essentially means "no holes." For example: the unit disk $\mathbb{D}$ is simply-connected; the vertical strip $\{z : 0 < \text{Re} z < 1\}$ is simply-connected; the punctured plane $\mathbb{C} \setminus \{0\}$ is not simply-connected; the set $\mathbb{C} \setminus \overline{\mathbb{D}}$ together with ∞ is simply-connected, but $\mathbb{C} \setminus \overline{\mathbb{D}}$ is not simply-connected; the open set inside a "figure 8" curve is not simply-connected because it is not connected.

Theorem 5.7 *A region $\Omega \subset \mathbb{C}$ is simply-connected if and only if every cycle in Ω is homologous to 0 in Ω. If Ω is not simply-connected then we can find a simple closed polygonal curve contained in Ω which is not homologous to 0.*

The point of Theorem 5.7 is that simply-connected is a geometric condition which is sufficient for Cauchy's theorem to apply.

Proof Suppose that Ω is simply-connected, that γ is a cycle contained in Ω and that $a \notin \Omega$. Because $\Omega^c = \mathbb{C}^* \setminus \Omega$ is connected, it must be contained in one component of the complement of γ in $\mathbb{C}^*$. Because $\infty \in \Omega^c$, a must be in the unbounded component of $\mathbb{C} \setminus \gamma$, and $n(\gamma, a) = 0$ by property (b).

Conversely, suppose that $\mathbb{C}^* \setminus \Omega = A \cup B$, where A and B are non-empty closed sets in $\mathbb{C}^*$ with $A \cap B = \emptyset$. Without loss of generality, $\infty \in B$. Since A is closed, a neighborhood of ∞ does not intersect A, and hence A is bounded. Pick $a_0 \in A$. We will construct a curve $\gamma_0 \subset \Omega$ such that $n(\gamma_0, a_0) \neq 0$, proving Theorem 5.7.

The construction is the same construction used to prove Runge's Theorem 4.23. Let

$$d = \text{dist}(A, B) = \inf\{|a - b| : a \in A, b \in B\} > 0.$$

If $B = \{\infty\}$, set $d = 1$. Pave the plane with squares of side $d/2$ such that a_0 is the center of one of the squares. Orient the boundary of each square in the positive, or counter-clockwise, direction (like all storms in the northern hemisphere). Shade each square S_j with $\overline{S_j} \cap A \neq \emptyset$. Let γ_0 denote the cycle obtained from $\cup \partial S_j$ after performing all possible cancellations. Then $\gamma_0 \subset \Omega$ because γ_0 does not intersect either A or B, and $n(\gamma_0, a_0) = 1$. Because γ_0 is a finite union of closed simple polygonal curves, at least one of these curves is not homologous to 0. □

Corollary 5.8 *Suppose f is analytic on a simply-connected region Ω. Then*

(i) *$\int_\gamma f(z)dz = 0$ for all closed curves $\gamma \subset \Omega$;*
(ii) *there exists a function F analytic on Ω such that $F' = f$;*
(iii) *if also $f(z) \neq 0$ for all $z \in \Omega$ then there exists a function g analytic on Ω such that $f = e^g$.*

Proof Corollary 5.8(i) follows from Cauchy's theorem and Theorem 5.7. To prove (ii), we follow the proof of Morera's Theorem 4.19. Fix $z_0 \in \Omega$. We can then define $F(z) = \int_{\sigma_z} f(\zeta)d\zeta$, where σ_z is any curve contained in Ω connecting z_0 to z. This definition does not depend on the choice of σ_z because the integral along any closed curve is zero. If $B \subset \Omega$ is a disk, then for $z \in B$ we can write $F(z)$ as an integral from z_0 to the center of B plus an integral along a horizontal then vertical line segment from the center of B to z. By the proof of Morera's Theorem 4.19, F is analytic and $F' = f$ on B and hence on all of Ω.

To prove (iii), note that f'/f is analytic on Ω, so that, by (ii), there is a function g analytic on Ω such that $g' = f'/f$. To compare f and e^g, set

$$h = \frac{f}{e^g} = fe^{-g}.$$

Then

$$h' = f'e^{-g} - fg'e^{-g} = f'e^{-g} - f'e^{-g} = 0.$$

This implies h is a constant. Adding a constant to g, we may suppose $e^{g(z_0)} = f(z_0)$, so that $h \equiv 1$ and $f = e^g$. This proves Corollary 5.8. □

Note that if Ω is not simply-connected then, by Theorem 5.7, there is a point $a \notin \Omega$ and a polygonal curve $\gamma \subset \Omega$ such that

$$\int_\gamma \frac{1}{\zeta - a} d\zeta \neq 0. \tag{5.5}$$

Then $f(z) = 1/(z - a)$ is analytic in Ω but cannot be the derivative of an analytic function, for otherwise the integral in (5.5) would be 0 by the fundamental theorem of calculus. We also cannot find a function g analytic on Ω with $e^{g(z)} = 1/(z-a)$, for otherwise $1/(z-a) = -g'(z)$ is the derivative of an analytic function.

The solutions to Corollary 5.8(ii) and (iii) are essentially unique, for if $F' = f = G'$ then $(F-G)' = 0$ and hence $F - G$ is constant. Likewise, if $e^g = f = e^h$ then $e^{g-h} = 1$ so that $g - h = 2\pi k i$ for some integer k, by continuity and the connectedness of Ω.

Definition 5.9 *If g is analytic in a region Ω and if $f = e^g$ then g is called a* **logarithm** *of f in Ω and is written $g(z) = \log f(z)$. The function g is uniquely determined by its value at one point $z_0 \in \Omega$.*

By Corollary 5.8(iii), if f is analytic and non-vanishing in a simply-connected region Ω then f has countably many logarithms, which differ by integer multiples of $2\pi i$. So, to specify $\log f$ uniquely as an analytic function, we must include in the definition its value at one point $z_0 \in \Omega$. For example, if Ω is simply-connected and $0 \notin \Omega$ then $\log z$ can be defined as an analytic function on Ω. Given $z_0 \in \Omega$, there are countably many choices for $\log z_0$ and hence for the function $\log z$.

A word of caution: the conclusion (iii) of Corollary 5.8 is that there is an analytic function g such that

$$f(z) = e^{g(z)}.$$

We are not claiming that a logarithm can be defined on $f(\Omega)$, the range of f, and then composed with f. The function g is *locally* the composition of a function $\log z$ and f, but there might not be a function h defined on all of $f(\Omega)$ such that $z = e^{h(z)}$. For example, the rational function $(z-1)/(z+1)$ maps the disk $\mathbb{D}$ onto the left half-plane $\{\mathrm{Re} z < 0\}$ and so the function $f(z) = e^{(z-1)/(z+1)}$ maps $\mathbb{D}$ onto $\mathbb{D} \setminus \{0\}$. Thus f satisfies the hypotheses of Corollary 5.8(iii) and indeed $g(z) = \log f(z) = (z-1)/(z+1)$ works. However, we cannot define $\log z$ as an analytic function on $f(\mathbb{D}) = \mathbb{D} \setminus \{0\}$ since $\arg z = \mathrm{Im} \log z$ will increase by 2π as a circle is traced counter-clockwise about 0.

Throughout this book, log denotes the natural log, not base 10. Some books call $\log z$ a "multiple-valued function" in $\mathbb{C} \setminus \{0\}$, which is a bit of a contradiction in terms. We will only consider $\log z$ on domains where it can be defined as a function, or we will define $\log f$ as an analytic function, as in Corollary 5.8(iii).

5.3 Removable Singularities

The main result in this section is in some sense a generalization of Exercise 2.5.

Corollary 5.10 (Riemann's removable singularity theorem) *Suppose f is analytic in $\Omega = \{z : 0 < |z-a| < \delta\}$ and suppose*

$$\lim_{z \to a} (z-a) f(z) = 0.$$

Then f extends to be analytic in $\{z : |z-a| < \delta\}$.

In other words, we can "remove" the "singularity" of f at a.

Proof Fix $z \in \Omega$ and choose ε and r so that $0 < \varepsilon < |z-a| < r < \delta$. Let C_ε and C_r denote the circles of radius ε and r centered at a, oriented in the counter-clockwise direction. The cycle $C_r - C_\varepsilon$ (see Figure 5.1) is homologous to 0 in Ω, so that, by Cauchy's integral formula,

$$f(z) = \frac{1}{2\pi i}\int_{C_r} \frac{f(\zeta)}{\zeta - z} d\zeta - \frac{1}{2\pi i}\int_{C_\varepsilon} \frac{f(\zeta)}{\zeta - z} d\zeta.$$

But

$$\left| \int_{C_\varepsilon} \frac{f(\zeta)}{\zeta - z} d\zeta \right| \le \max_{\zeta \in C_\varepsilon} |f(\zeta)| \frac{1}{|z-a| - \varepsilon} 2\pi\varepsilon.$$

But if $\zeta \in C_\varepsilon$ then $|f(\zeta)|\varepsilon = |f(\zeta)||\zeta - a| \to 0$ as $\varepsilon \to 0$ and hence

$$f(z) = \frac{1}{2\pi i}\int_{C_r} \frac{f(\zeta)}{\zeta - z} d\zeta. \tag{5.6}$$

By Lemma 4.30, the right-hand side of (5.6) is analytic in $\{z : |z-a| < r\}$. Thus if we define $f(a)$ as the value of the right-hand side of (5.6) when $z = a$, then this extension is analytic at a and we have extended f to be analytic in $\{z : |z-a| < \delta\}$. □

The most important special case of Riemann's removable singularity theorem is as follows: if f is bounded and analytic in a punctured neighborhood of a, then f extends to be analytic in a neighborhood of a.

We say that a compact set E has **one-dimensional Hausdorff measure equal to** 0 if for every $\varepsilon > 0$ there are finitely many disks D_j with radius r_j so that

$$E \subset \cup_j D_j$$

and

$$\sum_j r_j < \varepsilon.$$

Corollary 5.11 (Painlevé) *Suppose $E \subset \mathbb{C}$ is a compact set with one-dimensional Hausdorff measure 0. If f is bounded and analytic on $U \setminus E$, where U is open and $E \subset U$, then f extends to be analytic on U.*

Proof As in the proof of Runge's Theorem 4.23 and Theorem 5.7, we can find a cycle $\gamma \subset U \setminus E$ which is the boundary of a finite union of closed squares $\{S_j\}$ so that $n(\gamma, a) = 0$ or 1 for all $a \notin \gamma$, $n(\gamma, b) = 1$ for all $b \in \cup S_j \setminus \gamma \supset E$, and $n(\gamma, b) = 0$ for $b \notin \cup S_j$ and hence for all $b \in \mathbb{C} \setminus U$. Cover E by finitely many disks D_j of radius r_j so that $\sum r_j < \varepsilon$. We may assume each D_j intersects E so that, for small ε, each D_j is contained in $\cup S_j \setminus \gamma$. Let $V = \{z : n(\gamma, z) = 1\}$, let $\sigma = \partial\left(\cup D_j\right)$, and let $\Omega = V \setminus \cup\overline{D_j}$. Then $\gamma + \sigma = \partial\Omega$, which we parameterize so that $\partial\Omega$ has positive orientation. See Exercise 5.4. Then, as in (5.4), $\gamma + \sigma \sim 0$ in $U \setminus \cup\overline{D_j}$, so that, by Cauchy's theorem,

$$f(z) = \frac{1}{2\pi i}\int_{\gamma} \frac{f(\zeta)}{\zeta - z} d\zeta + \frac{1}{2\pi i}\int_{\sigma} \frac{f(\zeta)}{\zeta - z} d\zeta,$$

for $z \in V \setminus \cup\overline{D_j}$. Fix $z \in V \setminus \overline{\cup D_j}$. Then the second integral tends to 0 as $\varepsilon \to 0$ because $\ell(\sigma) \le \ell(\cup D_j) < 2\pi\varepsilon$ and because f is bounded, exactly as in the proof of Riemann's theorem. Thus

$$\frac{1}{2\pi i}\int_\gamma \frac{f(\zeta)}{\zeta - z}d\zeta$$

provides an analytic extension of f to E, by Lemma 4.30. □

Painlevé asked for geometric conditions on a compact set E to be removable for bounded analytic functions in 1888. **Removable** means that the second sentence of Corollary 5.11 holds. A major accomplishment in complex analysis within the last fifteen years was an answer to this question.

We can use the Cauchy integral formula and Riemann's theorem to give a formula for the inverse of a one-to-one analytic function. Suppose f is analytic and one-to-one on a region Ω. Let γ be a cycle in Ω such that $n(\gamma, a) = 0$ for all $a \in \Omega^c$. Fix w and suppose $f(z) - w$ has exactly one zero in $\{z : n(\gamma, z) = 1\}$. Then $(f(z) - w)/(z - f^{-1}(w))$ is a difference quotient for $f'(f^{-1}(w))$, and so

$$\lim_{z\to f^{-1}(w)} \frac{z - f^{-1}(w)}{f(z) - w} f'(z)z = f^{-1}(w).$$

By Riemann's removable singularity theorem,

$$h(z) = \frac{f'(z)z}{f(z) - w} - \frac{f^{-1}(w)}{z - f^{-1}(w)}$$

extends to be analytic in Ω, and, by Cauchy's theorem,

$$\int_\gamma h(z)dz = 0.$$

Since $n(\gamma, f^{-1}(w)) = 1$, we obtain

$$f^{-1}(w) = \frac{1}{2\pi i}\int_\gamma \frac{f'(z)z}{f(z) - w}dz. \tag{5.7}$$

We proved that f^{-1} is analytic in Corollary 4.17, but we can also use (5.7) to show that f^{-1} is analytic, by imitating the proof of Lemma 4.30.

5.4 Laurent Series

An **annulus** is the region between two concentric circles. If f is analytic on the annulus $A = \{z : r < |z - a| < R\}$ then, by Runge's Theorem 4.27, we can approximate f by a rational function with poles only at a. The **Laurent series** is another version of this result, similar to a power series expansion.

Theorem 5.12 (Laurent series) *Suppose f is analytic on $A = \{z : r < |z - a| < R\}$. Then there is a unique sequence $\{a_n\} \subset \mathbb{C}$ so that*

$$f(z) = \sum_{n=-\infty}^{\infty} a_n(z - a)^n,$$

where the series converges uniformly and absolutely on compact subsets of A. Moreover,

$$a_n = \frac{1}{2\pi i} \int_{C_s} \frac{f(\zeta)}{(\zeta - a)^{n+1}} d\zeta, \tag{5.8}$$

where C_s is the circle centered at a with radius s, $r < s < R$, oriented counter-clockwise.

In the course of the proof of Theorem 5.12 we shall see that the integrals in (5.8) do not depend on the choice of s, so long as $r < s < R$.

Proof Without loss of generality, $a = 0$. Set

$$f_s(z) = \frac{1}{2\pi i} \int_{C_s} \frac{f(\zeta)}{\zeta - z} d\zeta,$$

where $C_s = \{se^{it} : 0 \leq t \leq 2\pi\}$. Then, by Lemma 4.30, f_s is analytic off C_s. If $r < |z| < s_1 < s_2 < R$, then $C_{s_2} - C_{s_1} \sim 0$ with respect to A and $n(C_{s_2} - C_{s_1}, z) = 0$. By Cauchy's integral formula, $f_{s_2}(z) - f_{s_1}(z) = 0$. This says that $f_s(z)$ does not depend on s, so long as $r < |z| < s < R$.

Expanding $\frac{1}{\zeta - z}$ in a power series expansion about 0, and interchanging the order of summation and integration, as we have done before, we conclude that f_s has a power series expansion

$$f_s(z) = \sum_{n=0}^{\infty} a_n z^n,$$

valid in $|z| < s$, where a_n satisfies (5.8). Likewise, $f_s(z)$ does not depend on s so long as $r < s < |z| < R$. Expanding $\frac{1}{\zeta - z}$ in a power series expansion about ∞, i.e., in powers of $1/z$, and interchanging the order of summation and integration, we conclude that f_s has a power series expansion

$$f_s(z) = -\sum_{n=1}^{\infty} a_{-n} z^{-n},$$

valid in $|z| > s$, where a_{-n} satisfies (5.8).

If $r < s_1 < |z| < s_2 < R$, then $C_{s_2} - C_{s_1} \sim 0$ in A, so that, by Cauchy's integral formula,

$$f(z) = \frac{1}{2\pi i} \int_{C_{s_2} - C_{s_1}} \frac{f(\zeta)}{\zeta - z} d\zeta = f_{s_2}(z) - f_{s_1}(z) = \sum_{n=-\infty}^{\infty} a_n z^n.$$

The uniqueness statement is Exercise 5.5. □

One consequence of Theorem 5.12 is that an analytic function f on A can be written as $f = f_1 + f_2$, where f_1 is analytic in $|z| < R$ and f_2 is analytic in $|z| > r$.

For example, the function

$$r(z) = \frac{z}{z^2 + 4} + \frac{2}{(z - 3)^2} - \frac{1}{z - 4}$$

is analytic in $\mathbb{C}$ except at $\pm 2i, 3, 4$ and hence is analytic on four regions centered at 0: $A_1 = \{z : |z| < 2\}$, $A_2 = \{z : 2 < |z| < 3\}$, $A_3 = \{z : 3 < |z| < 4\}$ and $A_4 = \{z : 4 < |z|\}$. Of course, A_1 is a disk, so that r has a power series expansion in A_1. We can use

$$\frac{1}{z-a} = \frac{1}{-a(1-\frac{z}{a})} = \sum_{n=0}^{\infty} \frac{-1}{a^{n+1}} z^n,$$

provided $|z| < |a|$, and

$$\frac{1}{z-a} = \frac{1}{z(1-\frac{a}{z})} = \sum_{n=1}^{\infty} \frac{a^{n-1}}{z^n},$$

provided $|z| > |a|$. These formulae give the Laurent expansion of $1/(z-4)$ in each region. Differentiating a series of this form with $a = 3$ will give the series for $1/(z-3)^2$. Setting $a = -4$ and replacing z with z^2 will give the series for $1/(z^2+4)$, and multiplying by z will give the series for $z/(z^2+4)$. If a rational function is given as a ratio of two polynomials, p/q, then use the techniques of Section 2.2 to find its partial fraction expansion, and then use the series expansion for $1/(z-a)$ as above. Sometimes it might be easier to find the partial fraction expansion for $1/q$, then multiply the resulting series by the polynomial p and collect together similar powers of z.

Laurent series are useful for analyzing the behavior of an analytic function near an isolated singularity. We say that f has an **isolated singularity** at b if f is analytic in $0 < |z-b| < \varepsilon$ for some $\varepsilon > 0$ and $f(b)$ is not defined. Write

$$f(z) = \sum_{n=-\infty}^{\infty} a_n(z-b)^n.$$

Note the following:

(a) If $a_n = 0$ for $n < 0$ then f extends to be analytic at b with $f(b) = a_0$. In this case we say that f has a **removable singularity** at b.

(b) If $a_n = 0$ for $n < n_0$ with $n_0 > 0$ and $a_{n_0} \neq 0$, then we can write

$$f(z) = (z-b)^{n_0} \sum_{n=0}^{\infty} a_{n_0+n}(z-b)^n = a_{n_0}(z-b)^{n_0} + a_{n_0+1}(z-b)^{n_0+1} + \cdots.$$

In this case b is called a **zero of order** n_0.

(c) If $a_n = 0$ for $n < -n_0$ with $n_0 > 0$ and $a_{-n_0} \neq 0$ then we can write

$$f(z) = (z-b)^{-n_0} \sum_{n=0}^{\infty} a_{-n_0+n}(z-b)^n = \frac{a_{-n_0}}{(z-b)^{n_0}} + \frac{a_{-n_0+1}}{(z-b)^{n_0-1}} + \cdots.$$

In this case b is called a **pole of order** n_0, and $|f(z)| \to \infty$ as $z \to b$.

In each of the above cases there is a unique integer k so that

$$\lim_{z \to b} (z-b)^k f(z)$$

exists and is non-zero, and

$$(z-b)^k f(z)$$

extends to be analytic and non-zero in a neighborhood of b.

(d) If $a_n \neq 0$ for infinitely many negative n, then b is called an **essential singularity**. For example,

$$f(z) = e^{-\frac{1}{z^2}} = \sum_{n=0}^{\infty} \frac{(-1)^n}{n!} z^{-2n}$$

has an essential singularity at 0, even though it is infinitely differentiable on the real line.

If f is analytic in $\{z : |z| > R\}$, then $f(1/z)$ has an isolated singularity at 0, and we say that f has an **isolated singularity at** ∞. We classify this singularity at ∞ as a zero, pole or essential singularity if $f(1/z)$ has a zero, pole or (respectively) essential singularity at 0. In terms of the Laurent expansion of f in $|z| > R$,

$$f(z) = \sum_{n=-\infty}^{\infty} b_n z^n,$$

f has an essential singularity at ∞ if $b_n \neq 0$ for infinitely many positive n. The reader can supply the corresponding statement for zeros and poles and their orders. For example, a polynomial of degree n has a pole of order n at ∞.

Definition 5.13 *A zero or pole is called* **simple** *if the order is* 1.

Definition 5.14 *If f is analytic in a region Ω except for isolated poles in Ω then we say that f is* **meromorphic in** Ω. *A meromorphic function in $\mathbb{C}$ is sometimes just called* **meromorphic**.

If $f \not\equiv 0$ is meromorphic in Ω, then $1/f$ is meromorphic in Ω, and a zero of order k becomes a pole of order k, and a pole of order k becomes a zero of order k. If f and g are meromorphic in Ω then the order of a zero or pole, for the meromorphic function f/g at b, can be found by factoring the appropriate power of $z - b$ out of f and g then taking the difference of these powers.

The next result gives an idea of the behavior near an essential singularity.

Theorem 5.15 *If f is analytic in $U = \{z : 0 < |z-b| < \delta\}$, and if b is an essential singularity for f, then $f(U)$ is dense in $\mathbb{C}$.*

In other words, every (punctured) neighborhood of an essential singularity has a dense image.

Proof If *$f(U)$ is not dense in* $\mathbb{C}$, there exists $A \in \mathbb{C}$ and $\varepsilon > 0$ so that $|f(z) - A| > \varepsilon$ for all $z \in U$. Then

$$\frac{1}{f(z) - A}$$

is analytic and bounded by $1/\varepsilon$ on U. By Riemann's theorem, $1/(f(z) - A)$ extends to be analytic in $U \cup \{b\}$. Thus $f(z) - A$ is meromorphic in $U \cup \{b\}$ and hence f is meromorphic in $U \cup \{b\}$. The Laurent expansion for f then has at most finitely terms with a negative power of $z - b$, contradicting the assumption that b is an essential singularity. □

5.5 The Argument Principle

The next result is useful for locating zeros and poles of meromorphic functions.

Theorem 5.16 (argument principle) *Suppose f is meromorphic in a region Ω with zeros $\{z_j\}$ and poles $\{p_k\}$. Suppose γ is a cycle with $\gamma \sim 0$ in Ω, and suppose $\{z_j\} \cap \gamma = \emptyset$ and $\{p_k\} \cap \gamma = \emptyset$. Then*

$$n(f(\gamma),0) = \frac{1}{2\pi i}\int_\gamma \frac{f'(z)}{f(z)}dz = \sum_j n(\gamma,z_j) - \sum_k n(\gamma,p_k). \tag{5.9}$$

In the statement of the argument principle, if f has a zero of order k at z, then z occurs k times in the list $\{z_j\}$, and a similar statement holds for the poles. If γ is a simple closed curve with $n(\gamma,z) = 0$ or $= 1$ for all $z \notin \gamma$, then we say that z is **enclosed** by γ if $n(\gamma,z) = 1$. A consequence of Theorem 5.16, is that if γ is a simple closed curve in Ω, with $n(\gamma,z) = 0$ or $= 1$ for all $z \notin \gamma$, and if γ is homologous to 0 in Ω, then the number of zeros "enclosed" by γ minus the number of poles "enclosed" by γ is equal to the winding number of the image curve $f(\gamma)$ about zero.

Proof The first equality in (5.9) follows from the change of variables $w = f(z)$. Note that $\gamma \sim 0$ and $\gamma \subset \Omega$ imply that $n(\gamma,a) = 0$ if a is sufficiently close to $\partial\Omega$. Thus $n(\gamma,z_j) \neq 0$ for only finitely many z_j and for only finitely many p_j because there are no cluster points of $\{z_j\}$ or $\{p_k\}$ in Ω. This implies that the sums in (5.9) are finite. Set $\Omega_1 = \Omega \setminus \{z_j : n(\gamma,z_j) = 0\} \cup \{p_k : n(\gamma,p_k) = 0\}$. Then $\gamma \sim 0$ in Ω_1.

If b is a zero or pole of f, we can write

$$f(z) = (z-b)^k g(z),$$

where g is analytic in a neighborhood of b and $g(b) \neq 0$. Then

$$f'(z) = k(z-b)^{k-1}g(z) + (z-b)^k g'(z)$$

and

$$\frac{f'(z)}{f(z)} = \frac{k}{z-b} + \frac{g'(z)}{g(z)}.$$

Since $g(b) \neq 0$, g'/g is analytic in a neighborhood of b and hence $f'/f - k(z-b)^{-1}$ is analytic near b. Thus

$$\frac{f'(z)}{f(z)} - \sum \frac{1}{z-z_j} + \sum \frac{1}{z-p_k} \tag{5.10}$$

is analytic in Ω_1. In the sums, we repeat z_j and p_k according to their multiplicity. By Cauchy's theorem, integrating (5.10) over γ gives (5.9). □

Corollary 5.17 (Rouché) *Suppose γ is a closed curve in a region Ω with $\gamma \sim 0$ in Ω and $n(\gamma,z) = 0$ or $= 1$ for all $z \in \Omega \setminus \gamma$. If f and g are analytic in Ω and satisfy*

$$|f(z) + g(z)| < |f(z)| + |g(z)| \tag{5.11}$$

for all $z \in \gamma$, then f and g have the same number of zeros enclosed by γ.

Equation (5.11) says that strict inequality holds in the triangle inequality. The number of zeros of f and g are counted according to their multiplicity.

Proof The function $\frac{f}{g}$ is meromorphic in Ω and satisfies

$$\left|\frac{f}{g} + 1\right| < \left|\frac{f}{g}\right| + 1 \tag{5.12}$$

on γ. By (5.11), $f \neq 0$ and $g \neq 0$ on γ, so that the hypotheses of the argument principle are satisfied.

The left-hand side of (5.12) is the distance from $w = f(z)/g(z)$ to -1. But $|w-(-1)| = |w| + 1$ if and only if $w \in [0, \infty)$. See Figure 5.5. Thus the assumption (5.11) implies that $\frac{f}{g}(\gamma)$ omits the half-line $[0, \infty)$ and so 0 is in the unbounded component of $\mathbb{C} \backslash \frac{f}{g}(\gamma)$. Hence $n(\frac{f}{g}(\gamma), 0) = 0$. By the argument principle, the number of zeros of $\frac{f}{g}$ equals the number of poles, and so the number of zeros of f equals the number of zeros of g, counting multiplicity. □

Example 5.18 $f(z) = z^9 - 2z^6 + z^2 - 8z - 2$.

How many zeros does f have in $|z| < 1$? The biggest term is $-8z$, so, comparing f and $-8z$, we have

$$|f(z) + 8z| = |z^9 - 2z^6 + z^2 - 2| \leq 1 + 2 + 1 + 2 = 6 < |8z|$$

on $|z| = 1$. By Rouché's theorem, f and $8z$ have the same number of zeros in $|z| < 1$, namely one. How many zeros does f have in $|z| < 2$? In this case we compare f with z^9:

$$|f(z) - z^9| = |-2z^6 + z^2 - 8z - 2| \leq 128 + 4 + 16 + 2 = 150 < |z^9| = 512.$$

Therefore f and z^9 have the same number of zeros in $|z| < 2$, namely 9. Thus 8 of the zeros of f lie in $1 < |z| < 2$, and the remaining zero lies in $|z| < 1$.

Example 5.19 $g(z) = z^4 - 4z + 5$

How many zeros does g have in $|z| < 1$? Compare g with the constant function 5 on $|z| = 1$:

$$|z^4 - 4z + 5 - 5| = |z^4 - 4z| \leq 5 \leq |5| + |z^4 - 4z + 5|.$$

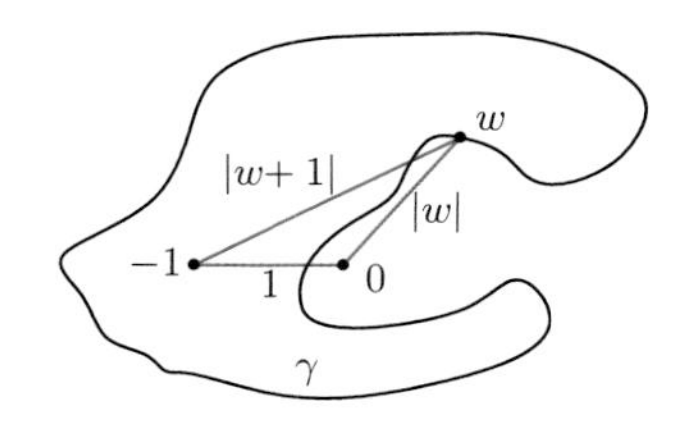

Figure 5.5 Proof of Rouché's theorem.

If equality holds in the first inequality, then $5 = |z^4 - 4z| = |z^3 - 4|$. But $|z| = 1$, so $z^3 = -1$, and $z^4 - 4z + 5 = -z - 4z + 5 = 5(-z + 1)$. Since $z \neq 1$, $|z^4 - 4z + 5| > 0$ and we have

$$|z^4 - 4z + 5 - 5| < |5| + |z^4 - 4z + 5|.$$

By Rouché's theorem, $z^4 - 4z + 5$ and 5 have the same number of zeros in $|z| < 1$, namely none.

A more elaborate process for locating zeros of polynomials is given in Exercise 5.13.

5.6 Exercises

A

5.1 Suppose γ is a closed curve and $a \notin \gamma$. Suppose R_a is a ray (half-line) from a to ∞ such that $\gamma \cap R_a$ consists of finitely many points. We say that the curve $\gamma(t)$ crosses R_a counter-clockwise (clockwise) at $\gamma(t_1) \in R_a$ if $\arg(\gamma(t) - a)$ is increasing (respectively, decreasing) as a function of t in a neighborhood of t_1. Prove that $n(\gamma, a)$ equals the number of counter-clockwise crossings of R_a minus the number of clockwise crossings. The set $\gamma \setminus R_a$ consists of finitely many pieces, but not all endpoints are necessarily "crossings."

5.2 Prove that homology is an equivalence relation on curves in a region Ω.

5.3 The proof of Corollary 5.8(iii) should remind you of the proof that the winding number is an integer. Explain this using the increase in the imaginary part of a continuous determination of $\log(z - a)$ along γ.

5.4 Suppose Ω is a bounded region whose boundary consists of finitely many disjoint piecewise differentiable simple closed curves. Suppose that $\partial\Omega$ is oriented so that the region lies on the left for each boundary component. This means that, except for finitely many t, $\gamma(t) + i\varepsilon(t)\gamma'(t) \in \Omega$ and $\gamma(t) - i\varepsilon(t)\gamma'(t) \notin \overline{\Omega}$ for sufficiently small $\varepsilon(t) > 0$. Prove $n(\partial\Omega, a) = 1$ if $a \in \Omega$ and $n(\partial\Omega, a) = 0$ if $a \in \mathbb{C} \setminus \overline{\Omega}$. Hint: If L is a short line segment from $a \in \Omega$ to $b \notin \overline{\Omega}$ that intersects γ in one point, use Exercise 5.1 to show $n(\gamma, a) = n(\gamma, b) + 1$. Then use the fact that $n(\partial\Omega, a)$ is constant in each component of $\mathbb{C} \setminus \partial\Omega$.

5.5 Prove the uniqueness of the Laurent series expansion; i.e., if

$$\sum_{n=-\infty}^{\infty} a_n z^n = \sum_{n=-\infty}^{\infty} b_n z^n$$

for $r < |z| < R$ then $a_n = b_n$ for all n. Convergence of the series on the region is part of the assumption. Hint: Liouville or multiply by z^k and integrate.

5.6 Note that in the proof of Laurent series expansions we proved that a function f which is analytic on $r < |z| < R$ can be written as $f = f_1 + f_2$, where f_1 is analytic in $|z| < R$, f_2 is analytic in $|z| > r$, and $f_2(z) \to 0$ as $|z| \to \infty$. Suppose that Ω is a bounded region in $\mathbb{C}$ such that $\partial\Omega$ is a finite union of disjoint (piecewise continuously differentiable) closed curves $\Gamma_j, j = 1, \ldots, n$. Suppose that f is analytic on $\overline{\Omega}$. Prove that $f = \sum f_j$, where f_j is analytic on the component of $\mathbb{C} \setminus \Gamma_j$ which contains Ω.

B

5.7 Prove that if a sequence of analytic polynomials converges uniformly on a region Ω then the sequence converges uniformly on a simply-connected region containing Ω. Hint: See Exercise 4.2(b).

5.8 We define the singularity at ∞ of $f(z)$ to be the singularity at 0 of $g(z) = f(1/z)$. Find the singularity at ∞ of the following functions. If the singularity is removable, give the value; if the singularity is a zero or pole, give the order.
(a) $\frac{z^2+1}{e^z}$; (b) $\frac{1}{e^{1/z}-1} - z$; (c) $e^{z/(1-z)}$; (d) $ze^{1/z}$; (e) $z^2 - z$; (f) $\frac{1}{z^3}e^{\frac{1}{z}}$.

5.9 Find the expansion in powers of z for

$$\frac{z}{(z^2+4)(z-3)^2(z-4)}$$

which converges in $3 < |z| < 4$.

5.10 Prove that the number of roots of the equation

$$z^{2n} + \alpha z^{2n-1} + \beta^2 = 0,$$

where α, β are real and non-zero and n is a natural number, which have positive real part is equal to n if n is even. If n is odd, the number of roots is $n - 1$ for $\alpha > 0$ and $n + 1$ for $\alpha < 0$. Hint: See what happens to $z^{2n} + \alpha z^{2n-1} + \beta^2$ as z traces the boundary of a large half-disk.

5.11 (a) How many zeros does $z^4 + z^3 + 5z^2 + 2z + 4$ have in the first quadrant? Hint: See where the image of the boundary of a large circular sector crosses $\mathbb{R}$.
(b) How many zeros does $p(z) = 3z^5 + 21z^4 + 5z^3 + 6z + 7$ have in $\overline{\mathbb{D}}$? How many zeros in $\{z : 1 < |z| < 2\}$?

5.12 Prove that all of the zeros of the polynomial

$$p(z) = z^n + c_{n-1}z^{n-1} + \ldots + c_1 z + c_0$$

lie in the disk centered at 0 with radius

$$R = \sqrt{1 + |c_{n-1}|^2 + \ldots + |c_1|^2 + |c_0|^2}.$$

C

5.13 This exercise gives an algorithm for counting the number of zeros of a polynomial in a disk. First translate and dilate the disk to the unit disk. Suppose $p(z) = a_n z^n + a_{n-1}z^{n-1} + \ldots + a_0$ is a polynomial with $a_n \neq 0$. Set

$$p^*(z) = z^n \overline{p\left(\frac{1}{\overline{z}}\right)} = \overline{a_0} z^n + \overline{a_1} z^{n-1} + \ldots + \overline{a_n}$$

and let $q(z) = \overline{a_0} p(z) - a_n p^*(z)$.
Case (a): if $|a_n| < |a_0|$, set $p_1 = q$.
Case (b): if $|a_n| > |a_0|$, set $p_1 = (a_0 p^* - \overline{a}_n p)/z$.
Case (c): if $|a_n| = |a_0|$ and $q \neq 0$ on $|z| = 1$, set $p_1 = q$.

Then the polynomial p_1 has smaller degree than p. Let $N(p)$ denote the number of zeros of p in $\mathbb{D} = \{z : |z| < 1\}$. Then prove

$$N(p) = \begin{cases} N(p_1), & \text{in cases (a) and (c)} \\ N(p_1) + 1, & \text{in case (b).} \end{cases}$$

Repeat the procedure until it stops. Try it for $p(z) = 2z^4 - 3z^3 + z^2 + z - 1$. If none of the cases applies, one possibility is to work with $p((1 - \varepsilon)z)$.

The following problem has been open since the 1930s and is of current interest in ergodic theory. Suppose p is a polynomial with integer coefficients

$$p(z) = a_n z^n + \ldots + a_0$$

with $|a_n| = |a_0| = 1$. Let $z_1, z_2, \ldots, z_k$ be the roots in $|z| < 1$. If p has at least one root in $|z| < 1$, how big can $\prod |z_j|$ be? (It is conjectured that

$$\prod_{p=1}^{k} |z_j| \leq 0.8501371\ldots$$

This value is achieved for the polynomial

$$p(z) = z^{10} + z^9 - z^7 - z^6 - z^5 - z^4 - z^3 + z + 1.$$

Eight zeros lie on the unit circle, one zero z_1 is outside the unit circle and the last is $1/z_1$, which is inside the unit disk.)

6 Elementary Maps

In this chapter we will study the mapping properties of the elementary functions and their compositions. The emphasis will be on the behavior of LFTs and the power, trigonometric and exponential functions related to familiar elementary functions of a real variable. These functions are all built from linear functions $a + bz$, $e^z = \sum_0^\infty z^n/n!$ and its locally defined inverse $\log z$ using algebraic operations and composition.

To facilitate our study, we will illustrate these functions using color pictures. Figure 6.1(b) shows a polar grid on the plane, where rays are colored using a standard color wheel in counter-clockwise order beginning along the negative reals: red, yellow, green, cyan, blue, magenta, red. Circles of radius $(1+\varepsilon)^n$, $n = -6, \ldots, 6$, are also plotted using a gray scale, increasing in darkness with the modulus except for the unit circle which is plotted in black with a thicker line width for emphasis. We will call this picture the **standard polar grid**. A "picture" of a complex-valued function f can be created by plotting points z using the same color that $f(z)$ has on the polar grid. For example, the Figure 6.1(a) shows the plot of a rational function. The rational function is a map from Figure 6.1(a) to the polar grid in Figure 6.1(b). Note that the colors near $z = 3$ cycle twice around in the same order as in the polar grid in Figure 6.1(b). This means that there is a zero of order 2 at $z = 3$. The colors near $2i$ and near $-2i$ cycle once in the opposite, or clockwise, order. This means that the function has poles of order one at $\pm 2i$. In fact, it is a picture of the function $(z-3)^2/(z^2+4)$. The preimage of the unit circle is black.

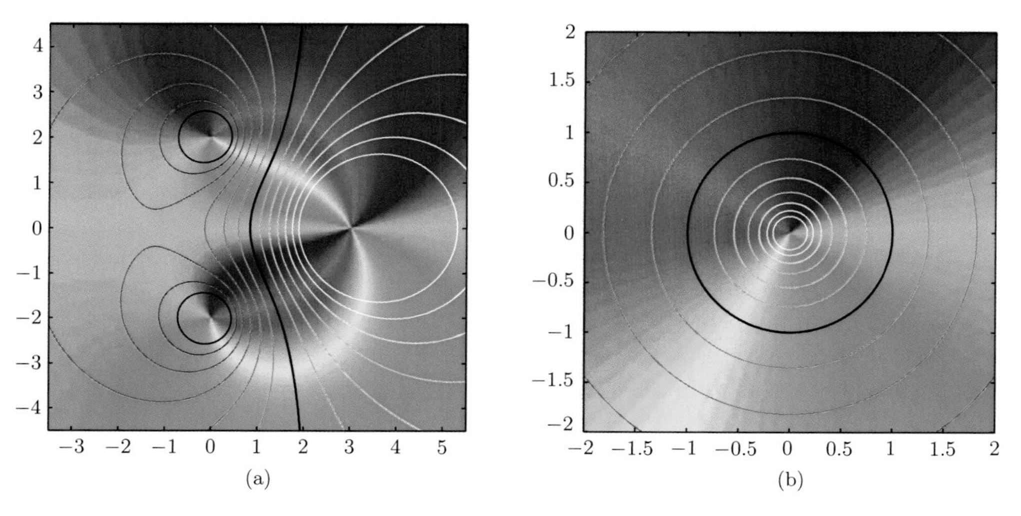

Figure 6.1 A rational function.

As we shall see, we can better understand some functions by plotting the color of z from the standard polar grid at the point $f(z)$, where z lies in a region on which f is one-to-one. This is of course a picture by the first method of the inverse function.

6.1 Linear Fractional Transformations

A **linear fractional transformation** (or **LFT**) is a non-constant rational function of the form

$$T(z) = \frac{az+b}{cz+d}, \tag{6.1}$$

where a, b, c, d are complex constants. An LFT is also called a **Möbius transformation**, though in some texts a Möbius transformation is an LFT that maps the unit disk onto itself. To avoid this confusion we will use the term LFT. The representation is not unique since the same function is obtained if a, b, c, d are replaced by ta, tb, tc, td with $t \neq 0$. Various normalizations are used, depending on the application. The four basic types of LFTs are:

$$\begin{array}{lll} \text{translation:} & T(z) = z + b, & \\ \text{rotation:} & T(z) = e^{i\theta}z, & \\ \text{dilation:} & T(z) = az, & \text{for } a > 0, \text{ and} \\ \text{inversion:} & T(z) = \frac{1}{z}. & \end{array}$$

The translation above shifts every point by the vector b. The rotation rotates the plane by an angle θ. The dilation expands (if $a > 1$) or contracts (if $a < 1$). The inversion is best understood by writing $z = re^{it}$. Then the argument of $1/z$ is $-t$ and the length of $1/z$ is the reciprocal of the length of z.

An LFT can be built out of these examples using composition. If $c = 0$ in (6.1) then

$$T(z) = \frac{a}{d}z + \frac{b}{d}. \tag{6.2}$$

In this case, T is a dilation by $\left|\frac{a}{d}\right|$, a rotation by $\arg\frac{a}{d}$, followed by a translation by $\frac{b}{d}$. If $c \neq 0$, then we can rewrite (6.1) as

$$T(z) = \frac{bc-ad}{c^2}\frac{1}{(z+\frac{d}{c})} + \frac{a}{c}. \tag{6.3}$$

In this case, T is a translation by $\frac{d}{c}$, an inversion, a dilation by $\left|\frac{bc-ad}{c^2}\right|$, a rotation by $\arg\frac{bc-ad}{c^2}$, followed by a translation by $\frac{a}{c}$.

Note that, by (6.2) and (6.3), T is non-constant if and only if $bc - ad \neq 0$.

Proposition 6.1 *The* LFTs *form a group under composition.*

Proof If S is an LFT, then, for constants $a, b, e^{i\theta}$, it is not hard to check that $S + b$, $e^{i\theta}S$, aS and $1/S$ are all LFTs, so, by (6.2) and (6.3), LFTs are closed under composition. The inverse of (6.1) is easily found to be

$$z = \frac{dw - b}{-cw + a}. \qquad \square$$

If T is given by (6.1) or (6.3) with $c \neq 0$, then T is meromorphic in $\mathbb{C}$ with a simple pole at $-\frac{d}{c}$. Moreover, T extends to be analytic at ∞ with $T(\infty) = \frac{a}{c}$. Thus T extends to be a one-to-one map of the extended plane $\mathbb{C}^*$ onto itself. The next theorem says that LFTs are the only such maps.

Theorem 6.2 *If f is analytic on $\mathbb{C} \setminus \{z_0\}$ and one-to-one then f is an* LFT.

Proof Using a translation, we may assume $z_0 = 0$. Then f has a Laurent series expansion

$$f(z) = \sum_{n=-\infty}^{\infty} a_n z^n.$$

If f has an essential singularity at 0, then, by Theorem 5.15, the image of every (punctured) neighborhood of 0 is dense in $\mathbb{C}$. In particular, if $B = \{z : |z - 1| < \frac{1}{2}\}$ then there is a $\zeta \notin B$ with $f(\zeta)$ in the open set $f(B)$. But then there is $z \in B$ with $f(z) = f(\zeta)$, contradicting the assumption that f is one-to-one. If f has a pole of order n at 0, then $1/f$ has a zero of order n at 0. Since $1/f$ is one-to-one, we must have $n \leq 1$ by Corollary 3.7. Applying the same argument to $f(1/z)$, we conclude that

$$f(z) = \frac{a}{z} + b + cz,$$

for some constants a, b, c. Set $w = f(z)$ then multiply by z. If $a \neq 0$ and $c \neq 0$ then, for each w,

$$cz^2 + (b - w)z + a = 0$$

has two roots in $\mathbb{C} \setminus \{0\}$, counting multiplicity, contradicting the assumption that f is one-to-one. Thus either $a = 0$ or $c = 0$, but not both since f is non-constant. In either case, f is an LFT. $\square$

In particular, Theorem 6.2 shows that if f is entire and one-to-one, then $f(z) = cz + b$, for some constants c and b. Thus the linear functions are precisely the analytic automorphisms of the plane. If f is an LFT, then f extends to be an automorphism of the extended plane $\mathbb{C}^*$, or, via stereographic projection, an automorphism of the Riemann sphere $\mathbb{S}^2$, and it is analytic in $\mathbb{C} \setminus \{z_0\}$, where z_0 is the simple pole. By Theorem 6.2, the LFTs are the only analytic automorphisms of the sphere in this sense. Exercise 3.8 characterized the analytic automorphisms of the disk $\mathbb{D}$. So we have now identified the analytic automorphisms of the disk, the plane and the sphere. As we will see in Section 15.4, subgroups of these automorphisms will allow us to transplant the study of analytic and meromorphic functions on any region, indeed on any Riemann surface, to the study of functions on the disk, plane or sphere.

The LFTs are characterized geometrically by the following theorem. A **"circle"** or **generalized circle** is a circle or a straight line. As we saw in Section 1.3, "circles" correspond precisely to the circles on the Riemann sphere. Lines lift to circles through the north pole. Similarly, a **"disk"** is a region (in $\mathbb{C}^*$) bounded by a "circle."

Theorem 6.3 LFTs *map "circles" onto "circles" and "disks" onto "disks."*

Proof We need only check this for the four basic types of LFTs. If $|z-c| = r$ then $|az-ac| = |a|r$ and $|(z-b)-(c-b)| = r$ so that rotations, dilations and translations map circles to circles. The equation of a straight line is given by

$$\text{Re}(c(z-b)) = 0,$$

since we can translate the line so it passes through 0 then rotate it to correspond to the imaginary axis. Rotations, dilations and translations map lines to lines exactly as in the case of circles. To check that inversions preserve "circles," suppose $|z-c| = r$ and set $w = 1/z$. Multiply out $|\frac{1}{w} - c|^2 = r^2$. If $r^2 = |c|^2$, then $\text{Re}(2cw) = 1$, the equation of a line. If $r^2 \neq |c|^2$, then by completing the square we obtain

$$\left| w + \frac{\overline{c}}{r^2 - |c|^2} \right|^2 = \left(\frac{r}{r^2 - |c|^2} \right)^2,$$

which is the equation of a circle. A similar reasoning for the image of a line is left to the reader to verify.

The equation of a disk is found by replacing the equals sign in the equation for a circle with $<$ or $>$, so that the proof of the statement for "disks" follows in a similar way. □

A useful restatement of Theorem 6.3 is that if T is an LFT and if D is a "disk" bounded by a "circle" C then $T(D)$ is a "disk" bounded by the "circle" $T(C)$.

The most common example is the one-to-one analytic map from the upper half-plane $\mathbb{H}$ onto the disk $\mathbb{D}$, sometimes called the Cayley transform,

$$C(z) = \frac{z-i}{z+i}.$$

See Figure 6.2. Note the order of the coloring around i and $-i$, the zero and pole of C. The absolute value $|C(z)|$ can be interpreted as the ratio of the distance from z to i over the distance from z to $-i$. Level curves (where C has constant modulus) and curves where C has constant argument are circles, as Theorem 6.3 guarantees. The real line is colored black. Indeed, the distance from $x \in \mathbb{R}$ to i is equal to the distance to $-i$, so that C maps $\mathbb{R} \cup \{\infty\}$ into $\partial\mathbb{D}$, and therefore by Theorem 6.3 onto $\partial\mathbb{D}$. Since $C(i) = 0$, the image of $\mathbb{H}$ must be $\mathbb{D}$ by Theorem 6.3. Another way to see that the image of $\mathbb{H}$ is $\mathbb{D}$ is to note that if $z \in \mathbb{H}$ then the distance from z to i is less than the distance from z to $-i$, so that $|C(z)| < 1$, and similarly $|C| > 1$ on the lower half-plane. Since C maps $\mathbb{C}^*$ onto $\mathbb{C}^*$, the image of $\mathbb{H}$ must be $\mathbb{D}$. The Cayley transform can be used, for example, to transform an integral on $\mathbb{R}$ to an integral on $\partial\mathbb{D}$ and vice versa.

Conformality can also be used to determine the image of an LFT. For example, the LFT given by $T(z) = (z-1)/(z+1)$ is real valued on $\mathbb{R}$ so it maps the unit circle to a "circle" which is orthogonal to $\mathbb{R}$ at the image of 1, namely 0, and passes through ∞, the image of -1. Thus it maps the unit circle to the imaginary axis, and, since $T(0) = -1$, it maps $\mathbb{D}$ onto the left half-plane $\{\text{Re}z < 0\}$.

The proof of the following theorem is useful for constructing LFTs.

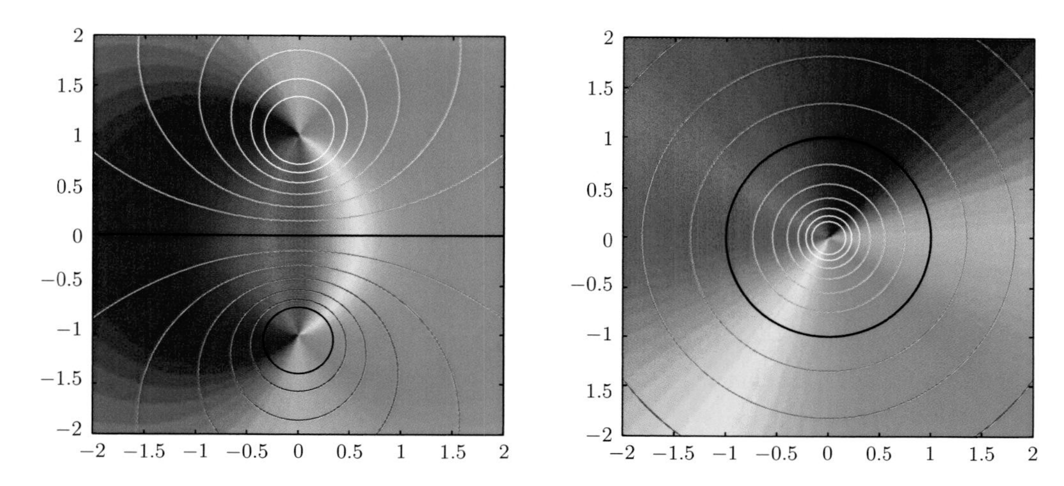

Figure 6.2 The Cayley transform.

Theorem 6.4 *Given z_1, z_2, z_3 distinct points in $\mathbb{C}^*$, and w_1, w_2, w_3 distinct points in $\mathbb{C}^*$, there is a unique* LFT, *T, such that*

$$T(z_i) = w_i, \tag{6.4}$$

for $i = 1, 2, 3$.

In the statement of Theorem 6.4, $\mathbb{C}^*$ is the extended plane, so that ∞ is included in the possibilities.

Proof First suppose that $w_1 = 0$, $w_2 = \infty$ and $w_3 = 1$. Set

$$T(z) = \left(\frac{z - z_1}{z - z_2}\right)\left(\frac{z_3 - z_2}{z_3 - z_1}\right).$$

Then $T(z_i) = w_i$, $i = 1, 2, 3$. For the general case, choose LFTs R and S so that $R(z_1) = S(w_1) = 0$, $R(z_2) = S(w_2) = \infty$ and $R(z_3) = S(w_3) = 1$, then

$$T = S^{-1} \circ R$$

satisfies (6.4) and is an LFT by Proposition 6.1. To prove that T is unique, suppose U is an LFT satisfying (6.4) then $V = S \circ U \circ R^{-1}$ is an LFT by Proposition 6.1 again, with $V(\infty) = \infty$, $V(0) = 0$ and $V(1) = 1$. It is easy then to check that $V(z) \equiv z$ and so $U = T$. □

One additional property of LFTs that is sometimes useful for determining their image is the following: if T is an LFT, and if z_1, z_2, z_3 are three points on the boundary of a "disk" D such that D lies to the left of ∂D as ∂D is traced from z_1 to z_2 to z_3, then $T(D)$ lies to the left of $T(\partial D)$ as it is traced from $T(z_1)$ to $T(z_2)$ to $T(z_3)$. Indeed, T is one-to-one and hence locally conformal, so that it preserves the angle, including direction, between the tangent and (inner) normal directions. For example, the unit disk $\mathbb{D}$ lies to the left of the unit circle as it is traced from 1 to i to -1. If $T(z) = (z - 1)/(z + 1)$ then $T(1) = 0$, $T(i) = i$ and $T(-1) = \infty$.

Thus the image of $\partial\mathbb{D}$ is the imaginary axis by Theorem 6.3, because the imaginary axis is the unique "circle" through $0, i, \infty$. The image $T(\mathbb{D})$ must lie to the left of the imaginary axis as it is traced from 0 to i to ∞, because angles are preserved, including direction. Thus $T(D)$ is the left half-plane $\{z : \text{Re}z < 0\}$.

See Penrose and Rindler [19] for the connection between Lorentz transformations and LFTs in the theory of relativity.

6.2 Exp and Log

In Exercise 2.6, we encountered the function

$$e^z = e^x e^{iy} = e^x(\cos y + i \sin y).$$

This function maps the horizontal line $y = c$ onto the ray $\arg z = c$ from 0 to ∞, and it maps each segment of length 2π in the vertical line $x = c$ onto the circle $|z| = e^c$. Figure 6.3(a) is mapped onto the standard polar grid by the map e^z.

By Exercise 2.7(a), $\frac{d}{dz}e^z = e^z$, which is non-zero, so e^z has a (local) inverse in a neighborhood of each point of $\mathbb{C} \setminus \{0\}$, called $\log z$. As we saw in Corollary 5.8(iii), $\log z$ can be defined as an analytic function on some regions which are not just small disks. For example, the function z is non-zero on the simply-connected region $\mathbb{C} \setminus (-\infty, 0]$. Then, $\log z$, with $\log 1 = 0$, is the function given by

$$\log z = \log |z| + i \arg z, \tag{6.5}$$

where $-\pi < \arg z < \pi$. Figure 6.3(b) is mapped onto the standard polar grid by this function. If instead we specified that $\log 1 = 2\pi i$, then (6.5) holds with $\pi < \arg z < 3\pi$. If $\Omega = \mathbb{C} \setminus (S \cup \{0\})$, where S is the spiral given in polar coordinates by $r = e^\theta$, $-\infty < \theta < \infty$, then Ω is simply-connected and $\text{Im} \log z$ is unbounded on Ω. In this case we can still specify, for example, $\log(-1) = \pi i$, and this uniquely determines the function $\log z$ on Ω.

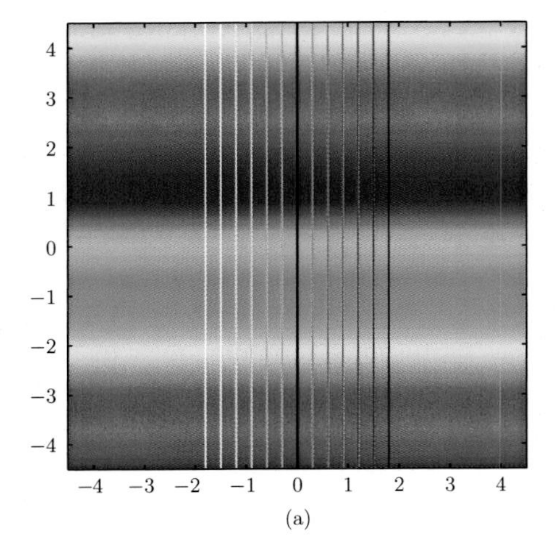

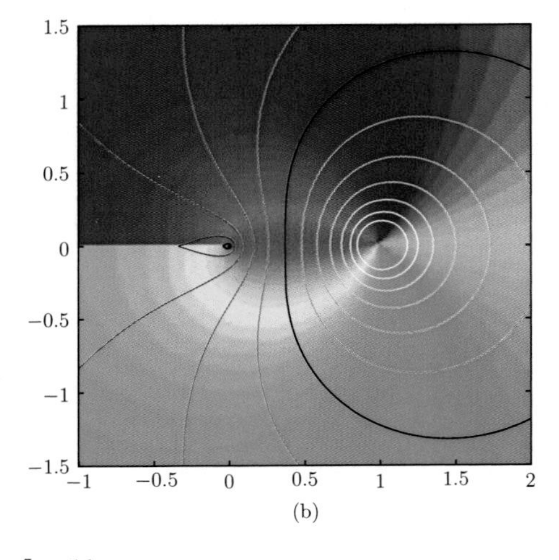

Figure 6.3 Maps e^z and $\log z$.

Figure 6.3(a) probably gives more insight into the log function than the inverse image of a polar grid in Figure 6.3(b). If Ω is any simply-connected region in the standard polar grid which omits the point $w = 0$, then $\log z$ maps it to a corresponding region in Figure 6.3(a). A horizontal strip of height 2π in Figure 6.3(a) is the image of a subset $\mathbb{C}\backslash R$ of the standard polar grid by the function $\log z$, where R is a ray from 0 to ∞. Any region contained in Figure 6.3(a) with the property that each vertical line intersects the region in a segment of length at most 2π is the image of a simply-connected region in the polar grid by the map $\log z$. The complement of the spiral $r = e^{\theta}$ in the polar grid is mapped by $\log z$ to the open strip lying between the lines $y = x$ and $y = x + 2\pi$ in Figure 6.3(a). A vertical shift by an integer multiple of 2π is also the image of the same region by $\log z$, but with a different choice of $\log(-1)$.

Note that, with the definition (6.5), $\log(zw) \neq \log(z) + \log(w)$ if $\arg(z) + \arg(w)$ is not in the interval $(-\pi, \pi)$; however, equality holds "modulo $2\pi i$."

6.3 Power Maps

If Ω is a simply-connected region not containing 0, and if $\alpha \in \mathbb{C}$, we define

$$z^{\alpha} = e^{\alpha \log z},$$

where $\log z$ can be specified by giving its value at one point $z_0 \in \Omega$. Then z^{α} is an analytic function on Ω.

For example, suppose $\Omega = \mathbb{C} \setminus (-\infty, 0]$, and define $\log 1 = 0$. If $z = re^{it}$, where $-\pi < t < \pi$, then $z^{1/4} = r^{1/4}e^{it/4}$. Figure 6.4(a) is the preimage of the standard polar grid, slit along $(-\infty, 0]$ by this map. The image of a sector of the form $\{z : |\arg z| < \beta\}$ by this map is the sector $\{z : |\arg z| < \beta/4\}$. Points z on the circle $|z| = r$ are mapped to points on the circle $|z| = r^{1/4}$. The map $z^{1/4}$ is locally conformal in Ω, but it is not conformal at 0. Indeed, angles are multiplied by $1/4$ at 0.

As with the logarithm, it might be easier to understand this map using Figure 6.4(b), which shows the image of the subset $\mathbb{C} \setminus (-\infty, 0]$ of the standard polar grid by this map. Note the range of colors and the location of the level lines. There are four possible definitions of $z^{1/4}$ on $\mathbb{C} \setminus (-\infty, 0]$, depending on the choice of $\log(1) = 2\pi ki$, where k is an integer. Each of the remaining three are rotations of Figure 6.4(b) by integer multiples of $2\pi/4$. If we put

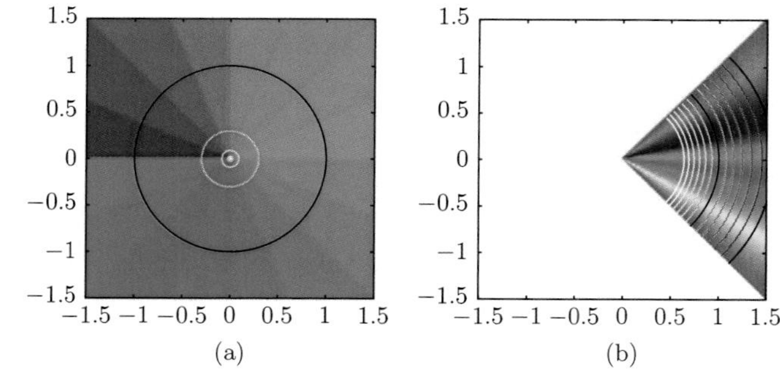

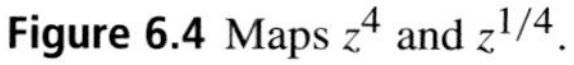
Figure 6.4 Maps z^4 and $z^{1/4}$.

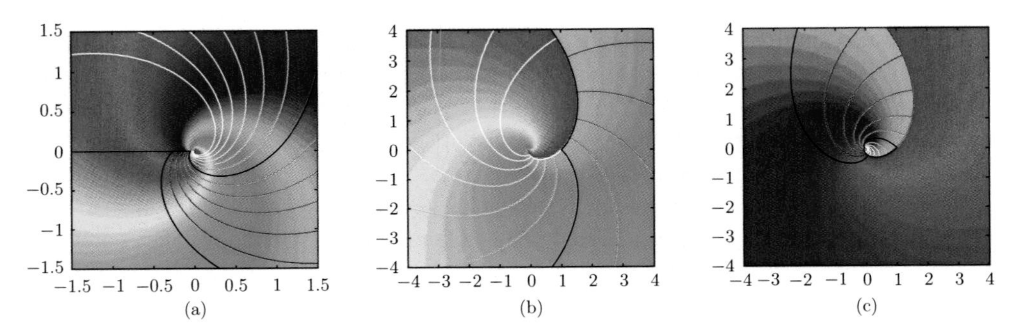

Figure 6.5 The map z^{1+i} and its inverse $z^{(1-i)/2}$.

the image of all four such functions together, we obtain Figure 6.4(c), which is of course the picture in the sense of Section 6.1 of the inverse function, z^4. Test your understanding by showing that there are exactly four possible definitions of $z^{1/4}$ in any simply-connected region not containing 0.

What does the image of $\mathbb{C} \setminus (-\infty, 0]$ by the map $z^{1/\pi}$, with $\log(1) = 0$, look like? There are in fact infinitely many definitions of $z^{1/\pi}$ on this region.

A more complicated function is

$$\varphi(z) = z^{1+i}$$

defined on $\Omega_0 = \mathbb{C} \setminus (-\infty, 0]$ with $\log(1) = 0$. Figure 6.5(a) illustrates this function by the method of Section 6.1, where we give z the color of $\varphi(z)$ in the standard polar grid.

But perhaps a better way to understand φ is to view it as a composition of three functions: $\log z$, $(1+i)z$ and e^z. The function $\log z$ is analytic on Ω_0 and has image equal to the horizontal strip $\Omega_1 = \{z : |\text{Im} z| < \pi\}$. The map $(1+i)z$ rotates Ω_1 counter-clockwise by an angle of $\pi/4$ and dilates by $\sqrt{2}$. The image of Ω_1 by this map is the strip

$$\Omega_2 = \{x + iy : x - 2\pi < y < x + 2\pi\}.$$

Each vertical line meets Ω_2 in a segment of length 4π. The function e^z maps a vertical segment of length 4π onto a circle, covered twice. The line $y = x$ is mapped onto a spiral S given in polar coordinates by $r = e^\theta$. The image of the parallel line $y = x + c$ is the rotation of S by the angle c. Thus e^z is an analytic two-to-one map of Ω_2 onto $\mathbb{C} \setminus \{0\}$, except that S is covered only once. So φ maps the rays $\arg z = \frac{c}{2}$ and $\arg z = \frac{c}{2} - \pi$ onto a rotation of S by the angle c, $0 < c < 2\pi$. Figures 6.5(b) and (c) illustrate this description of the map φ defined on the lower and upper (respectively) half-planes in the standard polar grid. Indeed, they are the illustrations by the method of Section 6.1 of the inverse functions $z^{\frac{1-i}{2}}$ on $\mathbb{C} \setminus (S \cup \{0\})$. Note how each picture is a continuation of the other across the spiral S.

Test your understanding by describing the image of circles centered at 0 by the map φ.

A related function is $\psi(z) = z^{1+i}$ with $\log(1) = 0$ defined on the region $\Omega_3 = \mathbb{C} \setminus (S \cup \{0\})$. From the description above, we can see that the image of Ω_3 by $\log z$ is the strip parallel to the line $y = x$ which meets each vertical line in a segment of length 2π, the middle half of Ω_2. The image of this region by the map $(1+i)z$ is a vertical strip Ω_4 of width 2π. The image of Ω_4 by the map e^z is the annulus $A = \{e^{-\pi} < |z| < e^\pi\}$, covered infinitely many times. Thus ψ maps $\mathbb{C} \setminus (S \cup \{0\})$ onto A, covering it infinitely many times.

This example highlights the need to define both the domain and the choice of $\log z$ to make sense of the function z^α. The function z^α has been used to explain a lithograph by M.C. Escher. See Exercise 6.9.

6.4 The Joukovski Map

The next function we will consider is

$$w(z) = \frac{1}{2}\left(z + \frac{1}{z}\right),$$

which is analytic in $\mathbb{C} \setminus \{0\}$. See Figure 6.6. By the quadratic formula,

$$z = w \pm \sqrt{w^2 - 1}, \tag{6.6}$$

so that w is two-to-one unless $w^2 = 1$. Since $w(z) = w(1/z)$, the two roots in (6.6) are reciprocals of each other, one inside $\mathbb{D}$ and one outside $\mathbb{D}$, or else complex conjugates of each other on $\partial\mathbb{D}$.

To understand the function w better, we view it as a one-to-one map on various subsets of $\mathbb{C}$. Fix $r > 0$ and write $z = re^{it}$, then

$$w = \frac{1}{2}\left(z + \frac{1}{z}\right) = \frac{1}{2}\left(r + \frac{1}{r}\right)\cos t + \frac{i}{2}\left(r - \frac{1}{r}\right)\sin t.$$

If we also write $w = u + iv$, then

$$\left(\frac{u}{\frac{1}{2}(r + \frac{1}{r})}\right)^2 + \left(\frac{v}{\frac{1}{2}(r - \frac{1}{r})}\right)^2 = 1,$$

which is the equation of an ellipse, unless $r = 1$. For each $r \neq 1$, the circles of radius r and $1/r$ are mapped onto the same ellipse. The circle of radius $r = 1$ is mapped onto the interval $[-1, 1]$. We leave as an exercise for the reader to show that the image of a ray from the origin

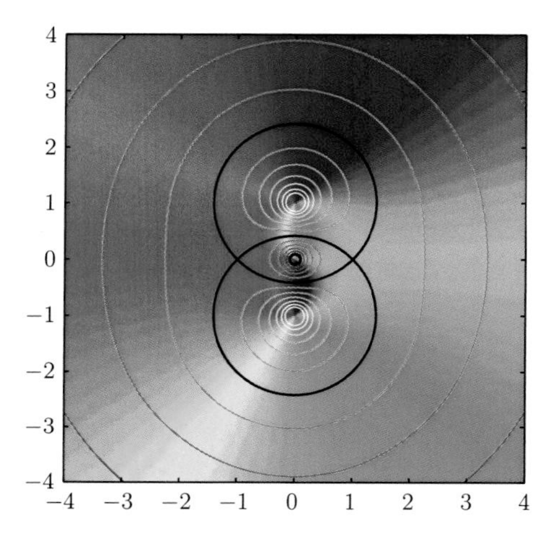

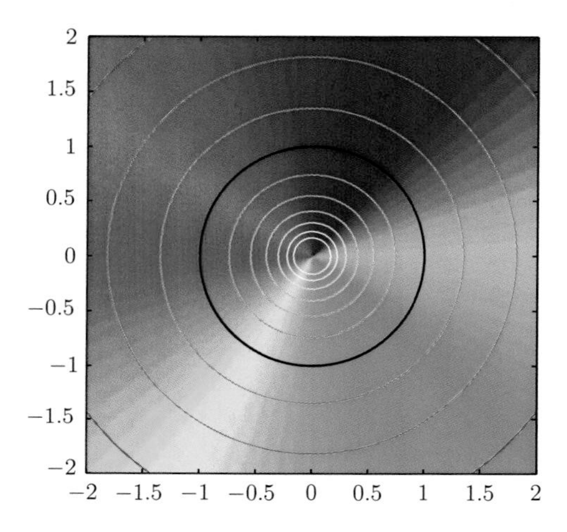

Figure 6.6 The map $(z + 1/z)/2$.

to ∞ which is not on a coordinate axis is a branch of a hyperbola which is perpendicular to each ellipse given above, by conformality.

The function w is a one-to-one analytic map of $\mathbb{C} \setminus \overline{\mathbb{D}}$ onto $\mathbb{C} \setminus [-1, 1]$, and a one-to-one analytic map of $\mathbb{D} \setminus \{0\}$ onto $\mathbb{C} \setminus [-1, 1]$. The function w is also analytic on $\mathbb{H} = \{z : \text{Im} z > 0\}$ and one-to-one, because $\text{Im}(1/z) < 0$ if $z \in \mathbb{H}$. The reader can check that the image of $\mathbb{H}$ by the map w is the region $\Omega_2 = \mathbb{C} \setminus \{(-\infty, -1] \cup [1, \infty)\}$. The inverse of w is given by (6.6) in each case, but we must make the correct choice for the square root on the image region. If U is any simply-connected region with $\pm 1 \notin U$ then we can define $\log(w^2 - 1)$ so as to be analytic in Ω by Corollary 5.8, and thus

$$w + \sqrt{w^2 - 1} = w + e^{\frac{1}{2}\log(w^2-1)}$$

is analytic and one-to-one on U with inverse function $\frac{1}{2}(z + 1/z)$.

For example, if $U = \mathbb{C}^* \setminus [-1, 1]$, write

$$\sqrt{w^2 - 1} = w\sqrt{1 - \frac{1}{w^2}}.$$

The function $1 - 1/w^2$ is analytic and non-zero on U. Define $\varphi(w) = \sqrt{1 - 1/w^2}$ so that $\varphi(\infty) = 1$. More concretely, the image of U by the map $1 - 1/w^2$ is contained in $\mathbb{C} \setminus (-\infty, 0]$ so if $\log \zeta = \log |\zeta| + i \arg \zeta$ with $-\pi < \arg \zeta < \pi$ then $\sqrt{1 - 1/w^2}$ is the composition of $\zeta = 1 - 1/w^2$ and $\exp(\frac{1}{2} \log \zeta)$, and therefore analytic. Then

$$\psi(w) = w\left(1 + \sqrt{1 - \frac{1}{w^2}}\right)$$

is an inverse to $(z + 1/z)/2$ defined on U. By the preceding discussion, there are two inverses to $(z + 1/z)/2$ on U. Since $\psi(w) \to \infty$ as $w \to \infty$, ψ must be the inverse to $(z + 1/z)/2$ with range $\mathbb{C} \setminus \overline{\mathbb{D}}$. See Figure 6.7(a). The inverse with range $\mathbb{D} \setminus \{0\}$ is

$$\psi_1(w) = w\left(1 - \sqrt{1 - \frac{1}{w^2}}\right),$$

which tends to 0 as $w \to \infty$. See Figure 6.7(b). Note that the pictures are not continuous across $(-1, 1)$, but are continuous across $(-\infty, -1) \cup (1, \infty)$. Readers are invited to test their understanding by finding the inverses defined on $\mathbb{C} \setminus \{(-\infty, -1] \cup [1, +\infty)\}$. See Figures 6.7(c) and (d), both of which are continuous across $(-1, 1)$ but not $(-\infty, -1) \cup (1, \infty)$. Note also the ellipses and hyperbolas where the functions have constant modulus and constant argument.

In Section 6.5 we will use the function $(z + 1/z)/2$ and the closely related function $(z - 1/z)/2$, which can be understood as a composition

$$\frac{1}{2}\left(z - \frac{1}{z}\right) = -i \cdot \frac{1}{2}\left((iz) + \frac{1}{(iz)}\right).$$

This function is the composition of rotation by $\pi/2$, followed by the map $\frac{1}{2}(z + 1/z)$, followed by rotation by $-\pi/2$. Thus, circles and lines through 0 are mapped to ellipses and orthogonal hyperbolas. The ellipses have semi-major axes along the imaginary axis. The unit circle is mapped to the interval $[-i, i]$.

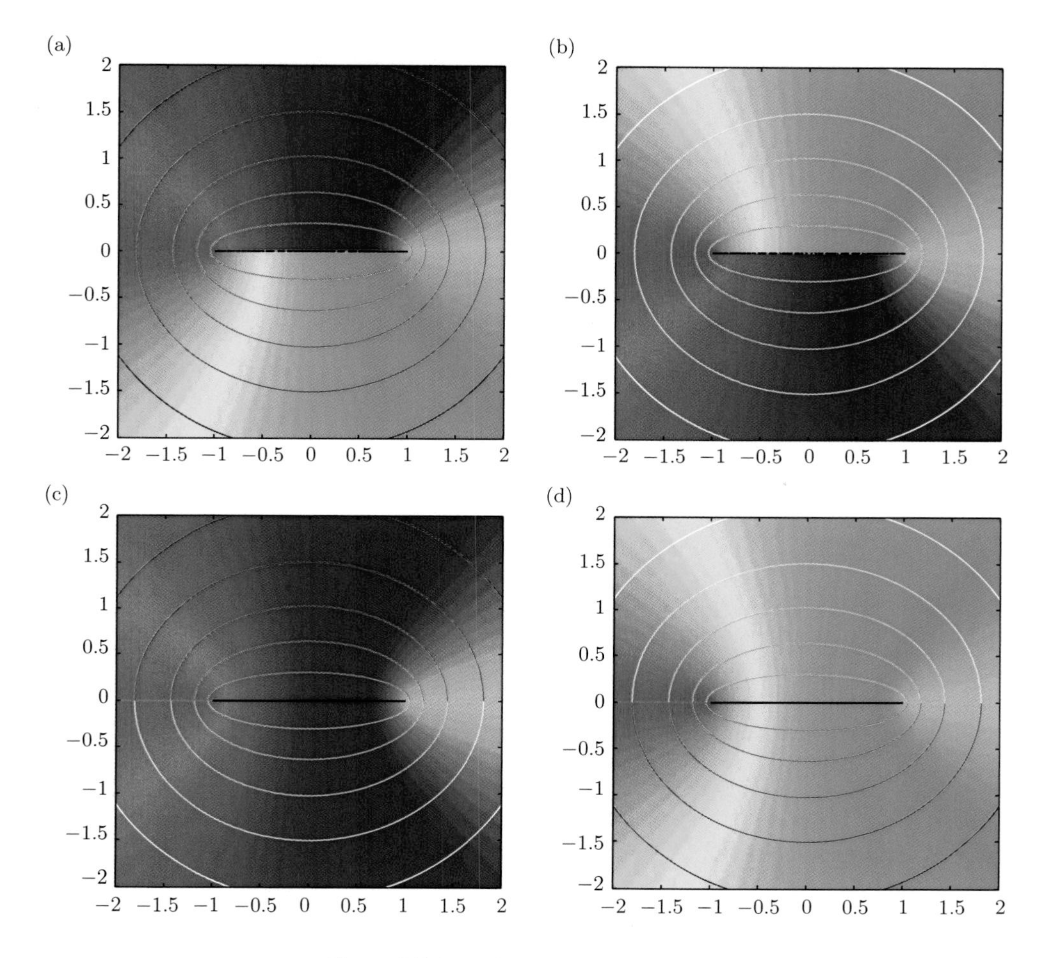

Figure 6.7 Some inverses of $(z + 1/z)/2$.

6.5 Trigonometric Functions

We define

$$\cos z = \frac{e^{iz} + e^{-iz}}{2} \text{ and } \sin z = \frac{e^{iz} - e^{-iz}}{2i}.$$

Then $\cos z$ and $\sin z$ are entire functions satisfying

$$\cos z + i \sin z = e^{iz},$$
$$(\cos z)^2 + (\sin z)^2 = 1,$$
$$\frac{d}{dz} \cos z = -\sin z \quad \text{and} \quad \frac{d}{dz} \sin z = \cos z.$$

These functions agree with their usual calculus definitions when z is real. However, we know by Liouville's theorem that they cannot be bounded in $\mathbb{C}$. The function $\cos z$ is best understood by viewing it as the composition of the maps iz, e^z and $\frac{1}{2}(z+1/z)$. See Figure 6.8. For example,

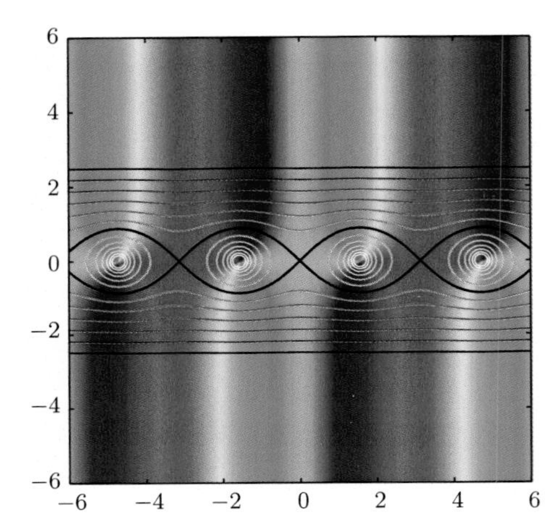

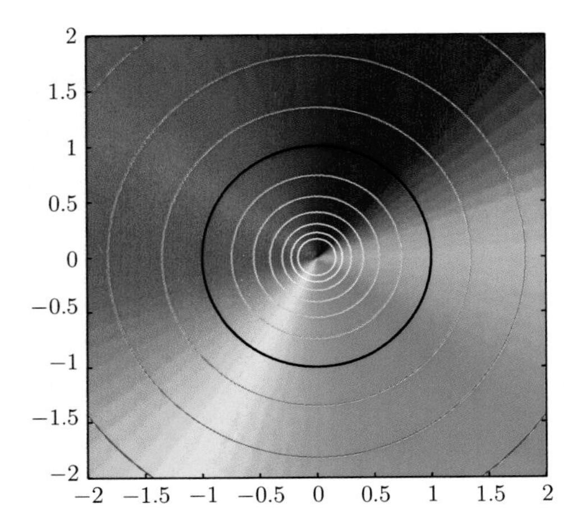

Figure 6.8 The map $\cos(z)$.

the vertical strip $\{z : |\mathrm{Re}z| < \pi\}$ is rotated to the horizontal strip $\{z : |\mathrm{Im}z| < \pi\}$ by the map iz. This horizontal strip is mapped onto $\mathbb{C} \setminus (-\infty, 0]$ by the map e^z. The composition e^{iz} maps vertical lines to rays from 0 to ∞ and maps horizontal lines to circles. Rays and circles are mapped by $\frac{1}{2}(z + 1/z)$ to branches of hyperbolas and ellipses, as we saw in Section 6.3.

Other trigonometric functions are defined using sin and cos, for example

$$\tan z = \frac{\sin z}{\cos z}.$$

Hyperbolic trigonometric functions are also defined using the exponential function:

$$\cosh z = \frac{e^z + e^{-z}}{2} \quad \text{and} \quad \sinh z = \frac{e^z - e^{-z}}{2}.$$

The inverse trigonometric functions can be found by working backward. For example, to find $\arccos z$, set $z = (e^{iw} + e^{-iw})/2$, multiply by e^{iw} and obtain a quadratic equation in e^{iw}, so that, by the quadratic formula,

$$e^{iw} = z \pm \sqrt{z^2 - 1}.$$

Thus

$$w = -i \log(z \pm \sqrt{z^2 - 1}).$$

If Ω is a simply-connected region such that $\pm 1 \notin \Omega$ then $f(z) = z + \sqrt{z^2 - 1}$ is an analytic function, as defined in Section 6.4. If $z + \sqrt{z^2 - 1} = 0$ then $z^2 = z^2 - 1$, which is impossible. Thus f is a non-vanishing function on Ω, and, by Corollary 5.8, we can define $F(z) = \log f(z)$ as an analytic function on Ω. It satisfies $\cos(F(z)) = z$, for $z \in \Omega$. Thus *cos*(z) has an analytic inverse function, $\arccos(z)$, on any simply-connected region which does not contain ± 1. The choice in the definitions of the square root and logarithmic functions are uniquely determined by specifying their values at a point. The second solution to the quadratic equation above, $z - \sqrt{z^2 - 1}$, is just a different choice for the analytic square root. To find the arccos with $\arccos(0) = \frac{\pi}{2}$, it may be better to write it in the form

$$\arccos(z) = w = -i\log(z + i\sqrt{1-z^2}),$$

with $\sqrt{1} = 1$ and $\log(i) = i\pi/2$.

6.6 Constructing Conformal Maps

In this section we will use the functions we have studied in this chapter to construct conformal maps. In modern usage, the phrase **conformal map** means a one-to-one analytic map. The entire function $f(z) = e^z$ is locally conformal everywhere, since its derivative is everywhere non-zero, but it is not a *conformal map* on $\mathbb{C}$ because it is not one-to-one. It is a conformal map on $\{z : |\text{Im}z| < \pi\}$, for example. The list below is not exhaustive, but rather is meant to illustrate the techniques that can be used at this point in the book.

$\mathbb{D}$ onto $\mathbb{D}$ with $f(z_0) = 0$ and $f'(z_0) > 0$ As we saw in Exercise 3.8, the conformal maps of $\mathbb{D}$ onto $\mathbb{D}$ are given by

$$f(z) = c\left(\frac{z-a}{1-\bar{a}z}\right),$$

where a and c are constants with $|a| < 1$ and $|c| = 1$. Set $a = z_0$ then $f(z_0) = 0$. Divide f by $z - z_0$ and let $z \to z_0$ to obtain $f'(z_0) = c/(1 - |z_0|^2)$. Setting $c = 1$ gives f.

Sector A conformal map of a sector $\Omega = \{z : a < \arg z < b\}$, with $0 < b - a \leq 2\pi$, onto $\mathbb{D}$ can be constructed in steps. The function

$$f(z) = z^\alpha = e^{\alpha \log z},$$

where $\alpha = \pi/(b-a)$, will map Ω onto a sector with angle at 0 equal to $\alpha(b-a) = \pi$, a half-plane. The choice of $\log z$ is already given in the description of Ω. A rotation $z \to e^{it}z$ will map the half-plane onto $\mathbb{H}$, and the Cayley transform $(z-i)/(z+i)$ will map $\mathbb{H}$ onto $\mathbb{D}$. It is usually sufficient to describe a conformal map as a composition of a sequence of simpler conformal maps.

Intersection or union of disks If Ω is the intersection of two disks, then, in order to map Ω onto $\mathbb{D}$, find the two points c, d where the bounding circles meet. The map

$$\frac{z-c}{z-d}$$

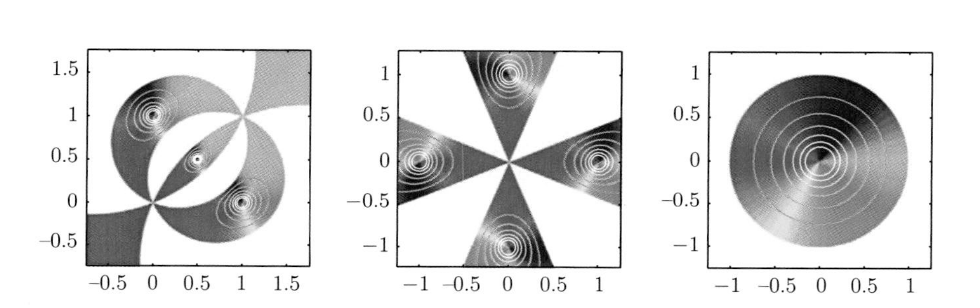

Figure 6.9 Four intersections of disks meeting at angle $\pi/4$.

will map each disk onto a "disk" with 0 and ∞ on its boundary, and hence the image of Ω is the intersection of two half-planes forming a sector at 0. Now apply the sector map constructed above. The same construction works for the union of two disks, whose interiors have a point in common. The region outside the union of two such disks is also the intersection of two "disks" in the extended plane if we add the point at ∞.

Figure 6.9 shows four regions, each of which has bounding circles meeting at 0 and $1+i$ at an angle of $\pi/4$. The point at ∞ is viewed as an interior point of one of the regions. The map $w = z/(z-(1+i))$ maps each region to a sector by Theorem 6.3. The map $\zeta = w^4$ will map each sector onto the right half-plane, and $T(z) = (z-1)/(z+1)$ maps the right half-plane onto $\mathbb{D}$. One way to see that the image of the right half-plane by the map T is $\mathbb{D}$ is to note that, for each point z on the imaginary axis, the distance from z to 1 equals the distance from z to -1, so that $|T(z)| = 1$. By Theorem 6.3, the right half-plane is mapped to a "disk" bounded by the unit circle. Since 1 is mapped to 0, the image must be the unit disk.

Half-plane with a slit The region $\Omega = \mathbb{H} \setminus I$, where I is the segment $[0, i]$ on the imaginary axis, can be mapped onto $\mathbb{H}$ by the map

$$\sqrt{z^2+1},$$

where $\sqrt{-1} = i$. Indeed, z^2 maps the half-line $\{re^{it} : r > 0\}$ onto the half-line $\{r^2e^{2it} : r > 0\}$, doubling the angle at 0. Thus the image of Ω is the slit plane $\mathbb{C} \setminus [-1, +\infty)$ and $z^2 + 1$ maps Ω onto $\mathbb{C} \setminus [0, +\infty)$. Similarly, the square root maps half-lines to half-lines, halving the angle at 0, so that $\sqrt{z^2+1}$ maps Ω onto $\mathbb{H}$. The choice of the square root is uniquely determined by the requirement that $\sqrt{-1} = i$. It can be given more explicitly as $\exp(\frac{1}{2}\log z)$, where $0 < \arg z = \operatorname{Im} \log z < 2\pi$. Note that the two "sides" of the slit $[0, i]$ correspond to the intervals $[-1, 0]$ and $[0, 1]$ in the closure of the upper half-plane. The interval $[0, +\infty)$ in the boundary of the slit plane corresponds to the interval $[1, +\infty)$ in the image region, the upper half-plane. In fact, the map $\sqrt{z^2+1}$ extends to be a one-to-one analytic map of $\mathbb{C} \setminus J$ onto $\mathbb{C} \setminus [-1, 1]$, where $J = i[-1, 1]$ is the union of I and its reflection about $\mathbb{R}$. See Exercise 6.4. We will use the inverse map given by $\sqrt{z^2-1}$ in Section 8.1.

Half-plane minus a circular arc Suppose $\Omega = \mathbb{H} \setminus A$, where A is an arc containing 0 lying on a circle C which is orthogonal to $\mathbb{R}$ at 0. Then Ω can be mapped to $\mathbb{H}$ by first applying the LFT $\sigma(z) = az/(1 - z/b)$, where b is the other point of intersection of C and $\mathbb{R}$ and $a > 0$. In fact, $b = |d|^2/\operatorname{Re} d$, where $d \in \mathbb{H}$ is the tip of the arc A. The map σ is real valued on $\mathbb{R}$ and has positive derivative at 0, so, by Theorem 6.3, σ maps $\mathbb{H}$ onto $\mathbb{H}$. Also by Theorem 6.3, the image of A must lie on a "circle" through $\sigma(0) = 0$ and ∞ which is perpendicular to $\mathbb{R}$ at 0, since LFTs are conformal. Thus the image of A is an interval $[0, ic]$ on the imaginary axis. We can choose $a > 0$ so that $c = 1$, and then the slit half-plane example applies. The circular slit is "opened up" to two intervals in $\mathbb{R}$ and the real line is mapped to the remaining portion of $\mathbb{R}$. See Figure 6.10.

Strip To map the strip $\{z : 0 < \operatorname{Re} z < 1\}$ onto $\mathbb{D}$, first apply the map $e^{\pi iz}$. The image of $\{\operatorname{Re} z = c\}$, $0 < c < 1$, is the ray $re^{i\pi c}$, $r > 0$, so the image of the strip is $\mathbb{H}$. Now apply the Cayley transform, $(z - i)/(z + i)$.

Half-strip To map the half-strip $\{z : 0 < \operatorname{Im} z < \pi,\ \operatorname{Re} z < 0\}$ onto $\mathbb{D}$, first apply the map e^z which has image $\mathbb{D} \cap \mathbb{H}$. Indeed, if $z = x + iy$ then $e^z = e^x e^{iy}$, so that the half-line

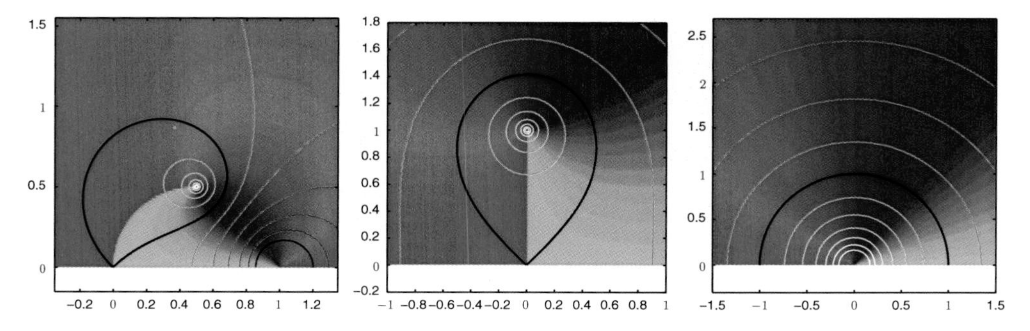

Figure 6.10 Mapping a circularly slit half-plane.

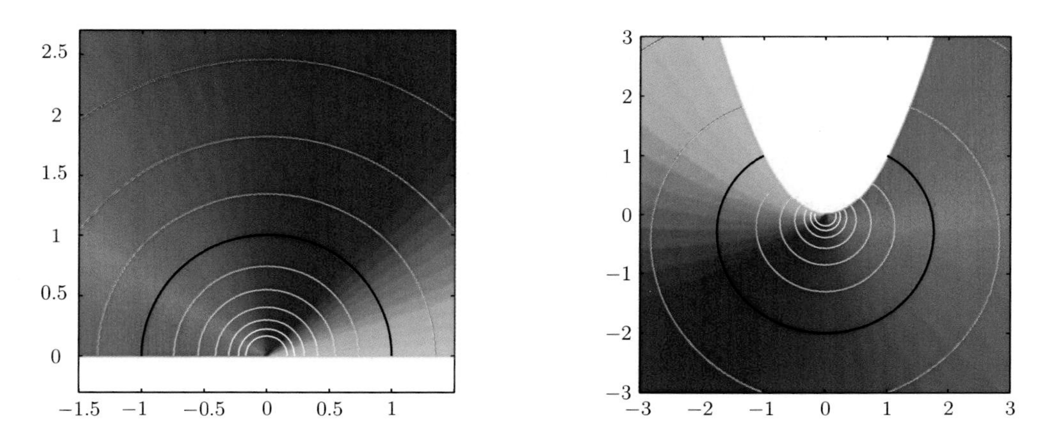

Figure 6.11 Below a parabola.

$\{x + iy : x < 0\}$ is mapped onto the portion of the ray $\arg z = y$ which lies in $|z| < 1$. Now apply the intersection of disks example, or use the map $\frac{1}{2}(z + 1/z)$ and a map of a half-plane to the disk.

Below a parabola To map $\mathbb{H}$ onto the region below the parabola $y = ax^2$ with $a > 0$, note that the image of a vertical line by the map z^2 is a parabola. To see this, fix $b > 0$. Write $u + iv = (b + iy)^2 = b^2 - y^2 + i2by$. Then $u = b^2 - (v/2b)^2$, which is the equation of a parabola that opens to the left. Thus, the image of the half-plane $\{z : \text{Re}z > b > 0\}$ by the one-to-one map z^2 is the region $P = \{u + iv : u > b^2 - (v/2b)^2\}$. To find the desired map, first apply $z \to -iz + b$, which maps $\mathbb{H}$ onto the half-plane $\text{Re}z > b$. Then apply z^2 to obtain the region P. Finally, apply the map $z \to -iz + ib^2$, which rotates the region clockwise by an angle of $\pi/2$ then shifts vertically by b^2 to obtain the region $\{x + iy : y < (x/2b)^2\}$. Choosing $b = 1/(2\sqrt{a})$ will give the desired map. See Figure 6.11.

Exterior of an ellipse To map $\mathbb{D}$ onto the exterior of an ellipse in $\mathbb{C}^*$, apply the map $z \to r/z$, $r > 1$, then $\frac{1}{2}(z + 1/z)$. See Section 6.4. In Figure 6.12, the domain, instead of the range, is given the standard (polar) coloring, so the inverse of the map is used to construct the picture.

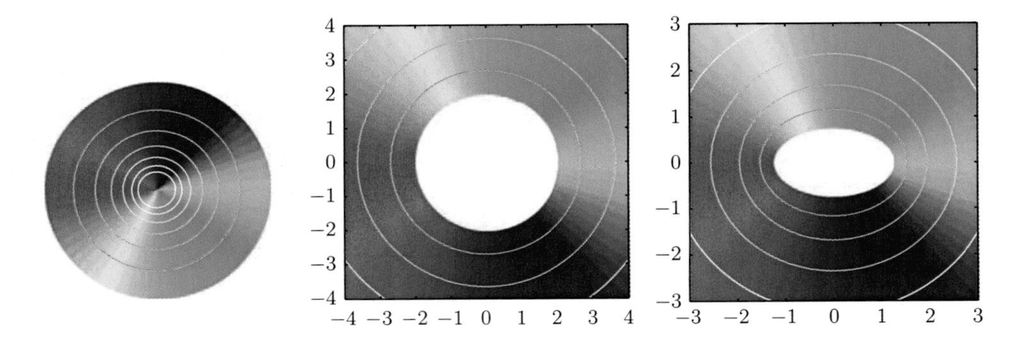

Figure 6.12 Exterior of an ellipse.

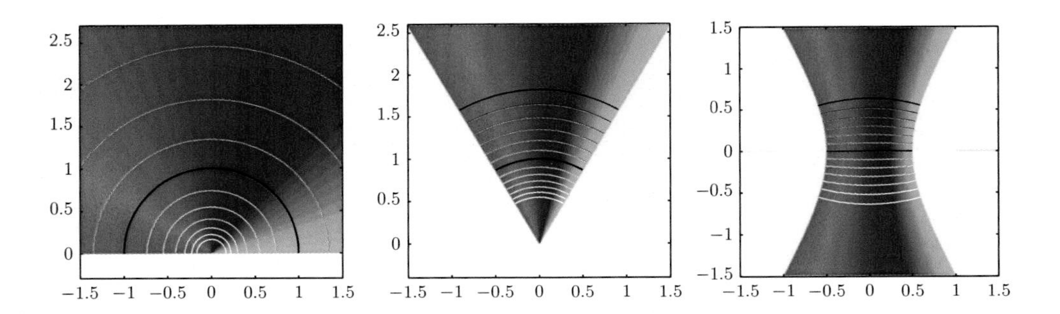

Figure 6.13 Region between the branches of a hyperbola.

Region between the branches of a hyperbola To map $\mathbb{H}$ onto the region between two branches of a hyperbola, first map $\mathbb{H}$ onto a sector symmetric about the y-axis using $e^{it}z^{\alpha}$ with $0 < \alpha < 1$ and $t = \pi(1 - \alpha)/2$. Then apply $\frac{1}{2}(z + 1/z)$. See Section 6.4. In Figure 6.13, as in Figure 6.12, we have used the standard coloring scheme on the domain, instead of the range.

See Exercise 6.11 for the construction of a conformal map to a region bounded by one branch of a hyperbola and Exercise 6.12 for the conformal map to a region above a parabola $y = ax^2$, $a > 0$. The only remaining regions bounded by a conic section are the interiors of ellipses. We will learn how to map the unit disk onto the interior of an ellipse in Chapter 8. See Exercise 8.8.

Many other examples can be constructed by using combinations of the above ideas. The conformal map in each of the examples above is a composition of a sequence of simpler conformal maps. The inverse map can be found by composing the inverses of the simpler functions in the reverse order. To find the conformal map of $\mathbb{D}$ onto a region Ω, it is usually easier to discover the map from Ω onto $\mathbb{D}$, then compute its inverse.

A natural question at this point is: how unique are these maps? A conformal map of Ω onto $\mathbb{D}$ can be composed with an LFT of the form given in the first example of this section and still map $\mathbb{D}$ onto $\mathbb{D}$.

Proposition 6.5 *If there exists a conformal map of a region Ω onto $\mathbb{D}$, then, given any $z_0 \in \Omega$, there exists a unique conformal map f of Ω onto $\mathbb{D}$ such that*

$$f(z_0) = 0 \text{ and } f'(z_0) > 0.$$

Proof If g is a conformal map of Ω onto $\mathbb{D}$, set $a = g(z_0)$. Then

$$h(z) = c\frac{z-a}{1-\bar{a}z}$$

maps $\mathbb{D}$ onto $\mathbb{D}$ if $|c| = 1$, and $f = h \circ g$ maps Ω onto $\mathbb{D}$ with $f(z_0) = 0$. Then

$$f'(z_0) = cg'(z_0)/(1 - |a|^2),$$

so that the proper choice of the argument of c will give $f'(z_0) > 0$. If k also maps Ω onto $\mathbb{D}$, with $k(z_0) = 0$ and $k'(z_0) > 0$, then $H = k \circ f^{-1}$ maps $\mathbb{D}$ onto $\mathbb{D}$ with $H(0) = 0$. By Schwarz's lemma, $|H(z)| \le |z|$ and $|H^{-1}(z)| \le |z|$ so that $|H(z)| = |z|$ and $H(z) = cz$, with $|c| = 1$. Since $H'(0) = k'(z_0)/f'(z_0) > 0$, we must have $c = 1$ and $k = f$. □

For example, to find a conformal map φ of $\mathbb{H}$ onto $\mathbb{D}$ such that $\varphi(z_0) = 0$ and $\varphi'(z_0) > 0$, first apply the Cayley transform, then apply an LFT of the form given in the first example of this section. Here it is actually easier to first apply the map

$$f(z) = \frac{z - z_0}{z - \overline{z_0}}$$

which maps $\mathbb{H}$ onto $\mathbb{D}$ because $|f(z)| = \text{dist}(z, z_0)/\text{dist}(z, \overline{z_0}) < 1$, when $z \in \mathbb{H}$, and is > 1 when $z \in \mathbb{C} \setminus \overline{\mathbb{H}}$ so that the image of $\mathbb{H}$ is $\mathbb{D}$ by Theorem 6.3. Then $f'(z_0) = \lim_{z \to z_0} f(z)/(z - z_0) = 1/(2i\text{Im}z_0)$. So $\varphi = if$ will work.

Another natural question is: what regions can be mapped conformally onto $\mathbb{D}$? Proposition 6.6 gives a necessary condition.

Proposition 6.6 *If φ is a conformal map of a region Ω onto $\mathbb{D}$, then Ω must be simply-connected.*

Proof Suppose γ is a closed curve contained in Ω, and suppose $a \in \mathbb{C} \setminus \Omega$. Let $f = \varphi^{-1}$, which is analytic on $\mathbb{D}$ by Corollary 4.17. Then $f'/(f - a)$ is analytic on $\mathbb{D}$ and

$$n(\gamma, a) = \frac{1}{2\pi i}\int_\gamma \frac{dw}{w-a} = \frac{1}{2\pi i}\int_{\varphi(\gamma)} \frac{f'(z)}{f(z)-a}dz = 0,$$

by Cauchy's Theorem 5.1 or Corollary 4.18. By Theorem 5.7, Ω is simply-connected. □

In Chapter 8 we will prove that every simply-connected region $\Omega \subset \mathbb{C}^*$, such that $\mathbb{C}^* \setminus \Omega$ contains at least two points, can be conformally mapped onto $\mathbb{D}$. Equivalently, any simply-connected region $\Omega \subset \mathbb{C}$ with $\Omega \ne \mathbb{C}$ can be conformally mapped onto $\mathbb{D}$.

We conclude this chapter by giving a picture of the "world's greatest function,"

$$I(z) = e^{\frac{z+1}{z-1}},$$

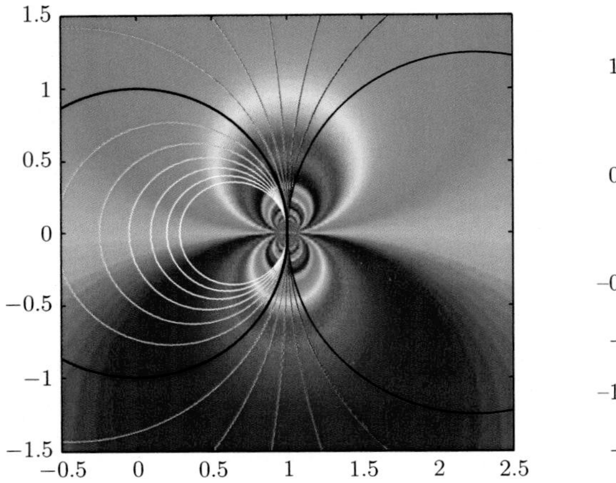

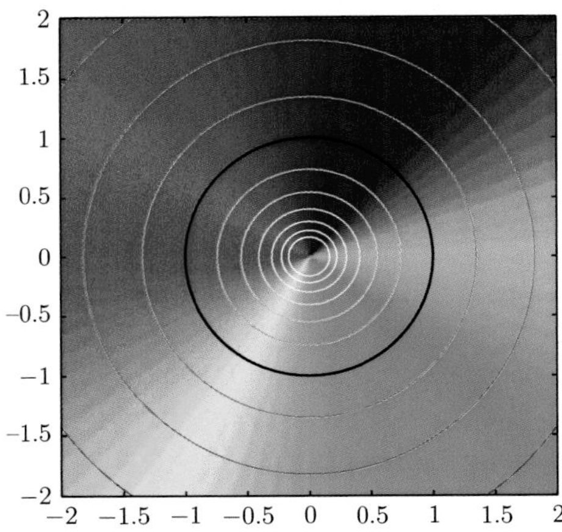

Figure 6.14 "World's greatest function," $\exp\,((z+1)/(z-1))$.

which we have encountered a couple of times so far. This function is a composition of an LFT which maps the unit disk onto the left half-plane followed by the exponential function. The preimage of the unit circle is black in Figure 6.14. Each level line is a circle through $z = 1$, where it has an essential singularity. The image of each level line covers the corresponding circle in the standard polar grid infinitely many times.

A Matlab program is given in Section A.2 of the Appendix for creating these color pictures. It also includes a 3D version, where the third coordinate is given by $\log |f|$. For more color pictures of analytic functions, see the beautiful text by E. Wegert [27].

6.7 Exercises

A

6.1 Provide details of the proof of the statement about "disks" in Theorem 6.3. Also provide the details for the image of a line by an inversion. Then check that the restatement of Theorem 6.3 given after the proof is correct.

6.2 The region $\Omega = \{z : \text{Im}z > 0 \text{ and } |z| > 1\}$ is the intersection of two disks. Map this region to a sector, then to $\mathbb{H}$. Find a conformal map of $\mathbb{H}$ onto $\mathbb{H}$ so that the composition of these maps fixes $1, -1$ and ∞. You should get $(z + 1/z)/2$. Why?

6.3 Find a conformal map f of $\Omega = \mathbb{C} \setminus (-\infty, 0]$ onto $\mathbb{D}$ such that $f(1) = 0$ and $f'(1) > 0$.

6.4 Prove that $\sqrt{z^2+1}$ extends to be a conformal map of $\mathbb{C} \setminus [-i, i]$ onto $\mathbb{C} \setminus [-1, 1]$. Hint: Consider the map on the lower and upper half-planes separately, but make a choice of the square roots so that the map extends continuously across $\mathbb{R} \setminus \{0\}$. An alternative approach is to define $\widetilde{J}(z) = -iJ(iz)$ on $|z| > 1$, where $J(z) = \frac{1}{2}(z + 1/z)$ is the Joukovski map. Then show $\sqrt{z^2+1} = J \circ \widetilde{J}^{-1}(z)$ and use the properties of J developed in Section 6.4.

6.5 Prove there is no one-to-one analytic mapping of $\{z : 0 < |z| < 1\}$ onto $\{z : 0 < |z| < \infty\}$.

6.6 Find the image of $\mathbb{H}$ by the map φ given by

$$\varphi(z) = z_0 + \frac{z_1 - z_0}{1 + z^2},$$

and show φ is one-to-one on $\mathbb{H}$ by writing it as a composition of the elementary maps z^2, $z + 1$, $1/z$, $z_0 + (z_1 - z_0)z$.

B

6.7 Construct a one-to-one analytic map f of the following regions Ω onto $\mathbb{D}$. If z_0 is given, find the map with $f(z_0) = 0$ and $f'(z_0) > 0$. You may leave your answer as an explicit sequence of maps, but you must show that each map does what you claim it does.

(a) $\Omega = \mathbb{D} \cap \{z : \text{Im}z > \frac{1}{2}\}$, $z_0 = \frac{3}{4}i$;
(b) $\Omega = \{z : |z| < 1\} \cap \{z : |z - 1| < 1\} \cap \{z : \text{Im}z > 0\}$;
(c) $\Omega = \mathbb{C}^* \setminus [0, 1]$, $z_0 = \infty$. Here $f'(\infty)$ means $\lim_{z\to\infty} z(f(z) - f(\infty))$.

6.8 Recall Exercise 3.9.

(a) Prove that if σ is a conformal map of $\mathbb{D}$ onto $\mathbb{D}$ then $\sigma((-1, 1))$ is a circular arc in $\mathbb{D}$ which is orthogonal to $\partial\mathbb{D}$. Here, as elsewhere, we view a diameter as part of a circle through ∞.
(b) Suppose that C is a circular arc in $\mathbb{D}$ which is orthogonal to $\partial\mathbb{D}$ at $a, b \in \partial\mathbb{D}$. Prove there is a conformal map of $\mathbb{D}$ onto $\mathbb{D}$ which maps the interval $(-1, 1)$ onto C so that the image of 0 is the closest point on C to 0. No computations are needed to solve (b).
(c) Show that the balls $\{z : \rho(z, z_0) < r\}$ in the pseudohyperbolic metric are Euclidean balls, though the center and radius in the two metrics are different.
(d) Prove that the hyperbolic metric on $\mathbb{D}$ is given by

$$\delta(z, w) = \inf \int_{\gamma_{z,w}} \frac{|d\zeta|}{1 - |\zeta|^2},$$

where the infimum is taken over all curves $\gamma_{z,w}$ in $\mathbb{D}$ connecting z and w. Hint: First prove it for $w = 0$, then use Exercise 3.9.
(e) Show that the **geodesic**, or the shortest curve, in the hyperbolic metric on $\mathbb{D}$ between z and w is the arc between z and w on the unique circle orthogonal to $\partial\mathbb{D}$ through z and w.
(f) In this non-Euclidean hyperbolic geometry, the "points" are the points in $\mathbb{D}$ and the "lines" are the circular arcs orthogonal to $\partial\mathbb{D}$. Two lines are parallel if they do not intersect. Prove that, in this geometry, Euclid's parallel axiom fails: through any point p not on a line L there are infinitely many lines which do not intersect L.

6.9 Draw a picture in the square $\{z = x + iy : 0 < x < \pi, 0 < y < \pi\}$. Reflect the picture about the line $y = 0$ and then reflect the union about $x = 0$. The picture can be extended to the plane by repeated reflections. Now apply the map $e^{\frac{1+i}{2}z}$ and describe the image. Use the Matlab program in Section A.2 of the Appendix to visualize this map and its inverse. A similar idea was used by M.C. Escher in one of his lithographs. See de Smit *et al.* [7].

6.10 (Rain = complement of the Seattle Umbrella) Map $\mathbb{C} \setminus \overline{\mathbb{D}}$ onto

$$\Omega = \mathbb{C}^* \setminus (\{z : |z| \leq 1 \text{ and } \mathrm{Im} z \geq 0\} \cup \{it : -1 \leq t \leq 0\})$$

so that ∞ is mapped to ∞. Extra credit: put a handle of your choice on the umbrella.

6.11 To map a half-plane onto a region bounded by one branch of a hyperbola, note that, for $0 < \alpha < 1$, the map $\varphi(z) = \frac{1}{2}(z + 1/z)$ maps a region of the form

$$S_\alpha = \left\{z : |z| > 1, -\frac{\pi\alpha}{2} < \arg z < \frac{\pi\alpha}{2}\right\}$$

onto the region H_α to the right of a hyperbola, but slit from $\cos\frac{\pi\alpha}{2}$ to 1:

$$H_\alpha = \left\{z = x + iy : \left(\frac{x}{2\cos\frac{\pi\alpha}{2}}\right)^2 - \left(\frac{y}{2\sin\frac{\pi\alpha}{2}}\right)^2 > 1\right\} \setminus \left[\cos\frac{\pi\alpha}{2}, 1\right].$$

When $\alpha = 1$, the image region is the right half-plane slit along $(0, 1]$; call it H_1. We can map S_1 onto S_α using the map z^α. Thus $f = \varphi \circ z^\alpha \circ \varphi^{-1}$ maps H_1 onto H_α. This map extends to be a one-to-one analytic map of the right half-plane onto the region to the right of the hyperbola. Supply details for this outline.

6.12 To map $\mathbb{H}$ onto the region above the parabola $y = ax^2$, $a > 0$, note that z^2 maps the half-strip $S_b = \{z : -b < \mathrm{Re} z < b, \mathrm{Im} z > 0\}$ onto the interior of a parabola, slit along the segment $[0, b^2)$:

$$P_b = \left\{z = x + iy : x < b^2 - \left(\frac{y}{2b}\right)^2\right\} \setminus [0, b^2).$$

As in Exercise 6.11, find a conformal map of a slit half-plane onto P_b which extends continuously across the slit, using the inverse of $\frac{1}{2}(z + 1/z)$, $\log z$ and z^2, composing with linear maps as needed. Supply details for this outline.

6.13 The function $g(z) = \sqrt{\log(2/(1 - z))}$ can be defined so that it is analytic and one-to-one on $\mathbb{D}$. Draw a picture of the image by considering a sequence of simpler maps. Show that $\mathrm{Im} g$ is continuous on $\partial\mathbb{D}$ but that $\mathrm{Re} g$ is not continuous. In fact, $\mathrm{Re} g$ is not bounded. (For Fourier series enthusiasts, the imaginary part of the power series for g gives a Fourier series of a continuous function whose conjugate Fourier series is not even bounded.)

C

The following two exercises are not hard, if you get the right idea.

6.14 Suppose C and D are tangent circles, one inside the other. Find a circle C_1 which is tangent to both C and D, then for $n \geq 2$ find C_n tangent to C, D and C_{n-1}, with $\{C_j, j = 1, \ldots, n\}$ bounding disjoint open disks. Prove that this process can be continued indefinitely and that the points of tangency of the C_n all lie on a circle. See Figure 6.15.

6.15 Suppose C and D are non-intersecting circles, one inside the other. Choose C_1 tangent to both C and D. Again choose C_j tangent to C, D and C_{j-1}, with $\{C_j, j = 1, \ldots, n\}$ bounding disjoint open disks. Under some circumstances, the chain of circles may "close up." In other words, C_n is also tangent to C_1 for some $n > 1$. Show that, for each $n \geq 3$, there

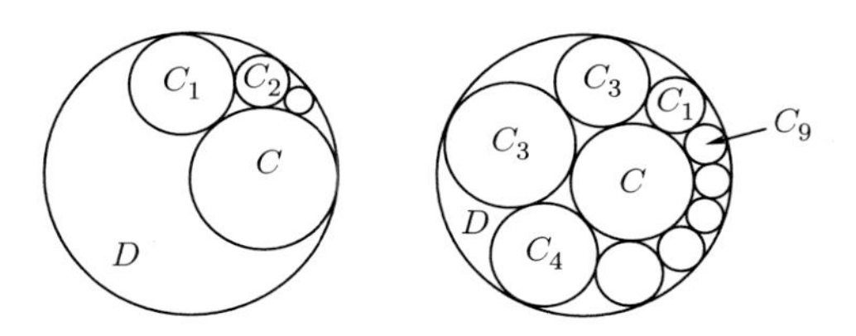

Figure 6.15 Tangent circles.

is a choice of C and D so that the chain always closes up. Given circles C and D, show that the property of "closing up" does not depend on the choice of C_1. See Figure 6.15.

It is known that, given a collection of disks in a simply-connected region whose interiors are disjoint, there is collection of disks in the unit disk with the same pattern of tangencies. If the disks are small enough, the map from one set of disks to the other approximates the conformal map between the region and the unit disk. See K. Stephenson [26] for an introduction to this subject.

6.16 (Mercator projection) Let π^{-1} denote the map from the unit sphere to the complex plane given by the inverse of stereographic projection. Cut the plane along $(-\infty, 0]$ and apply the map $L(z) = i \log z$. By Exercise 1.11, the composition $L \circ \pi^{-1}$ is a map of the (slit) sphere onto a vertical strip which preserves angles between curves. In other words, if two curves on the unit sphere meet at an angle θ when viewed from inside the sphere, then the images of these curves in the strip meet at the same angle. The map $M(w) = \overline{-i \log \pi^{-1}(w)}$, $w \in \mathbb{S}^2$, then preserves angles between curves when viewed as if we are standing on the outside of the sphere or above the strip. The north pole corresponds to the top of the strip, the south pole corresponds to the bottom of the strip, and vertical lines correspond to great circles through the north and south poles, called longitude lines. A horizontal line intersects the strip in a segment which corresponds to a circle with constant angle of elevation, called a latitude line. The map M, called the Mercator projection, was of great commercial value because it allowed ships to sail beyond the sight of land. This is because a line between two points on the strip (the map) corresponds to a path on the sphere with constant compass heading, by conformality. The local stretch in distances, however, does depend on the direction.

Show that the ratio of the vertical stretch to the horizontal stretch on a latitude line with elevation ϕ is given by $\sec \phi$, so that the height of the latitude line on the strip (map) is given by

$$\int_0^{\phi} \sec t \, dt. \tag{6.7}$$

Mercator, perhaps unaware of the connection with stereographic projection, deduced (6.7) geometrically via Riemann sums, which he used to estimate the value. It was an unsolved problem for 40 years to find this integral in terms of elementary functions. Use the map M to find the integral. Calculus students typically learn how to do the integral with a trick, unaware of the great importance of the integral.

PART II

7 Harmonic Functions

Harmonic functions arise in applications as solutions of Laplace's equation

$$\Delta u \equiv \frac{\partial^2 u}{\partial x^2} + \frac{\partial^2 u}{\partial y^2} = 0.$$

They are useful for the study of analytic functions in two ways: if f is analytic then $\text{Re}f$ is harmonic and, more importantly, $\log|f|$ is harmonic where $f \neq 0$. The most useful property of harmonic functions is the mean-value property, and its consequence, the maximum principle. To avoid difficulties with proving differentiability properties, we take the mean-value property as definition, then give the connection with Laplace's equation in Theorem 7.12.

7.1 The Mean-Value Property and the Maximum Principle

In this section, we define harmonic and subharmonic functions, prove the maximum principle, and then show how to find the values of a harmonic function on a disk from its values on the bounding circle.

Definition 7.1 *A continuous real-valued function u is* **harmonic** *on a region $\Omega \subset \mathbb{C}$ if for each $z \in \Omega$ there is an $r_z > 0$ (depending possibly on z) such that*

$$u(z) = \frac{1}{2\pi}\int_0^{2\pi} u(z + re^{it})dt \tag{7.1}$$

for all $r < r_z$.

Equation (7.1) is called the **mean-value property**. The mean-value property says that the value of u at the center of the disk is equal to the average of its values on the boundary of the disk. This is interesting because it says that if you know u on the circle, then you can recover its value inside the disk, at least at the center. We learn later how to recover values at other points of the disk. Harmonicity is a local property because we only require the mean-value equality on sufficiently small disks, not necessarily all disks contained in Ω. It is true though that the mean-value property holds for all circles which bound a closed disk contained in Ω, if u is harmonic. See Corollary 7.6.

Definition 7.2 *A continuous function u with values in $[-\infty, +\infty)$ is* **subharmonic** *on a region Ω if for each $z \in \Omega$ there is an $r_z > 0$ (depending possibly on z) such that*

$$u(z) \le \frac{1}{2\pi} \int_0^{2\pi} u(z + re^{it})dt \tag{7.2}$$

for all $r < r_z$.

In some texts, the continuity assumption in Definition 7.2 is replaced by upper semicontinuity, though we do not need this extra flexibility. See Exercise 7.14. Equation (7.2) is called the **mean-value inequality**. Note that we allow a subharmonic function to take the value $-\infty$ but not $+\infty$. The mean-value inequality is trivially satisfied for those points z where $u(z) = -\infty$. If u is continuous and $u(z) > -\infty$ then u is bounded below in some neighborhood N of z, so that the integral in (7.2) is defined for sufficiently small $r > 0$. We can also define the integral of u on any circle in Ω as the decreasing limit of the integrals of $u_n(z) = \max(-n, u(z))$.

The reader can easily verify that if u and $-u$ are subharmonic then u is harmonic. If u_1 and u_2 are harmonic then $A_1u_1 + A_2u_2$ is harmonic, for real numbers A_1, A_2, and $|A_1u_1|$ is subharmonic. If u is subharmonic then Au is subharmonic provided $A > 0$. If u_1 and u_2 are subharmonic then $u(z) = \max(u_1(z), u_2(z))$ is subharmonic.

For example, an analytic function f has the mean-value property, as can be seen by expanding in a (local) power series and interchanging the order of integration and summation. Then, by taking real and imaginary parts, $u = \text{Re}f$ and $v = \text{Im}f$ are harmonic. It also follows that $|f|$ is subharmonic. Harmonicity, like analyticity, is a local property, so that $\log|f|$ is harmonic on $\Omega \setminus \{f = 0\}$ since $\log|f|$ is the real part of $\log f$, which can be defined to be analytic in a neighborhood of any point where $f \neq 0$. This also shows that $\log|f|$ is subharmonic on Ω. Of these examples, perhaps the most important for the study of analytic functions is the subharmonicity of $\log|f|$. We will use subharmonic functions to study harmonic functions as well as analytic functions.

The next result is perhaps the most important elementary result in complex analysis.

Theorem 7.3 (maximum principle) *Suppose u is subharmonic on a region* Ω. *If there exists* $z_0 \in \Omega$ *such that*

$$u(z_0) = \sup_{z \in \Omega} u(z), \tag{7.3}$$

then u is constant.

A subharmonic function can have a local maximum, but the function will be constant in a neighborhood of any point where the local maximum occurs, by Theorem 7.3. For example, $u(z) = \max(0, \log|z|)$ is subharmonic in $\mathbb{C}$ and has a local maximum at each point of $\mathbb{D}$.

Proof Suppose (7.3) holds. Set $E = \{z \in \Omega : u(z) = u(z_0)\}$. Since u is continuous, E is closed in Ω. By (7.3), the set E is non-empty. We need only show E is open, since Ω is connected. If $z_1 \in E$, then by (7.2)

$$\frac{1}{2\pi} \int_0^{2\pi} [u(z_1) - u(z_1 + re^{it})]dt \le 0, \tag{7.4}$$

for $r < r_{z_1}$. But the integrand in (7.4) is continuous and non-negative by (7.3) and hence identically 0 for all t and all $r < r_{z_1}$. This proves E is open and hence equal to Ω. □

Another form of the maximum principle is as follows:

Corollary 7.4 *If u is a non-constant subharmonic function in a bounded region Ω and if u is continuous on $\overline{\Omega}$ then*

$$\max_{z\in\overline{\Omega}} u(z)$$

occurs on $\partial\Omega$ but not in Ω.

The reader should verify the alternative form: if u is continuous and subharmonic on Ω then

$$\limsup_{z\to\partial\Omega} u(z) = \sup_{\Omega} u(z).$$

If Ω is unbounded, then ∞ must also be viewed as part of $\partial\Omega$. The function $u(z) = \mathrm{Re}z$ is harmonic on $\Omega = \{z : \mathrm{Re}z > 0\}$ and satisfies $u = 0$ on $\partial\Omega \cap \mathbb{C}$ but u is not bounded by 0.

Theorem 7.5 (Schwarz) *If g is real valued and continuous on $\partial\mathbb{D}$, set*

$$u(z) = \frac{1}{2\pi}\int_0^{2\pi} \frac{1-|z|^2}{|e^{it}-z|^2} g(e^{it})dt,$$

for $z \in \mathbb{D}$. Then u is harmonic in $\mathbb{D}$ and

$$\lim_{z\to\zeta} u(z) = g(\zeta), \tag{7.5}$$

for all $\zeta \in \partial\mathbb{D}$.

Proof The function

$$G(z) = \frac{1}{2\pi}\int_0^{2\pi} \frac{e^{it}+z}{e^{it}-z} g(e^{it})dt$$

is analytic on $\mathbb{D}$, as can be seen by expanding the kernel

$$\frac{e^{it}+z}{e^{it}-z} = \frac{1+e^{-it}z}{1-e^{-it}z} = 1 + 2\sum_{1}^{\infty} e^{-int}z^n \tag{7.6}$$

and interchanging the order of summation and integration. The identity

$$\frac{1-|z|^2}{|e^{it}-z|^2} = \mathrm{Re}\left(\frac{e^{it}+z}{e^{it}-z}\right) \tag{7.7}$$

shows that $u = \mathrm{Re}G$ and hence u is harmonic. Note also that if $g \equiv 1$, then $G \equiv 1$, since $\int e^{-int}dt = 0$ if $n \neq 0$. Thus for all $z \in \mathbb{D}$

$$\frac{1}{2\pi}\int_0^{2\pi} \frac{1-|z|^2}{|e^{it}-z|^2} dt = 1. \tag{7.8}$$

To prove (7.5) fix t_0 and $\varepsilon > 0$ then choose $\delta > 0$ so that $|g(e^{it}) - g(e^{it_0})| < \varepsilon$ if $t \in I_\delta = \{t : |t - t_0| < \delta\}$. Then, using (7.8),

$$\begin{aligned} |u(z) - g(e^{it_0})| &= \left| \frac{1}{2\pi} \int_0^{2\pi} \frac{1 - |z|^2}{|e^{it} - z|^2} (g(e^{it}) - g(e^{it_0}))dt \right| \\ &\leq \varepsilon \frac{1}{2\pi} \int_{I_\delta} \frac{1 - |z|^2}{|e^{it} - z|^2} dt + M(z) \int_{\partial\mathbb{D}\setminus I_\delta} |g(e^{it}) - g(e^{it_0})| \frac{dt}{2\pi}, \end{aligned}$$

where

$$M(z) = \sup_{\{t : |t - t_0| \geq \delta\}} \frac{1 - |z|^2}{|e^{it} - z|^2}.$$

The first term is at most ε by (7.8). Moreover, $M(z) \to 0$ as $z \to e^{it_0}$ because $|e^{it} - z|$ is bounded below for $|t - t_0| \geq \delta$. Thus $u(z) \to g(e^{it_0})$ as $z \to e^{it_0}$. □

The proof of Schwarz's theorem shows that we need assume only that g is integrable on $\partial\mathbb{D}$ and continuous at ζ for (7.5) to hold. The kernel

$$P_z(t) = \frac{1}{2\pi} \frac{1 - |z|^2}{|e^{it} - z|^2}$$

is called the **Poisson kernel** and $u = PI(g) \equiv \int P_z(t) g(e^{it}) dt$ is called the **Poisson integral of g**. If B is a disk and g is continuous on ∂B, then $PI_B(g)$ denotes the harmonic function on B which equals g on ∂B, and is called the Poisson integral of g on B. The reader can give a formula for $PI_B(g)$ by transplanting the problem to the unit disk using a linear map. Note that by (7.8) if g is integrable and satisfies $m \leq g \leq M$ then $m \leq PI_B(g) \leq M$. The Poisson kernel for the upper half-plane $\mathbb{H}$ is given by

$$P_w^{\mathbb{H}}(s) = \frac{1}{\pi} \frac{v}{(u - s)^2 + v^2} = \frac{1}{\pi} \operatorname{Im} \left(\frac{1}{s - w} \right),$$

where $w = u + iv$, $v > 0$ and $s \in \mathbb{R}$. This can be proved using the change of variables $z = (w - i)/(w + i)$, $e^{it} = (s - i)/(s + i)$ in Schwarz's Theorem 7.5.

The next result shows how to find the values of a harmonic function in the disk from its values on the boundary.

Corollary 7.6 *If u is harmonic on $\mathbb{D}$ and continuous on $|z| \leq 1$, then for $z \in \mathbb{D}$*

$$u(z) = \frac{1}{2\pi} \int_0^{2\pi} \frac{1 - |z|^2}{|e^{it} - z|^2} u(e^{it}) dt. \tag{7.9}$$

Proof Let $U(z)$ denote the right-hand side of (7.9). Then by Schwarz's theorem $u - U$ is harmonic on $\mathbb{D}$, continuous on $\overline{\mathbb{D}}$ and equal to 0 on $\partial\mathbb{D}$. By the maximum principle applied to $u - U$ and $U - u$, we conclude $u = U$. □

If u is harmonic in a region Ω and if D is a closed disk contained in Ω, then, by a linear change of variables and Corollary 7.6, the mean-value property holds for ∂D. See Exercise 7.1.

The next corollary shows how to recapture an analytic function from the boundary values of its real part.

Corollary 7.7 *If u is harmonic on $|z| < 1$ and continuous on $|z| \le 1$, then*

$$f(z) = \frac{1}{2\pi}\int_0^{2\pi} \frac{e^{it}+z}{e^{it}-z} u(e^{it})dt$$

is the unique analytic function on $\mathbb{D}$ with $\mathrm{Re}f = u$ *and* $\mathrm{Im}f(0) = 0$.

The function $(e^{it}+z)/(e^{it}-z)$ is called the **Herglotz kernel** and the integral is called the **Herglotz integral**.

Proof By the first part of the proof of Schwarz's Theorem 7.5, f is analytic. The real part of f is then equal to u by (7.7) and Corollary 7.6. If g is another analytic function with $\mathrm{Re}g = u$, then $f - g$ is purely imaginary and hence not an open mapping. Thus $f - g$ is constant. Finally note that $f(0) = \int u(e^{it})dt$ is real, so that if $g(0)$ is real then $g = f$. □

It follows from Corollary 7.7 that a harmonic function on a region Ω is the real part of an analytic function on each disk contained in Ω. One consequence is that it is not hard to prove that if u is harmonic on a disk and if f is analytic then $u \circ f$ is harmonic. Indeed, harmonicity is local and on a disk $u = \mathrm{Re}g$ for some analytic function g, by Corollary 7.7. Hence $u \circ f = \mathrm{Re}(g \circ f)$ is harmonic. If $u = \mathrm{Re}z$ and $f = z^2$ then $f \circ u$ is not harmonic. The function $u = \log|z|$ is harmonic on $\Omega = \{z : 0 < |z| < \infty\}$ and is the real part of an analytic function on each disk that does not contain 0, but it is not the real part of an analytic function on all of Ω because $\arg z$ is not continuous on Ω.

An application to analytic functions is given in Corollary 7.8.

Corollary 7.8 (jump theorem) *Suppose f is an integrable function such that $f : \partial\mathbb{D} \to \mathbb{C}$. Let*

$$F(z) = \int_{|\zeta|=1} \frac{f(\zeta)}{\zeta - z}\frac{d\zeta}{2\pi i}.$$

Then F is analytic on $\mathbb{C} \setminus \partial\mathbb{D}$ and, for $|\zeta| = 1$,

$$\lim_{z\to\zeta}\left[F(z) - F\left(\frac{1}{\bar{z}}\right)\right] = f(\zeta)$$

at all points of continuity ζ of f.

The function F is called the **Cauchy integral** or **Cauchy transform** of f. The jump theorem says that the analytic function F jumps by $f(\zeta)$ as z crosses the unit circle at ζ. Note that if f is analytic on $\overline{\mathbb{D}}$ then $F = f$ in $\mathbb{D}$ and $F = 0$ in $|z| > 1$ by Cauchy's integral formula, Theorem 5.2.

Proof We already proved that F is analytic off $\partial\mathbb{D}$ in Lemma 4.30. To prove the corollary, just manipulate the integrals:

$$F(z) - F\left(\frac{1}{\overline{z}}\right) = \int_{|\zeta|=1} f(\zeta)\left(\frac{1}{\zeta - z} - \frac{1}{\zeta - 1/\overline{z}}\right)\frac{d\zeta}{2\pi i}$$
$$= \int_{|\zeta|=1} f(\zeta)\frac{z - 1/\overline{z}}{(\zeta - z)(\zeta - 1/\overline{z})}\frac{d\zeta}{2\pi i}$$
$$= \int_0^{2\pi} f(e^{it})\frac{1 - |z|^2}{|e^{it} - z|^2}\frac{dt}{2\pi}.$$

Applying Schwarz's Theorem 7.5 to the real and imaginary parts of f completes the proof. □

If g is integrable on $\partial\mathbb{D}$, set $a_n = \frac{1}{2\pi}\int_0^{2\pi} g(e^{it})e^{-int}dt$. Then

$$\sum_{n=-\infty}^{\infty} a_n e^{int}$$

is called the **Fourier series of g**. Note that $|a_n| \leq \frac{1}{2\pi}\int_0^{2\pi}|g(e^{it})|dt$. By (7.6) and (7.7)

$$\frac{1 - |z|^2}{|e^{it} - z|^2} = 1 + \sum_{n=1}^{\infty} e^{-int}z^n + \sum_{n=1}^{\infty} e^{int}\,\overline{z}^n.$$

Interchanging the order of summation and integration, the harmonic "extension" of g to $\mathbb{D}$ is given by

$$G(z) \equiv PI(g)(z) = a_0 + \sum_{n=1}^{\infty} a_n z^n + \sum_{n=1}^{\infty} a_{-n}\overline{z}^n.$$

In other words, G is found from the Fourier series of g by replacing e^{it} with z and e^{-it} with $\overline{z}$. The interplay between convergence of the Fourier series of g and the harmonic functions ReG and ImG is delicate. One connection is given by Schwarz's Theorem 7.5, and another by Abel's limit theorem, Theorem 3.13, which implies that if the Fourier series for g converges at e^{it_0} then G has that series sum as its non-tangential limit at e^{it_0}.

Fourier series were first developed by Fourier because he wanted to solve the following problem. Find a harmonic function u which satisfies $0 < u < 1$ in the half-strip $S = \{x + iy : x > 0,\ 0 < y < \pi\}$ with $u(0, y) = 1$, $0 < y < \pi$, and $u(x, 0) = u(x, \pi) = 0$ for $x > 0$. If a thin plate of homogeneous material in the shape of a long strip is heated to one temperature on the top and bottom, and another temperature on the end, and allowed to reach a state of equilibrium, then the resulting temperature is harmonic on the strip. Fourier's idea was that $e^{-nx}\sin ny = \text{Im}(-e^{-nz})$ is harmonic and equal to 0 on the top and bottom edges of S. So, set $u(z) = \sum_{n=1}^{\infty} c_n e^{-nx}\sin ny$ and choose c_n so that

$$1 = u(0, y) = \sum_{n=1}^{\infty} c_n \sin ny. \tag{7.10}$$

The coefficients c_k can be determined by integrating both sides of (7.10) against $\sin ky$ by orthogonality. The right-hand side of (7.10) is called a **Fourier sine series**. Alternatively, because the right-hand side of (7.10) is odd, define $F(e^{iy}) = 1$ for $y \in (0, \pi)$ and $F(e^{iy}) = -1$ for $y \in (-\pi, 0)$. Then the coefficients c_n can be found by adding the n and $-n$ terms of the Fourier series for F. There are convergence issues though. We will not prove these facts, but instead give another solution to Fourier's problem in Section 7.3.

7.2 Cauchy–Riemann and Laplace Equations

If $f(z) = u(z) + iv(z)$ we will sometimes use the notation $f(x, y) = u(x, y) + iv(x, y)$, where $z = x + iy$ with x, y real and $u(x, y)$ and $v(x, y)$ are real valued. If f is analytic then, by the definition of the complex derivative,

$$\begin{aligned} f'(z) &= \lim_{h\to 0} \frac{f(x+h, y) - f(x, y)}{h} = f_x(x, y) = u_x(x, y) + iv_x(x, y) \\ &= \lim_{k\to 0} \frac{f(x, y+k) - f(x, y)}{ik} = \frac{1}{i} f_y(x, y) = v_y(x, y) - iu_y(x, y). \end{aligned}$$

Thus

$$u_x = v_y \text{ and } u_y = -v_x. \tag{7.11}$$

Equations (7.11) are called the **Cauchy–Riemann equations.**

There is another useful notation for first-order derivatives. Write $z = x + iy$ and define

$$f_z \equiv \frac{\partial f}{\partial z} \equiv \frac{1}{2}\left(f_x - if_y\right)$$

and

$$f_{\bar{z}} \equiv \frac{\partial f}{\partial \bar{z}} \equiv \frac{1}{2}\left(f_x + if_y\right).$$

Then it is an easy exercise to verify the chain rule:

$$(f \circ g)_z = f_z \circ g\, g_z + f_{\bar{z}} \circ g\, \overline{g}_z$$

and

$$(f \circ g)_{\bar{z}} = f_z \circ g\, g_{\bar{z}} + f_{\bar{z}} \circ g\, \overline{g}_{\bar{z}}.$$

As a mnemonic device, think of z and $\bar{z}$ as independent variables (even though one is the complex conjugate of the other). View f as $f(z, \bar{z})$ and view the composition of f and g as $f(g(z, \bar{z}), \overline{g(z, \bar{z})})$. Then formally differentiate using the two-variable chain rule. It may also help to write $\frac{1}{i}$ instead of $-i$ and $-\frac{1}{i}$ instead of i since the "denominators" contain $z = x + iy$ and $\bar{z} = x - iy$, respectively.

The Cauchy–Riemann equations can be restated in this terminology as

$$f_{\bar{z}} = 0.$$

The Cauchy–Riemann equations characterize analytic functions by the following theorem.

Theorem 7.9 *Suppose u and v are real valued and continuously differentiable on a region Ω. Then u and v satisfy the Cauchy–Riemann equations (7.11) if and only if $f = u + iv$ is analytic on Ω.*

Proof We have already proved that if $f = u + iv$ is analytic then u and v satisfy the Cauchy–Riemann equations. Conversely, if u and v satisfy the Cauchy–Riemann equations, then, by Taylor's theorem applied to the real-valued functions u and v,

$$\begin{aligned} f(x+h, y+k) - f(x, y) = hu_x(x, y) + ku_y(x, y) + i(hv_x(x, y) + kv_y(x, y)) \\ + \varepsilon(h, k), \end{aligned}$$

where $\varepsilon(h,k)/\sqrt{|h|^2+|k|^2} \to 0$ as $|h|, |k| \to 0$. Dividing by $h+ik$ and applying the Cauchy–Riemann equations, we obtain

$$\lim_{\sqrt{|h|^2+|k|^2}\to 0} \frac{f(x+h,y+k)-f(x,y)}{h+ik} = u_x(x,y) - iu_y(x,y).$$

So $f'(z)$ exists and is continuous, and therefore f is analytic. □

A harmonic function on a region Ω is the real part of an analytic function on each disk contained in Ω by Corollary 7.7. So harmonic functions have continuous partial derivatives of all orders. A harmonic function is not always the real part of an analytic function on all of Ω. However, the Cauchy–Riemann equations allow us to associate another analytic function with a harmonic function as in the following theorem.

Theorem 7.10 *A function u is harmonic on a region Ω if and only if $u_x - iu_y$ exists and is analytic on Ω. If Ω is simply-connected then u is harmonic on Ω if and only if $u = \text{Re}f$ for some f analytic on Ω.*

Proof If u is harmonic on Ω and if B is a disk contained in Ω then, by Corollary 7.7, $f = u+iv$ for some analytic function f on B. Moreover, f' is analytic and $f' = u_x + iv_x = u_x - iu_y$ by the Cauchy–Riemann equations. This proves that $u_x - iu_y$ exists and is analytic on each B and hence on Ω. Conversely if $g = u_x - iu_y$ exists and is analytic on Ω, then g has a power series expansion on any disk $B \subset \overline{B} \subset \Omega$. Integrating the series term by term gives an analytic function f with $f' = g$. If $w = \text{Re}f$ then, by the Cauchy–Riemann equations, $w_x = u_x$ and $w_y = u_y$, so that $u = w + c$ on B, where c is a constant. Since $w = \text{Re}f$ is harmonic, u must also be harmonic on B and hence on all of Ω.

If u is harmonic on a simply-connected region Ω, then, by Corollary 5.8(ii) and the third sentence in this proof, there is an analytic function f on all of Ω such that $f' = u_x - iu_y$. By the Cauchy–Riemann equations, $w = \text{Re}f$ and u have the same partial derivatives on Ω and so $u = \text{Re}(f+c)$ for some constant c.

Finally, if f is analytic on Ω, then the mean-value property holds for f and hence for $u = \text{Re}f$. So u is harmonic. □

If $f = u + iv$ is analytic in a region Ω then v is called a **harmonic conjugate** of u in Ω. Because $-if$ is analytic, $-u$ is a harmonic conjugate of v. If v_1 and v_2 are harmonic conjugates of u on a region Ω then $v_1 - v_2$ is constant because the difference of two analytic functions is an open map, if non-constant, and hence cannot be purely imaginary on an open set.

If f is analytic then f has continuous partial derivatives of all orders. By the Cauchy–Riemann equations applied to $u = \text{Re}f$ and $v = \text{Im}f$,

$$u_{xx} = (v_y)_x = (v_x)_y = (-u_y)_y = -u_{yy}, \tag{7.12}$$

and hence $u_{xx} + u_{yy} = 0$. Similarly, $v_{xx} + v_{yy} = 0$.

Definition 7.11 *The Laplacian of u is the second-order derivative given by $\Delta u = u_{xx} + u_{yy}$. We say that u* **satisfies Laplace's equation** *on a region Ω if u has continuous second-order partial derivatives (including the mixed partials) and $\Delta u = 0$ on Ω.*

The next result relates Laplace's equation to harmonicity, and shows that both are essentially the same as the maximum principle.

Theorem 7.12 *Suppose u is real valued and continuous on a region Ω. Then the following are equivalent:*

(i) *u is harmonic on Ω,*
(ii) *u satisfies Laplace's equation on Ω,*
(iii) *if B is an open disk with $B \subset \overline{B} \subset \Omega$ and if v is harmonic on B, then $u - v$ and $v - u$ satisfy the maximum principle on B.*

Proof We have already seen that (i) implies (ii). By the maximum principle (i) implies (iii). If (iii) holds and if $B \subset \overline{B} \subset \Omega$, then let v be the Poisson integral of $u|_{\partial B}$ on B. Then v is harmonic on B and $u - v$ and $v - u$ are equal to 0 on ∂B by Schwarz's Theorem 7.5. By (iii), $u = v$ on B. This proves u is harmonic on each $B \subset \Omega$, and hence (i) holds.

Finally we show that (ii) implies (i). Set $g = u_x - iu_y$. Now if R is a rectangle with sides parallel to the axes contained in Ω, we claim that $\int_{\partial R} g(\zeta)d\zeta = 0$. To see this, note that

$$\int_{\partial R} (u_x - iu_y)(dx + idy) = \int_{\partial R} u_x dx + u_y dy + i \int_{\partial R} u_x dy - u_y dx.$$

By the fundamental theorem of calculus applied to each segment in ∂R, the first integral is zero. Also by the fundamental theorem of calculus, integrating along horizontal lines in R and vertical lines in R, the second integral can be rewritten as

$$\int_R (u_{xx} + u_{yy})dxdy,$$

which is also equal to 0 by (ii). (We have just followed a proof of the simplest form of Green's theorem.) By Morera's theorem, $g = u_x - iu_y$ is analytic on Ω, and by Theorem 7.10, u is harmonic. □

See Exercise 7.13 for weaker conditions sufficient for analyticity and harmonicity.

For an application of the Cauchy–Riemann equations, recall that if f is analytic and $f'(z_0) \neq 0$ then f is locally conformal. That is, f preserves angles between curves passing through z_0, in both magnitude and direction. See Section 3.2. Corollary 7.13 gives a converse to this statement.

Corollary 7.13 *If f is continuously differentiable (with respect to x and y) on a region Ω, and if f preserves angles between curves at each point of Ω, then f is analytic in Ω and $f' \neq 0$ on Ω.*

Proof Suppose $z_0 \in \Omega$ and $\theta \in [0, 2\pi]$. Set $\gamma(t) = z_0 + te^{i\theta}$ and $w(t) = f(\gamma(t))$. Because f preserves angles between curves at z_0, the angle between $w(t)$ and $\gamma(t)$ at $t = 0$, $\arg(w'(0)/\gamma'(0))$, does not depend on θ. By the chain rule,

$$\begin{aligned} w'(t) &= f_z \gamma'(t) + f_{\bar{z}} \overline{\gamma'(t)} \\ &= f_z e^{i\theta} + f_{\bar{z}} e^{-i\theta}, \end{aligned}$$

so that

$$\frac{w'(0)}{\gamma'(0)} = f_z + f_{\bar{z}}\, e^{-2i\theta},$$

which traces a circle with radius $|f_{\bar{z}}(z_0)|$ as θ varies from 0 to π. Because $\arg(w'(0)/\gamma'(0))$ is constant, $f_{\bar{z}}(z_0) = 0$. By Theorem 7.9, f is analytic. An analytic function does not preserve angles where $f' = 0$, as described in Section 3.2, and the corollary follows. □

As a consequence of the proof of Corollary 7.13, note that the local stretching of $\gamma(t)$ at $t = 0$ is given by $|w'(0)/\gamma'(0)|$. So if instead of preserving angles we assume that the local stretch is the same in all directions and non-zero at each point $z_0 \in \Omega$, then either $f_{\bar{z}} = 0$ or $f_z = 0$ at z_0. Because f is continuously differentiable, either f or $\bar{f}$ is analytic in a neighborhood of each point of Ω. The closed sets $\{w \in \Omega : f_z(w) = 0\}$ and $\{w \in \Omega : f_{\bar{z}}(w) = 0\}$ are disjoint because the local stretch is non-zero at each point. Because Ω is connected, either f or $\bar{f}$ is analytic in Ω. Moreover, the non-zero local stretch at z_0 is given by $|f'(z_0)|$ or $|(\bar{f})'(z_0)|$.

7.3 Hadamard, Lindelöf and Harnack

The maximum principle can be used to give more information than is immediately apparent from its statement.

Theorem 7.14 (Hadamard's three-circles theorem) *Suppose f is analytic in the annulus $A = \{z : r < |z| < R\}$. Let $m = \limsup_{|z|\to r} |f(z)|$ and $M = \limsup_{|z|\to R} |f(z)|$, and suppose $m, M < \infty$. If $z \in A$, then*

$$|f(z)| \leq M^{\omega(z)} m^{1-\omega(z)},$$

where $\omega(z) = \log(|z|/r)/\log(R/r)$.

Theorem 7.14 is called the three-circles theorem because $\omega(z)$ depends only on $t = |z|$, so the theorem relates the max on the circle of radius t to the max on the two bounding circles. Another way to state the conclusion is that if $M(s) = \sup_{|z|=s} |f(z)|$ then $\log M(s)$ is a convex function of $\log s$.

Proof The function $\omega \log M + (1 - \omega) \log m$ is harmonic on A, equal to $\log M$ on $|z| = R$ and equal to $\log m$ on $|z| = r$. Thus $u = \log|f| - (\omega \log M + (1 - \omega)\log m)$ is subharmonic on A and $\limsup_{z\to\partial A} u(z) \leq 0$. By the maximum principle, $u \leq 0$ in A. □

Another way to solve Fourier's problem in Section 7.1 is to note that

$$\varphi(z) = (z + 1/z) \circ e^z = e^z + e^{-z}$$

is a conformal map of the half-strip $S = \{x + iy : x > 0,\ 0 < y < \pi\}$ onto the upper half-plane $\mathbb{H}$, which maps the vertical end of the strip to the interval $(-2, 2)$. If $-\infty < a < b < \infty$ then the function $\theta(z) = \operatorname{Im}\log \frac{b-z}{a-z}$ is harmonic in $\mathbb{H}$ and equal to the angle θ formed at $z \in \mathbb{H}$ between line segments from z to b and from z to a. See Figure 7.1.

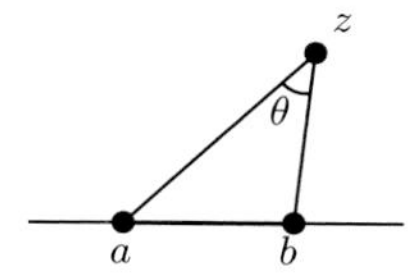

Figure 7.1 The angle $\theta(z)$ as a harmonic function.

The angle θ approaches π as $z \to (a,b)$ and approaches 0 as $z \to \mathbb{R} \setminus [a,b]$. Thus

$$u(z) = \frac{1}{\pi} \arg \left(\frac{2-z}{-2-z} \right) \circ (e^z + e^{-z})$$

solves Fourier's problem.

It is natural to ask if this function is the only solution. The answer is provided by an extension to the maximum principle, called **Lindelöf's maximum principle**.

Theorem 7.15 (Lindelöf) *Suppose Ω is a region and suppose $\{\zeta_1, \ldots, \zeta_n\}$ is a finite subset of $\partial\Omega$, not equal to all of $\partial\Omega$. If u is subharmonic on Ω with $u \le M < \infty$ on Ω and if*

$$\limsup_{z \in \Omega \to \zeta} u(z) \le m,$$

for all $\zeta \in \partial\Omega \setminus \{\zeta_1, \ldots, \zeta_n\}$, then $u \le m$ on Ω.

In the statement of Lindelöf's theorem, if Ω is unbounded, then we view ∞ as a boundary point, which may or may not be one of the exceptional points $\{\zeta_j\}$.

Proof First suppose that Ω is bounded and let $d = \operatorname{diam}(\Omega)$. For $\varepsilon > 0$ set

$$u_\varepsilon(z) = u(z) + \varepsilon \sum_{j=1}^{n} \log \left| \frac{z - \zeta_j}{d} \right|. \tag{7.13}$$

Then u_ε is subharmonic in Ω, $u_\varepsilon \le u$ and $u_\varepsilon \to -\infty$ as $z \to \zeta_j$, for $j = 1, \ldots, n$. Thus $\limsup_{z \to \partial\Omega} u_\varepsilon(z) \le m$, and so by the maximum principle $u_\varepsilon \le m$ on Ω. Fix z and let $\varepsilon \to 0$ in (7.13) to obtain $u(z) \le m$.

If Ω is not bounded, we may suppose that $\zeta_j \ne \infty$ for $j = 1, \ldots, n$ by composing with an LFT if necessary. Given $\varepsilon > 0$, we can choose R so that $R > \max_j |\zeta_j|$ and $u(z) \le m + \varepsilon$ for $z \in \Omega \cap \{|z| > R\}$. Now apply the bounded case to $u - \varepsilon$ on $\Omega \cap \{|z| < R\}$ to conclude that $u - \varepsilon \le m$ on $\Omega \cap \{|z| < R\}$ and hence on Ω. Let $\varepsilon \to 0$ to conclude $u \le m$ on Ω. □

Lindelöf's maximum principle shows that there is a unique bounded harmonic function solving Fourier's problem since the difference of two such solutions is bounded and harmonic with boundary values equal to 0 except at three boundary points, 0, πi and ∞. Similarly, if $I_j = (a_j, b_j)$, $j = 1, \ldots, n$, are disjoint intervals on $\mathbb{R}$ then

$$u(z) = \sum_{j=1}^{n} \frac{c_j}{\pi} \arg \left(\frac{b_j - z}{a_j - z} \right)$$

is the unique bounded harmonic function on $\mathbb{H}$ with boundary values c_j on I_j and 0 on $\mathbb{R} \setminus \cup \bar{I}_j$.

Another consequence of Lindelöf's maximum principle is the **three-lines** version of Hadamard's theorem, which plays an important role in complex interpolation theory of operators. It follows from the proof of Hadamard's three-circles theorem and Lindelöf's maximum principle.

Corollary 7.16 *If u is subharmonic on the strip $S_0 = \{z = x + iy : 0 < x < 1\}$ set $m_0 = \limsup_{\text{Re}z \to 0} u(z)$ and $m_1 = \limsup_{\text{Re}z \to 1} u(z)$. If $u \le M < \infty$ on S_0 then $u(z) \le m_1 x + m_0(1-x)$.*

For example, suppose g and h are continuous functions on a compact set X and suppose $d\mu$ is a positive measure on X. Then the function

$$F(z) = \int_X |g|^{pz}|h|^{q(1-z)}d\mu$$

is analytic on S_0 for $p, q > 0$. This follows from Morera's theorem after interchanging the order of integration. Thus

$$\log\left|\int_X |g|^{pz}|h|^{q(1-z)}d\mu\right|$$

is subharmonic on S_0, and bounded above. By Corollary 7.16, for $0 < x < 1$,

$$\log\left|\int_X |g|^{pz}|h|^{q(1-z)}d\mu\right| \le x\log\int_X |g|^p d\mu + (1-x)\log\int_X |h|^q d\mu.$$

If $1/p + 1/q = 1$ with $p > 1$ then set $z = x = 1/p$ and exponentiate to obtain

$$\int_X |gh|d\mu \le \left(\int_X |g|^p d\mu\right)^{\frac{1}{p}}\left(\int_X |h|^q d\mu\right)^{\frac{1}{q}},$$

which is called **Hölder's inequality**.

Positive harmonic functions cannot grow too quickly in the unit disk, as the following result, known as **Harnack's inequality**, shows.

Theorem 7.17 (Harnack) *Suppose u is a positive harmonic function on $\mathbb{D}$. Then*

$$\left(\frac{1-r}{1+r}\right)u(0) \le u(z) \le \left(\frac{1+r}{1-r}\right)u(0),$$

for $|z| = r < 1$.

Proof We may assume u is harmonic on $\overline{\mathbb{D}}$ by replacing u with $u(sz)$, $s < 1$, and then letting $s \to 1$. Because u then satisfies the Poisson integral formula (7.9), we first estimate the Poisson kernel:

$$\frac{1-r}{1+r} = \frac{1-r^2}{(1+r)^2} \le \frac{1-|z|^2}{|e^{it}-z|^2} \le \frac{1-r^2}{(1-r)^2} = \frac{1+r}{1-r}.$$

Then, because u is positive and the mean-value property holds,

$$\left(\frac{1-r}{1+r}\right)u(0) \le \int_0^{2\pi} \frac{1-|z|^2}{|e^{it}-z|^2}u(e^{it})\frac{dt}{2\pi} \le \left(\frac{1+r}{1-r}\right)u(0).$$

□

A similar estimate can be used to show that two positive harmonic functions on $\mathbb{D}$ vanishing on a interval of $\partial\mathbb{D}$ must approach 0 at the same rate.

Theorem 7.18 (boundary Harnack inequality) *Suppose u and v are positive harmonic functions on $\mathbb{D}$ which extend to be continuous and equal to 0 on a closed arc $I \subset \partial\mathbb{D}$. Let $U_\delta = \{z \in \mathbb{D} : \operatorname{dist}(z, \partial\mathbb{D} \setminus I) > \delta > 0\}$. Then for $z \in U_\delta$*

$$\frac{\delta^2}{4}\left(\frac{u(0)}{v(0)}\right) \le \frac{u(z)}{v(z)} \le \frac{4}{\delta^2}\left(\frac{u(0)}{v(0)}\right).$$

Proof Fix $z \in U_\delta$ with $|z| < r < 1$ and set $\delta_r = \operatorname{dist}(z/r, \partial\mathbb{D} \setminus I)$. By the Poisson integral formula,

$$\begin{aligned}\frac{u(z)}{1-|z/r|^2} &= \frac{1}{2\pi}\int_{\partial\mathbb{D}} \frac{u(re^{it})}{|e^{it}-z/r|^2}dt \\ &\le \frac{1}{2\pi}\int_{\partial\mathbb{D}\setminus I} \frac{u(re^{it})}{\delta_r^2}dt + \frac{1}{2\pi}\int_I \frac{u(re^{it})}{|e^{it}-z/r|^2}dt \\ &\le \frac{u(0)}{\delta_r^2} + \frac{1}{2\pi}\int_I \frac{u(re^{it})}{|e^{it}-z/r|^2}dt.\end{aligned}$$

Similarly,

$$\begin{aligned}\frac{v(z)}{1-|z/r|^2} &= \frac{1}{2\pi}\int_{\partial\mathbb{D}} \frac{v(re^{it})}{|e^{it}-z/r|^2}dt \\ &\ge \frac{1}{2\pi}\int_{\partial\mathbb{D}\setminus I} \frac{v(re^{it})}{4}dt + \frac{1}{2\pi}\int_I \frac{v(re^{it})}{|e^{it}-z/r|^2}dt \\ &= \frac{v(0)}{4} - \frac{1}{2\pi}\int_I \frac{v(re^{it})}{4}dt + \frac{1}{2\pi}\int_I \frac{v(re^{it})}{|e^{it}-z/r|^2}dt.\end{aligned}$$

Taking the ratio and letting $r \to 1$, we obtain the right-hand inequality, since $\delta_r \to \operatorname{dist}(z, \partial\mathbb{D}\setminus I) > \delta > 0$, and $u(re^{it})$ and $v(re^{it})$ converge uniformly to 0 on I. The left-hand inequality is proved by reversing the roles of u and v. □

Corollary 7.19 *Let K be a compact subset of a region Ω. Then there exists a constant C depending only on Ω and K such that if u is positive and harmonic on Ω then for all $z, w \in K$*

$$\frac{1}{C}u(w) \le u(z) \le Cu(w). \tag{7.14}$$

The point of Corollary 7.19 is that the constant C can be taken to be independent of the function u.

Proof If B is a disk, let $2B$ be the disk with the same center as B and twice the radius. Suppose B is a disk such that $2B \subset \Omega$. Let φ be a linear map of $\mathbb{D}$ onto $2B$, then by Harnack's inequality applied to $u \circ \varphi$ we have that (7.14) holds for $z \in B$ and w equal to the center of B, with $C = 3$. Thus (7.14) holds for all $z, w \in B$ with $C = 9$. Cover K by a finite collection of disks $\mathcal{B} = \{B_j\}$ with $2B_j \subset \Omega$. Add more disks if necessary so that $\cup B_j$ is connected. If

$B_j, B_k \in \mathcal{B}$ with $B_j \cap B_k \neq \emptyset$ then (7.14) holds on $B_j \cup B_k$ with $C = 81$. Because there are only finitely many disks and because their union is connected, (7.14) holds on $\cup\{B_j : B_j \in \mathcal{B}\}$, and therefore on K, with a constant C depending only on the number of disks in $\mathcal{B}$, and not on u. □

We end this section with a simple but powerful consequence. We will use Theorem 7.20 in Sections 8.2 and 13.1.

Theorem 7.20 (Harnack's principle) *Suppose $\{u_n\}$ are harmonic on a region Ω such that $u_n(z) \leq u_{n+1}(z)$ for all $z \in \Omega$. Then either*

(i) $\lim_{n\to\infty} u_n(z) \equiv u(z)$ *exists and is harmonic on* Ω, *or*
(ii) $\lim_{n\to\infty} u_n(z) = +\infty$,

where convergence is uniform on compact subsets of Ω. In case (ii), *this means that given $K \subset \Omega$ compact and $M < \infty$ there is an $n_0 < \infty$ such that $u_n(z) \geq M$ for all $n \geq n_0$ and $z \in K$.*

Proof By assumption, if $n > m$ then $u_n - u_m \geq 0$, and by the maximum principle $u_n - u_m$ is strictly positive or identically 0. Fix $z_0 \in K$, where K is compact. By Corollary 7.19 there is a constant C such that

$$\frac{1}{C}(u_n(z_0) - u_m(z_0)) \leq u_n(z) - u_m(z) \leq C(u_n(z_0) - u_m(z_0)),$$

for all $z \in K$. Thus $\{u_n(z_0)\}$ is Cauchy if and only if $\{u_n\}$ is uniformly Cauchy on K. Thus if the increasing sequence $\{u_n(z_0)\}$ converges, then $\{u_n\}$ converges uniformly on compact subsets of Ω. Similarly if $u_n(z_0) \to \infty$ then $u_n(z) \to \infty$ uniformly on compact subsets of Ω. The limit function u is harmonic by the mean-value property (see Exercise 7.1). □

The assumption that the sequence u_n is increasing is essential. Note that one needs only to prove the sequence is bounded at one point to prove uniform convergence on compact subsets.

7.4 Exercises

A

7.1 (a) If u is harmonic on a region Ω and if B is a closed disk contained in Ω then the average of u on the boundary of B is equal to the value of u at the center of B. Hint: Use Corollary 7.6 and a linear change of variables. It is also possible to prove this using Exercise 7.5 and Theorem 7.12.

(b) Use (a) to show that if $\{u_n\}$ are harmonic on a region Ω and converge uniformly on compact subsets of Ω to a function u then u is harmonic. This approach using the mean-value property is easier than working with Laplace's equation.

7.2 Suppose v is subharmonic on $\mathbb{D}$ and continuous on $\overline{\mathbb{D}}$. Suppose u is harmonic on $\mathbb{D}$ and continuous on $\overline{\mathbb{D}}$ such that $v \leq u$ on $\partial\mathbb{D}$. Then $v \leq u$ in $\mathbb{D}$. Use this to prove that

$$\int_0^{2\pi} v(re^{it})dt \leq \int_0^{2\pi} v(se^{it})dt,$$

for $r < s \leq 1$. Conclude that if v is subharmonic on a region Ω then v satisfies the mean-value inequality for each circle that bounds a closed disk contained in Ω. In particular, if f is analytic on $\mathbb{D}$ then

$$\int_0^{2\pi} |f(re^{it})|^p dt \leq \int_0^{2\pi} |f(se^{it})|^p dt$$

for $r < s < 1$ and $p > 0$. A similar inequality holds for $v(z) = \log|f(z)|$. Hint: Apply the Poisson integral formula to $\max(-n, v)$.

7.3 Show that the one-variable analogs of harmonic and subharmonic are linear and convex in two ways: using the second derivative and using the maximum principle on intervals.

7.4 (a) Prove that in polar coordinates (r, θ)

$$r^2 \Delta u = \left(r\frac{\partial}{\partial r}\right)\left(r\frac{\partial}{\partial r}\right)u + \left(\frac{\partial}{\partial \theta}\right)\left(\frac{\partial}{\partial \theta}\right)u.$$

(b) Prove that if u is harmonic in $\mathbb{C} \setminus \{0\}$ and if u depends only on r (not θ) then $u = a\log r + b$, where a and b are constants.

B

7.5 (a) Prove the following version of Green's theorem: suppose that Ω is a bounded region, bounded by finitely many C^2 simple closed curves. If u and v are twice continuously differentiable on $\overline{\Omega}$ then

$$\int_{\partial\Omega} \left(u\frac{\partial v}{\partial \eta} - v\frac{\partial u}{\partial \eta}\right)|dz| = \int_{\Omega} (u\Delta v - v\Delta u)dxdy,$$

where $\eta(\zeta)$ is the unit normal at $\zeta \in \partial\Omega$ pointing out of the region Ω, $|dz|$ is arc-length measure on $\partial\Omega$ and $dxdy$ is area measure on Ω. If you already know a proof of some version of Green's theorem, you may use it to derive this version.

(b) Prove that (ii) implies (i) in Theorem 7.12 by directly verifying that the mean-value property holds. Hint: In $\mathbb{D} \setminus B(0, \varepsilon)$ set $v = \log|z|$, apply (a) and let $\varepsilon \to 0$.

(c) Similarly, show that if v is twice continuously differentiable and $\Delta v \geq 0$ then v is subharmonic because the mean-value inequality holds.

(d) If g is differentiable on $\mathbb{R}$ and $g'' > 0$ then prove that $v(re^{it}) = g(\log r)$ is subharmonic.

7.6 Prove that the following are equivalent for a real-valued twice continuously differentiable function v on a region Ω:

(a) v is subharmonic,
(b) $\Delta v \geq 0$,
(c) if B is a disk with $B \subset \overline{B} \subset \Omega$ and if u is harmonic on $\overline{B}$ then $v - u$ satisfies the maximum principle.

Hint: See Exercise 7.5 and the proof of Theorem 7.12.

7.7 (a) Prove a "Schwarz's lemma" for harmonic functions. If u is real valued and harmonic in $\mathbb{D}$ with $|u| \leq 1$ and $u(0) = 0$, then

$$|u(z)| \leq \frac{4}{\pi} \arctan |z|.$$

When can equality hold?

(b) Formulate and prove an upper bound for $u(z)$, where u is harmonic on $\mathbb{D}$ with $u(0) = 0$ and $u \leq 1$.

Hint: Use a conformal map.

7.8 If $u(x, y)$ is harmonic in the unit disk $\mathbb{D} = \{z = x + iy : |z| < 1\}$, define

$$f(z) = 2u\left(\frac{z}{2}, \frac{z}{2i}\right) - u(0, 0).$$

In other words, formally replace the real variables x and y in the two-variable Taylor series for u by the complex variables α and β then set $\alpha = z/2$ and $\beta = z/2i$. Where will the resulting series converge? Prove that f is well defined (i.e., that this substitution makes sense) and analytic in $\mathbb{D}$ and $\mathrm{Re} f = u$. Note that this procedure does not work for the functions $u = \log(x^2 + y^2)$ and $\mathrm{Re}\frac{1}{z}$ which are harmonic in $\mathbb{D} \setminus \{0\}$. However, if u is harmonic in a neighborhood of (x_0, y_0) then, for $z_0 = x_0 + iy_0$,

$$f(z) = 2u((z - z_0)/2 + x_0, (z - z_0)/(2i) + y_0) - u(x_0, y_0)$$

satisfies $\mathrm{Re} f = u$. Try this procedure for a few familiar functions such as $e^x \cos y$ and $\mathrm{Re}(z^n)$.

7.9 Suppose u is harmonic in $\mathbb{C}$ and satisfies

$$|u(z)| \leq M|z|^k,$$

when $|z| > R$. Show that u is the real part of a polynomial of degree at most k.

7.10 (a) Suppose u is harmonic in a punctured ball $B(z_0, r) \setminus \{z_0\}$. Prove

$$\lim_{z \to z_0} \frac{u(z)}{\log |z - z_0|} = 0 \tag{7.15}$$

if and only if u extends to be harmonic in $B(z_0, r)$. This is the "harmonic" version of Riemann's removable singularity theorem. In particular, if u is bounded and harmonic on the punctured ball, then u extends to be harmonic on $B(z_0, r)$.

(b) Suppose u is non-constant and harmonic in $\mathbb{C}$. Prove

$$\lim_{r \to \infty} \frac{M(r)}{\log r} = \infty,$$

where $M(r) = \sup_{|z|=r} u(z)$.

7.11 Suppose v is twice continuously differentiable and subharmonic in $\mathbb{C}$. Prove that

$$\lim_{r \to \infty} \frac{M_1(r)}{\log r}$$

exists (possibly infinite), where $M_1(r)$ is the mean value of v on $|z| = r$.

C

7.12 Suppose u is harmonic in $\mathbb{C}$ and satisfies

$$u(z) \leq A|z| + B,$$

where A and B are constants. Prove $u = ax + by + c$, where a, b and c are constant. Note that the assumption is about u not $|u|$.

7.13 (**Weyl's lemma** for analytic functions.) A continuous function f is **weakly-analytic** on a region Ω provided

$$\int_{\Omega} f \frac{\partial \varphi}{\partial \bar{z}} dA = 0$$

for all compactly supported continuously differentiable functions φ on Ω. Here dA denotes area measure. Prove that a continuous function is weakly-analytic if and only if it is analytic. Fill in the details of the following outline.

(a) If R is a rectangle and ψ is continuously differentiable on $\overline{R}$ then

$$\int_R \psi_{\bar{z}} \, dA = -i \int_{\partial R} \psi(z) dz.$$

Apply this to $\psi = f\varphi$ on a union of squares to deduce that an analytic function is weakly-analytic.

(b) Set $\varphi(z) = \frac{3}{\pi}\left(1 - |z|^2\right)^2$ on $|z| < 1$ and 0 elsewhere. Then show that φ is continuously differentiable and $\int_{\mathbb{C}} \varphi \, dA = 1$.

(c) Suppose f is weakly-analytic on Ω. Set $\varphi_\delta(z) = \frac{1}{\delta^2}\varphi(\frac{z}{\delta})$ and

$$f_\delta(z) = \int_{\mathbb{C}} f(w)\varphi_\delta(z - w) dA(w).$$

If R is a rectangle with $R \subset \overline{R} \subset \Omega$ prove that for δ sufficiently small

$$\int_{\partial R} f_\delta(z) dz = 0.$$

Hint: Use (a) and $\frac{\partial}{\partial \bar{w}}\varphi_\delta(z - w) = -\frac{\partial}{\partial \bar{z}}\varphi_\delta(z - w)$.

(d) Prove $|f_\delta(z) - f(z)| \leq \varepsilon \int_\Omega \varphi_\delta(z - w) dA(w)$, for δ sufficiently small.

(e) Conclude that f_δ is analytic and converges uniformly to f on compact subsets of Ω, and so f is analytic on Ω.

(f) Formulate and prove a similar result for harmonic functions using the Laplacian instead of $\partial/\partial\bar{z}$.

7.14 In the definition of subharmonic, the continuity assumption can be replaced by upper semi-continuity. Let φ_δ be as in Exercise 7.13. Suppose v is upper semi-continuous on a region Ω and suppose v satisfies the mean-value inequality on sufficiently small disks contained in Ω. Define

$$v_\delta(z) = \int_{\mathbb{C}} v(w)\varphi_\delta(z - w) dA(w),$$

for $z \in \Omega$ and $\delta < \text{dist}(z, \partial\Omega)$.

(a) Prove that v_δ is subharmonic and decreases to v (eventually) on compact subsets of Ω.

(d) Use part (a) or check the proofs directly using the definition of upper semi-continuity to show that the results of this chapter remain valid for this weaker definition of subharmonic.

8 Conformal Maps and Harmonic Functions

Conformal maps can be used to transfer problems on a complicated region to related problems on a simpler region. Using the inverse of the conformal map, the solutions on the simpler region can be transferred back to the original region. For example, suppose Ω is a simply-connected region such that a conformal map φ of $\mathbb{D}$ onto Ω extends to be a one-to-one and continuous map of $\overline{\mathbb{D}}$ onto $\overline{\Omega}$. If f is a continuous function on $\partial\Omega$, then we can find a harmonic function u on Ω which extends to be continuous on $\overline{\Omega}$ and equal to f on $\partial\Omega$. To accomplish this, let $g = f \circ \varphi$ on $\partial\mathbb{D}$ and let v be the Poisson integral of g. Then $u = v \circ \varphi^{-1}$ is the desired solution.

In this chapter we give the **geodesic zipper algorithm** for computing a conformal map of a simply-connected region whose boundary contains a prescribed (finite) collection of points, onto the upper half-plane $\mathbb{H}$. Lindelöf's maximum principle for harmonic functions is used to bound the constructed curve between these points. We then use the geodesic zipper algorithm and Harnack's principle to prove the Riemann mapping theorem, which says that every simply-connected region, other than $\mathbb{C}$, can be conformally mapped onto the unit disk. In Section 8.3, we explore symmetric regions and how symmetry of the region is reflected in symmetry of conformal maps. Finally, we give the Schwarz–Christoffel formula for the conformal map of the upper half-plane onto a region bounded by a polygonal curve.

8.1 The Geodesic Zipper Algorithm

Given distinct points $z_0, z_1, \ldots, z_n \in \mathbb{C}$, we can construct a simple closed curve γ whose image, also denoted by the symbol γ, contains these points and conformal maps of the half-plane $\mathbb{H}$ onto the two regions in $\mathbb{C}^*$ bounded by γ as follows. Let γ_1 be the line segment from z_0 to z_1 and let $\Omega_1 = \mathbb{C}^* \setminus \gamma_1$. Then

$$\varphi_1(z) = z_0 + \frac{z_1 - z_0}{1 + z^2} \tag{8.1}$$

defines a conformal map of $\mathbb{H}$ onto Ω_1 such that $\varphi_1(\infty) = z_0$ and $\varphi_1(0) = z_1$. See Exercise 6.6. If $z_2 \notin \gamma_1$, let σ_2 be the unique circular arc (image) in $\mathbb{H}$ from 0 to $\varphi_1^{-1}(z_2)$ which is orthogonal to $\mathbb{R}$. As described in Section 6.6 (see Figure 6.10), we can find a conformal map φ_2 of $\mathbb{H}$ onto $\mathbb{H} \setminus \sigma_2$ using the map $\sqrt{z^2 - 1}$ of $\mathbb{H}$ onto $\mathbb{H} \setminus [0, i]$, followed by an LFT which maps $\mathbb{H}$ onto $\mathbb{H}$ and maps the vertical segment $[0, i]$ onto σ_2. Note that $\varphi_2(0) = \varphi_1^{-1}(z_2)$. Then $\gamma_2 \equiv \varphi_1(\sigma_2)$ is a simple arc (image) from z_1 to z_2 and $\varphi_1 \circ \varphi_2$ is a conformal map of $\mathbb{H}$ onto $\Omega_2 \equiv \mathbb{C}^* \setminus (\gamma_1 + \gamma_2)$. Similarly if $z_3 \notin \gamma_1 \cup \gamma_2$, let σ_3 be the unique circular arc (image) in $\mathbb{H}$ from 0 to $\varphi_2^{-1} \circ \varphi_1^{-1}(z_3)$ which is orthogonal to $\mathbb{R}$ and let φ_3 be the conformal map of $\mathbb{H}$ onto

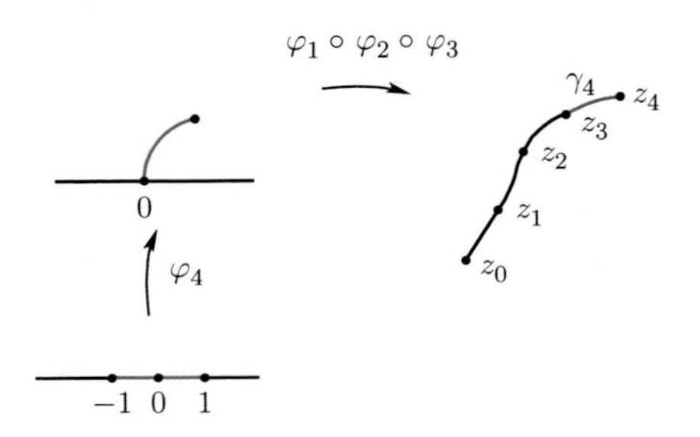

Figure 8.1 Construction of γ_4 and φ_4.

$\mathbb{H} \setminus \sigma_3$ with $\varphi_3(0) = \varphi_2^{-1} \circ \varphi_1^{-1}(z_3)$. Then $\gamma_3 = \varphi_1 \circ \varphi_2(\sigma_3)$ is a simple arc (image) from z_2 to z_3 and $\varphi_1 \circ \varphi_2 \circ \varphi_3$ is a conformal map of $\mathbb{H}$ onto $\Omega_3 \equiv \mathbb{C}^* \setminus (\gamma_1 + \gamma_2 + \gamma_3)$. Repeating this idea, we obtain a conformal map $\varphi = \varphi_1 \circ \varphi_2 \circ \cdots \circ \varphi_n$ of $\mathbb{H}$ onto $\mathbb{C}^* \setminus (\gamma_1 + \gamma_2 + \cdots + \gamma_n)$, where $\gamma_1 + \gamma_2 + \cdots + \gamma_2$ is (the image of) a simple curve from z_0 to z_n passing through $z_1, \ldots, z_{n-1}$. See Figure 8.1.

Next, let σ_{n+1} be the semi-circle (or half-line) in $\mathbb{H}$ from 0 to $\varphi^{-1}(z_0) \in \mathbb{R} \cup \{\infty\}$ which is orthogonal to $\mathbb{R}$. Let $\gamma_{n+1} = \varphi(\sigma_{n+1})$. Then $\gamma \equiv \gamma_1 + \gamma_2 + \cdots + \gamma_{n+1}$ is (the image of) a simple closed curve containing $z_0, z_1, \ldots, z_n$. The semicircle (or half line) σ_{n+1} divides $\mathbb{H}$ into two regions, D^+ and D^-, and so $\varphi(\mathbb{H} \setminus \sigma_{n+1})$ is the union of two disjoint simply-connected regions Ω^+ and Ω^- with $\mathbb{C}^* = \Omega^+ \cup \Omega^- \cup \gamma$. Finally we can find an LFT, τ, of $\mathbb{H}$ onto $\mathbb{H}$ so that $\tau(\sigma_{n+1})$ is the positive imaginary axis. Using the maps $\sqrt{z}$ and $-\sqrt{-z}$, with $\sqrt{1} = 1$, followed by τ^{-1} we obtain conformal maps of $\mathbb{H}$ onto D^+ and D^-. Composing with φ yields conformal maps φ^+ and φ^- of $\mathbb{H}$ onto Ω^+ and Ω^-, respectively. Each map in the full composition has an explicit inverse map, and composing the inverse maps in the reverse order produces conformal maps of Ω^+ and Ω^- onto $\mathbb{H}$.

This process can be applied to any finite sequence of distinct points $z_0, \ldots, z_n$ unless the points are out of order, in the sense that some z_j already belongs to $\gamma_1 + \cdots + \gamma_{j-1}$. When this happens, z_j can be skipped.

The map $\sqrt{z^2 - 1}$ extends to be analytic and one-to-one in a neighborhood of each of the intervals in $\mathbb{R} \setminus \{-1, 0, 1\}$. LFTs are conformal everywhere. It follows that φ^+ and φ^- extend to be one-to-one continuous maps of $\mathbb{H} \cup \mathbb{R} \cup \{\infty\}$ onto the closures of Ω^+ and Ω^-, respectively. This is perhaps clearer if the map φ is examined on D^+ and D^- separately.

The curve γ also has a continuously turning (unit) tangent. This follows because if σ is a curve whose image is orthogonal to $\mathbb{R}$ at 0, then $z^2 - 1$ applied to σ has an image which is tangential to $\mathbb{R}$ at $z = -1$. Because $\sqrt{z}$ is analytic in a neighborhood of -1, the image α of σ by the map $\sqrt{z^2 - 1}$ is tangential to the interval $[0, i]$ at $z = i$, and thus α meets $[0, i]$ at an angle of π. Because conformal maps preserve angles between curves, any subsequent composition with conformal maps on $\mathbb{H}$ will preserve this angle of intersection. Because γ is built from compositions of $\sqrt{z^2 - 1}$ and LFTs, it must have a continuously turning (unit) tangent. The tangent is also continuous at z_0 because σ_{n+1} is orthogonal to $\mathbb{R}$ not only at 0 but also at its other endpoint on $\mathbb{R} \cup \{\infty\}$, and that angle is preserved by $\varphi_2 \circ \varphi_3 \circ \cdots \circ \varphi_n$. But φ_1 then doubles the angle to π because of the presence of z^2 in (8.1). It might be easier to understand orthogonality and the behavior of z^2 when the point in question is at ∞ by using the LFT $-1/z$ of $\mathbb{H}$ onto $\mathbb{H}$, which interchanges ∞ and 0. Because the (unit) tangent direction is continuous, the curve γ is continuously differentiable when it is parameterized by arc-length.

The constructed curve γ can be described in the language of hyperbolic geometry.

Definition 8.1 *If $\Omega \subset \mathbb{C}^*$ is simply-connected, then γ is called a* **hyperbolic geodesic** *in Ω if there is a conformal map ψ of the upper half-plane $\mathbb{H}$ onto Ω such that $\gamma = \psi(I)$, where I is the positive imaginary axis. The image of an arc contained in a hyperbolic geodesic will be called a* **geodesic arc**.

For example, a subarc in $\mathbb{H}$ of a circle orthogonal to $\mathbb{R}$ is a geodesic arc using an LFT and Theorem 6.3. See Exercises 3.8, 3.9 and 6.8 for more information about the hyperbolic geometry on $\mathbb{D}$.

Definition 8.2 *An (open) simple arc γ is called* **analytic** *if there exists a function g which is one-to-one and analytic in a neighborhood N of the open interval $(0, 1)$ with $g(t) = \gamma(t)$, $0 < t < 1$. A simple closed curve σ is called analytic if there exists a one-to-one analytic function g defined in a neighborhood of $\partial\mathbb{D}$ with $g(e^{it}) = \sigma(e^{it})$, $0 \le t \le 2\pi$. When the parameterization of an arc or closed curve is not specified, we say that it is analytic if it is analytic for some parameterization.*

See Exercise 8.2(f) for an equivalent local definition of an analytic arc. By definition, then, a geodesic arc is the image of an analytic arc. In particular, it has a parameterization that is infinitely differentiable.

The image of the curve $\gamma = \gamma_1 + \cdots + \gamma_{n+1}$ consists of a line segment from z_0 to z_1 followed by the images of arcs $\gamma_j, j = 2, \ldots, n+1$. The image of the arc γ_j is, by construction, a hyperbolic geodesic in the region $\Omega_j = \mathbb{C}^* \setminus (\gamma_1 + \cdots + \gamma_{j-1})$ and therefore γ_j is analytic except at its endpoints. We have proved the following:

Theorem 8.3 *Given distinct points $z_0, z_1, \ldots, z_n \in \mathbb{C}$, the geodesic algorithm constructs a continuously differentiable and piecewise analytic simple closed curve γ whose image contains these points, together with conformal maps of $\mathbb{H}$ onto the two regions in $\mathbb{C}^* \setminus \gamma$, as well as the inverses of these maps.*

Given a large number of data points $z_0, \ldots, z_n$ on a simple closed curve we can use this algorithm on a computer to find conformal maps from $\mathbb{H}$ onto the two regions bounded by the piecewise geodesic curve. A natural question, then, is: what does the curve look like between the data points? Is it close to the closed curve from which we took the data points? The proof of Theorem 8.4 is a typical use of harmonic functions and the maximum principle to estimate analytic functions.

Theorem 8.4 (Jørgensen) *Suppose Δ is a closed disk contained in a simply-connected region Ω. If J is a hyperbolic geodesic in Ω, then $J \cap \Delta$ is connected and $J \cap (\partial\Delta)$ consists of at most two points.*

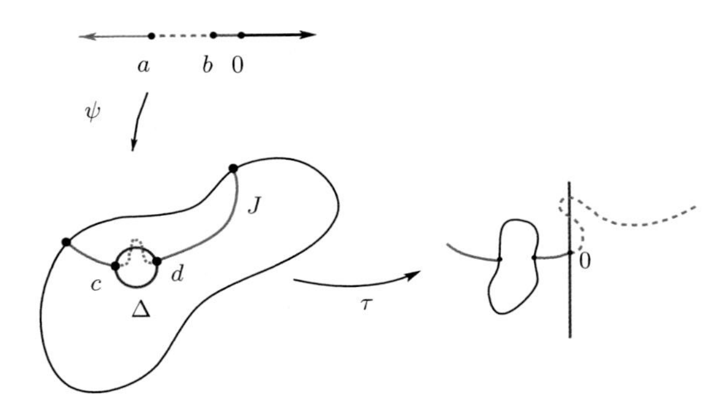

Figure 8.2 Proof of Jørgensen's theorem.

Jørgensen's theorem says that closed disks contained in Ω are **strictly convex** in the hyperbolic geometry on Ω.

Proof Because J is a hyperbolic geodesic, there is a conformal map ψ of $\mathbb{C} \setminus [0,\infty)$ onto Ω such that $\psi((-\infty,0)) = J$. Here we composed the conformal map in Definition 8.2 with $\sqrt{z}$. We may suppose that $J \cap \Delta$ consists of more than one point. Let c be the first intersection of J with Δ and let d be the last intersection. So there are $a < b < 0$ with $\psi(a) = c \in \partial\Delta$, $\psi(b) = d \in \partial\Delta$, $\psi((-\infty,a)) \cap \Delta = \emptyset$ and $\psi((b,0)) \cap \Delta = \emptyset$. By Theorem 6.3, we can choose a constant $\lambda \in \mathbb{C}$ so that the LFT $\tau(z) = \lambda(z-c)/(z-d)$ maps the interior of Δ onto the right half-plane, $\{z : \text{Re}z > 0\}$. Then $\tau(\partial\Omega \cup J \setminus \Delta) \subset \{z : \text{Re}z < 0\}$. See Figure 8.2.

Set

$$f(z) = \lambda \left(\frac{\psi(z) - c}{\psi(z) - d}\right)\left(\frac{z-b}{z-a}\right).$$

Then f is bounded and analytic in $U = \mathbb{C} \setminus ((-\infty,a] \cup [b,\infty))$. Moreover, if $x \in (-\infty,a) \cup (b,\infty)$ then

$$\limsup_{z \in U \to x} \text{Re}f(z) = \left(\frac{x-b}{x-a}\right) \limsup_{z \in U \to x} \text{Re}\ \tau(\psi(z)) < 0.$$

By Theorem 7.15, Lindelöf's maximum principle, $\text{Re}f(z) < 0$ in U. But if $a < x < b$, then $(x-b)/(x-a) < 0$ so that

$$\text{Re}\left(\lambda \frac{\psi(x) - c}{\psi(x) - d}\right) = \text{Re}\ \tau(\psi(x)) > 0.$$

Thus $\psi(x) \in \text{interior}(\Delta)$, proving Theorem 8.4. □

It follows from Jørgensen's theorem that the geodesic arc between $c, d \in \Omega$ lies in the intersection of all closed disks $\Delta \subset \Omega$ with $c, d \in \partial\Delta$. This is a "lens" shaped region given by the intersection of two disks.

Corollary 8.5 *Suppose D is an open disk contained in a simply-connected region $\Omega \subset \mathbb{C}^*$ and suppose J is a geodesic in Ω which is tangent to ∂D at $z_0 \in \Omega$; then $J \cap D = \emptyset$.*

Proof If $z_1 \in J \cap D$ then we can find a slightly smaller closed disk $\Delta \subset D \cup \{z_0\} \subset \Omega$ tangent to J at z_0 and with z_1 in the interior of Δ. By Jørgensen's theorem, the portion of J between z_0 and z_1 is contained in Δ. Because Δ is a compact subset of Ω, we can rotate Δ slightly about the point z_0 and remain inside Ω, yet still contain both z_0 and z_1. Then the portion of J between z_0 and z_1 must be contained in the intersection of these rotations by Jørgensen's theorem. This is impossible if J is tangent to ∂D at z_0. □

Corollary 8.5 can be helpful in locating a geodesic. If $z_0 \in \Omega$ and if J is a geodesic in Ω containing z_0, then let D_1 and D_2 be the largest possible disjoint open disks contained in Ω and tangent to J at z_0. Then J does not intersect $D_1 \cup D_2$.

Using Jørgensen's theorem we can build approximations to conformal maps on a simply-connected region.

Definition 8.6 *A* **disk-chain** $D_0, D_1, \ldots, D_n$ *is a sequence of pairwise disjoint open disks such that* ∂D_j *is tangent to* ∂D_{j+1}, *for* $j = 0, \ldots, n-1$. *A* **closed disk-chain** *is a disk-chain such that* ∂D_n *is also tangent to* ∂D_0.

Any simple closed polygon P, for example, can be covered by a closed disk-chain of disks with arbitrarily small radii and centers on P, for example by placing disks centered at each vertex, then covering each of the remaining open line segments with disks centered on the segments. Another method for constructing a disk-chain is to cover the plane with a hexagonal grid of disks with radius $\varepsilon > 0$. Then select a sequence of these disks to form a disk-chain. Yet another method is to use the curves γ constructed in the proof of Runge's Theorem 4.23, for a bounded simply-connected region Ω. Pave the plane with squares of side length d and shade the squares whose closure is contained in Ω. Let σ be the boundary of a connected component of the interior of the union of the closed shaded squares. Because σ is the union of edges of squares with sides of length d, disks of diameter d centered at the endpoints of each square edge in σ then form a closed disk-chain. See Figure 8.3.

If $D_0, D_1, \ldots, D_n$ is a closed disk-chain, set

$$z_j = \partial D_j \cap \partial D_{j+1},$$

for $j = 0, \ldots, n$, where $D_{n+1} \equiv D_0$. The next theorem gives one geometric method for describing the contours in the constructed curve γ from the geodesic zipper algorithm.

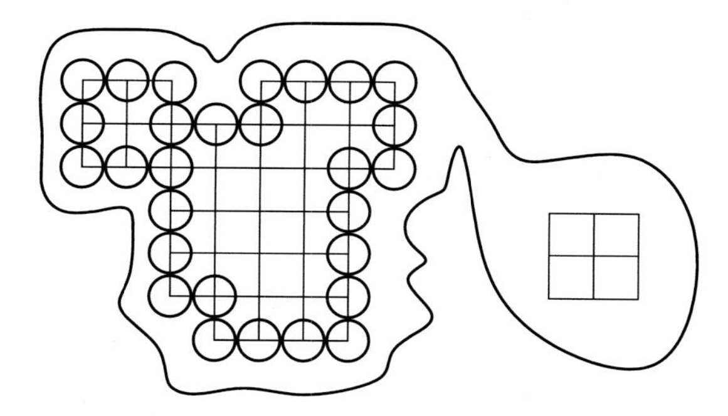

Figure 8.3 A disk-chain.

Theorem 8.7 (Marshall-Rohde) *If $D_0, D_1, \ldots, D_n$ is a closed disk-chain, then the geodesic zipper algorithm applied to the data $z_0, z_1, \ldots, z_n$ produces explicit formulae for the conformal maps $\varphi^{\pm}$ of the upper half-plane $\mathbb{H}$ onto the two regions bounded by a continuously differentiable and piecewise analytic curve γ with*

$$\gamma \subset \bigcup_{0}^{n} (D_j \cup z_j).$$

The algorithm also produces explicit formulae for the inverses of $\varphi^{\pm}$.

We emphasize, though, that the geodesic zipper algorithm does not require the construction of a disk-chain. Disk-chains give a simple geometric method for locating the constructed curve. See also Exercise 8.10.

Proof Set

$$U_j = \{z_0\} \cup \bigcup_{k=1}^{j} (D_k \cup \{z_k\}),$$

for $j = 1, 2, \ldots, n$.

Since γ_1 is a straight-line segment, $\gamma_1 \subset U_1$ and γ_1 is not tangent to ∂D_1 at z_1. We proceed by induction. Suppose now that $\gamma_1 + \cdots + \gamma_{j-1} \subset U_{j-1}$. By Corollary 8.5, γ_{j-1} is not tangent to ∂D_{j-1} at z_{j-1}. Because $\gamma_1 + \cdots + \gamma_j$ has a continuous tangent and D_j is tangent to D_{j-1} at z_{j-1}, the initial portion of the curve γ_j near z_{j-1} is contained in D_j. If $j < n+1$, find a slightly smaller closed disk $\Delta \subset D_j \cup \{z_j\}$ so that $z_j \in \partial\Delta$ and $\Delta \cap \gamma_j$ contains at least two points. Then Δ is a closed disk in Ω_{j-1}. See Figure 8.4. By Jørgensen's theorem, $\gamma_j \subset U_j$. If $j = n+1$, note that both γ_n and γ_0 are not tangent to ∂D_{n+1}. Since $\gamma_1 + \cdots + \gamma_{n+1}$ has a continuous tangent, we can apply Jørgensen's theorem on slightly smaller closed disks $\Delta \subset D_{n+1}$ to conclude that $\gamma_{n+1} \subset D_{n+1} \cup \{z_n, z_0\}$. Theorem 8.7 follows. □

Figure 8.5 shows conformal maps of grids on the disk and its complement to the interior and exterior of a simple closed curve using the geodesic zipper algorithm. Given a simply-connected region Ω in the plane, the geodesic zipper algorithm can be used to approximate a conformal map of $\mathbb{D}$ onto Ω, by using lots of points on the boundary. Exercise 8.10 shows that, for a given number of points, a more judicious choice of these points can lead to more accurate approximations than using, for instance, approximately equally spaced points on the boundary or the points of tangency of disks constructed from the approximation of Ω by a union of squares.

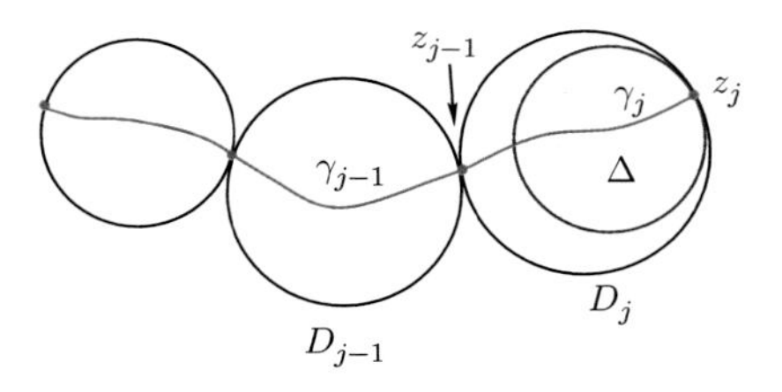

Figure 8.4 A smaller disk $\Delta \subset D_j$.

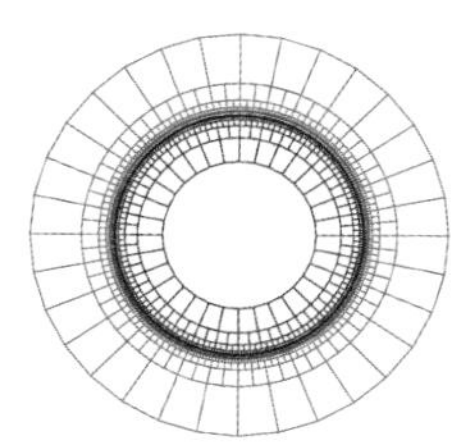
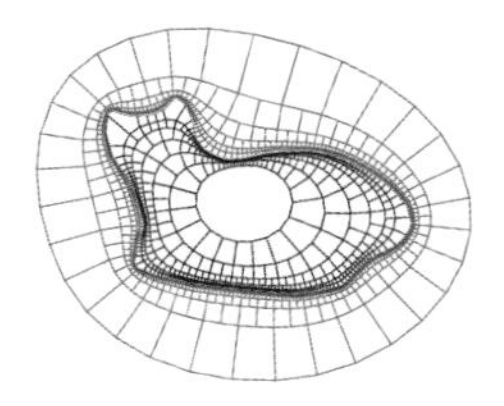

Figure 8.5 Conformal mapping using the geodesic zipper algorithm.

8.2 The Riemann Mapping Theorem

Our next major goal is to prove the Riemann mapping theorem, that every simply-connected region contained in the plane (except the plane itself) can be mapped, one-to-one and analytically, onto the unit disk. The original approach to this problem, due to Riemann, was to minimize a certain integral (with side conditions) over a collection of functions. This method had the advantage of a physical interpretation, but it *assumed* that there was a minimizer. Weierstrass then showed that there are similar minimization problems with no minimizer. Existence proofs for this and other related problems dominated mathematical thinking in analysis for the next 50 or so years. Around 1900, Hilbert finally patched up the difficulties in Riemann's approach, though the proof does not apply to all simply-connected regions. See J. Gray [11] for a discussion of some of the history.

We give a somewhat constructive proof of the Riemann mapping theorem in this section. Other proofs will be given in Chapters 10 and 13. The idea of the proof here is to construct, using the geodesic zipper algorithm, conformal maps from an increasing sequence of regions onto the disk, then prove that the sequence converges to a conformal map of the union of these regions onto the disk.

We first prove some preliminary results. The next two results, due to Hurwitz, are typical applications of the argument principle and will be key to finding solutions to many extremal problems.

Theorem 8.8 (Hurwitz) *Suppose $\{g_n\}_{n=1}^{\infty}$ is a sequence of analytic functions on a region Ω and suppose $g_n(z) \neq 0$ for all $z \in \Omega$ and all n. If g_n converges uniformly to g on compact subsets of Ω, then either g is identically zero in Ω or g is never equal to 0 in Ω.*

Proof The limit function g is analytic on Ω by Weierstrass's theorem. In particular, if g is not identically zero, then the zeros of g are isolated in Ω. Moreover, again by Weierstrass's theorem, g_n' converges to g' uniformly on compact subsets of Ω. If $\Delta \subset \Omega$ is a closed disk with $g \neq 0$ on $\partial\Delta$, then g_n'/g_n converges to g'/g uniformly on $\partial\Delta$. By the argument principle, for n sufficiently large, the number of zeros of g_n in Δ is the same as the number of zeros of g in Δ. Because g_n is never zero, the theorem follows. □

Corollary 8.9 (Hurwitz) *If $\{g_n\}_1^{\infty}$ is a sequence of one-to-one and analytic functions on a region Ω, and if g_n converges to g uniformly on compact subsets of Ω, then either g is one-to-one and analytic on Ω or g is constant.*

Proof Fix $w \in \Omega$ and apply Hurwitz's Theorem 8.8 to $g - g(w)$ on $\Omega \setminus \{w\}$. □

The following lemma says that, to prove convergence of a sequence of analytic functions, it is enough to prove convergence of their real parts, together with the full sequence at one point.

Lemma 8.10 *Suppose f_n is analytic on a region Ω_n with $\Omega_n \subset \Omega_{n+1} \subset \Omega = \cup_n \Omega_n$. Suppose: $z_0 \in \Omega_1$; $\{\text{Re} f_n\}$ converges uniformly on compact subsets of Ω; and $\{f_n(z_0)\}$ converges. Then $\{f_n\}$ converges uniformly on compact subset of Ω.*

In the statement of Lemma 8.10, uniform convergence of f_n on a compact set $K \subset \Omega$ means that f_n is defined on K for $n \geq n_0 = n_0(K)$ and $\{f_n\}_{n_0}^{\infty}$ converges uniformly on K.

Proof If f is analytic on $\overline{\mathbb{D}}$, then, by Corollary 7.7 for $|z| \leq r < 1$,

$$|f(z) - i\text{Im} f(0)| = \left| \int_0^{2\pi} \frac{e^{it} + z}{e^{it} - z} \text{Re} f(e^{it}) dt \right| \leq \frac{1+r}{1-r} \sup_{|z|=1} |\text{Re} f(z)|. \tag{8.2}$$

If g_n is analytic on $\overline{\mathbb{D}}$ and $\{\text{Re} g_n\}$ is a Cauchy sequence in the uniform norm on $\overline{\mathbb{D}}$ then, by (8.2) applied to $g_n - g_m$, we conclude $\{g_n - i\text{Im} g_n(0)\}$ is Cauchy in the uniform norm on $|z| \leq r$.

Now suppose $\{\text{Re} f_n\}$ converges uniformly on compact subsets of Ω. If $z_1 \in \Omega$, and if $B(z_1) = \{z : |z - z_1| \leq \frac{1}{2} \text{dist}(z_1, \partial\Omega)\}$, then, by a linear change of variables,

$$\{f_n - i\text{Im} f_n(z_1)\} \text{ is a Cauchy sequence on } B(z_1). \tag{8.3}$$

Let $E = \{z_1 \in \Omega : \{f_n(z_1)\} \text{ converges}\}$. Then E is open by (8.3). If $z_1 \in \Omega \setminus E$ then, by (8.3) and the convergence of $\{\text{Re} f_n(z_1)\}$, we have $B(z_1) \subset \Omega \setminus E$. By hypothesis, $z_0 \in E$ and Ω is connected so $E = \Omega$. Because any compact subset of Ω can be covered by finitely many balls of the form $B(z_1)$, $\{f_n\}$ converges uniformly on compact subsets of Ω. □

Next we combine these results with Harnack's principle and Schwarz's lemma. See [11] for the history of contributions to this result by Harnack, Schwarz, Carathéodory, and others.

Theorem 8.11 (Carathéodory) *Suppose Ω is a simply-connected region with $z_0 \in \Omega \neq \mathbb{C}$. Suppose also that Ω_n are simply-connected regions with*

$$z_0 \in \Omega_n \subset \Omega_{n+1} \subset \bigcup_{n=1}^{\infty} \Omega_n = \Omega,$$

and suppose f_n is a conformal map of Ω_n onto $\mathbb{D}$, with $f_n(z_0) = 0$ and $f_n'(z_0) > 0$, $n = 1, 2, \ldots$. Then $\{f_n\}$ converges uniformly on compact subsets of Ω to a conformal map f of Ω onto $\mathbb{D}$, with $f(z_0) = 0$ and $f'(z_0) > 0$.

Proof We first show that we may suppose $\Omega \subset \mathbb{D}$ and $z_0 = 0$, without loss of generality, by constructing a conformal map g_0 of Ω into $\mathbb{D}$, with $g_0(z_0) = 0$ and $g_0'(z_0) > 0$. Indeed, if g_n is a conformal map of $g_0(\Omega_n)$ onto $\mathbb{D}$ with $g_n(0) = 0$ and $g_n'(0) > 0$ then $f_n = g_n \circ g_0$.

By assumption, there is a point $z_1 \in \mathbb{C} \setminus \Omega$. Because Ω is simply-connected we can define $\log(z - z_1)$ so as to be analytic on Ω, by Corollary 5.8. Set

$$g_1(z) = \sqrt{z - z_1} \equiv e^{\frac{1}{2}\log(z-z_1)}.$$

If $g_1(z) = g_1(w)$ then, by squaring, $z - z_1 = w - z_1$ and hence $z = w$. Thus g_1 is one-to-one. The same argument shows that if $z \neq w$ then $g_1(z) \neq -g_1(w)$. If $g_1(z_2) \neq 0$ then $g_1(\Omega)$ covers a neighborhood of $g_1(z_2)$ because g_1 is open, and hence $g_1(\Omega)$ omits a neighborhood of $-g_1(z_2)$. Thus

$$g_2(z) = \frac{C}{g_1(z) + g_1(z_2)}$$

is analytic and one-to-one on Ω and bounded by 1 for C sufficiently small. The function

$$g_3(z) = \frac{g_2(z) - g_2(z_0)}{1 - \overline{g_2(z_0)}g_2(z)}$$

is the composition of g_2 with an LFT, and hence one-to-one with $g_3(z_0) = 0$. Now set $g_0(z) = \frac{|g_3'(z_0)|}{g_3'(z_0)} g_3(z)$. Henceforth we will assume $\Omega \subset \mathbb{D}$ and $z_0 = 0$.

Because Ω_n is simply-connected and $z/f_n(z)$ is a non-vanishing function on Ω_n, we can define $\log(z/f_n(z))$ as an analytic function on Ω_n with value $\log(1/f_n'(0)) \in \mathbb{R}$ at 0. Then $u_n(z) \equiv \operatorname{Re}\log(z/f_n(z)) = \log|z/f_n(z)|$ is harmonic on Ω_n. By Schwarz's lemma applied to $f_{n+1} \circ f_n^{-1}$, we have that $u_n(z) \leq u_{n+1}(z)$ for all $z \in \Omega_n$. Moreover, $u_n(0) = \log|1/f_n'(0)| \leq 0$ by Schwarz's lemma applied to f_n^{-1}, because $\Omega_n \subset \mathbb{D}$. Applying Harnack's principle, we conclude that u_n converges uniformly on compact subsets of Ω to a harmonic function u. The value of $\log(z/f_n(z))$ at 0 is equal to $u_n(0)$, which converges to $u(0)$. By Lemma 8.10, $\log(z/f_n(z))$ converges uniformly on compact subsets of Ω, and hence $f_n(z) = z\exp(-\log(z/f_n(z)))$ converges uniformly on compact subsets of Ω to an analytic function f. The limit function f is bounded by 1 on Ω with $f(0) = 0$ and $f'(0) = e^{-u(0)} > 0$. By Hurwitz's theorem (Corollary 8.9), f is a one-to-one conformal map.

It remains to prove that f maps Ω onto $\mathbb{D}$. Set $g_n = f \circ f_n^{-1}$. Then g_n is analytic on $\mathbb{D}$ with $|g_n| \leq 1$ and $g_n'(0) \to 1$. If $\zeta_0 \in \mathbb{D} \setminus f(\Omega)$ then $g_n \neq \zeta_0$ so that

$$T(z) = \frac{z - \zeta_0}{1 - \overline{\zeta_0}z}$$

is a non-vanishing analytic function on $g_n(\mathbb{D})$, which is simply-connected by Proposition 6.6. Then, as in the first part of this proof, T has an analytic square root $\sqrt{T} = \exp(\frac{1}{2}\log T)$ on $g_n(\mathbb{D})$ which is one-to-one. Let

$$S(z) = \frac{z - \sqrt{T}(0)}{1 - \overline{\sqrt{T}(0)}z}.$$

Thus $h_n = S \circ \sqrt{T} \circ g_n$ is an analytic function mapping $\mathbb{D}$ into $\mathbb{D}$, vanishing at 0 with

$$|h_n'(0)| = \frac{1}{1 - |\zeta_0|}\frac{1}{2|\zeta_0|^{\frac{1}{2}}}(1 - |\zeta_0|^2)g_n'(0) = \left(\frac{1 + |\zeta_0|}{2|\zeta_0|^{\frac{1}{2}}}\right) g_n'(0) > 1,$$

for n sufficiently large. This contradicts Schwarz's lemma and hence f maps Ω onto $\mathbb{D}$. $\square$

Theorem 8.12 *If $\Omega \subset \mathbb{C}$ is a simply-connected region with $\Omega \neq \mathbb{C}$ and $z_0 \in \Omega$, then we can find conformal maps f_n constructed using the geodesic zipper algorithm, with regions $\Omega_n = f_n^{-1}(\mathbb{D})$ such that $\Omega_n \subset \Omega_{n+1}$ and $\cup_n \Omega_n = \Omega$. Moreover, we can choose f_n so that $f_n(z_0) = 0$ and $f_n'(z_0) > 0$.*

Proof By the first part of the proof of Theorem 8.11, we may suppose that $\Omega \subset \mathbb{D}$ and $z_0 = 0$. If $\delta > 0$ then, as in the proof of Runge's Theorem 4.23, pave the plane with squares of side length δ. If S is a square, $3S$ will denote the square with the same center as S but edge length three times as long. Shade each square S with the property that $3S \subset \Omega$. Let U be the component of the interior of the union of the closed shaded squares with $0 \in U$. The set U is non-empty if δ is sufficiently small. Then ∂U is the union of edges of squares of side length δ. Let $C_\delta = D_0 \cup D_1 \cup \cdots \cup D_m$ be the closed disk-chain of disks of diameter δ centered at the endpoints of each square edge in ∂U. See Figure 8.3. Then

$$\delta/2 \leq \text{dist}(\zeta, \partial\Omega) \leq 4\delta, \tag{8.4}$$

for each $\zeta \in C_\delta$. Let Ω_δ be the simply-connected region constructed via the geodesic zipper algorithm, as in Theorem 8.7, with $\partial\Omega_\delta \subset C_\delta$, and let f_δ be the corresponding conformal map of Ω_δ onto $\mathbb{D}$, which we can normalize so that $f_\delta(0) = 0$ and $f_\delta'(0) > 0$.

Choose a sequence $\delta_n > 0$ with $\delta_{n+1} < 8\delta_n$. Then, by (8.4), $\Omega_{\delta_n} \subset \Omega_{\delta_{n+1}}$ and $z_0 \in \Omega_{\delta_1}$ for δ_1 sufficiently small. If $z_1 \in \Omega$ then there is a polygonal curve $\gamma \subset \Omega$ from z_0 to z_1. If $4\delta_n < \text{dist}(\gamma, \partial\Omega)$ then $\gamma \subset \Omega_n$ and hence $\Omega = \cup_n \Omega_n$. □

The reason for reducing to the case where $\Omega \subset \mathbb{D}$ is to guarantee C_δ is a finite union of disks. Finally we deduce the Riemann mapping theorem.

Theorem 8.13 (Riemann mapping theorem) *Suppose Ω is a simply-connected region with $z_0 \in \Omega \neq \mathbb{C}$. Then there is a unique one-to-one analytic map f of Ω onto $\mathbb{D}$ with $f(z_0) = 0$ and $f'(z_0) > 0$.*

Proof Theorem 8.13 follows from Theorem 8.12, Theorem 8.11 and Proposition 6.5. □

8.3 Symmetry and Conformal Maps

In this section we will explore the connection between symmetry in a simply-connected region and symmetry of a conformal map from the region to the disk $\mathbb{D}$.

We say that a region Ω is **symmetric about** $\mathbb{R}$ provided $\bar{z} \in \Omega$ if and only if $z \in \Omega$. If Ω is symmetric about $\mathbb{R}$ and if f is analytic on Ω, then $\overline{f(\bar{z})}$ is analytic on Ω. Indeed, the power series expansion for $\overline{f(\bar{z})}$ based at $\bar{b}$ has coefficients which are the complex conjugates of the coefficients in the power series for f based at b. If f is also real valued on an interval in $\Omega \cap \mathbb{R}$, then

$$f(z) \equiv \overline{f(\bar{z})} \tag{8.5}$$

by the uniqueness theorem, Corollary 2.9. Geometrically, (8.5) says that f maps symmetric points to symmetric points. For one-to-one functions we have the following related result:

Proposition 8.14 *If f is a conformal map of a simply-connected symmetric region Ω onto $\mathbb{D}$ such that $f(x_0) = 0$ and $f'(x_0) > 0$ for some $x_0 \in \mathbb{R} \cap \Omega$, then $f(z) = \overline{f(\overline{z})}$.*

Proof Apply the uniqueness conclusion in the Riemann mapping theorem to the functions f and $\overline{f(\overline{z})}$. □

Proposition 8.14 says that if the region is symmetric then we can take the mapping function to be symmetric. We would like to use this idea to help construct conformal maps.

Theorem 8.15 (Schwarz reflection principle) *Suppose Ω is a region which is symmetric about $\mathbb{R}$. Set $\Omega^+ = \Omega \cap \mathbb{H}$ and $\Omega^- = \Omega \cap (\mathbb{C} \setminus \overline{\mathbb{H}})$. If v is harmonic on Ω^+, continuous on $\Omega^+ \cup (\Omega \cap \mathbb{R})$ and equal to 0 on $\Omega \cap \mathbb{R}$, then the function defined by*

$$V(z) = \begin{cases} v(z), & \text{for } z \in \Omega \setminus \Omega^- \\ -v(\overline{z}), & \text{for } z \in \Omega^- \end{cases}$$

is harmonic on Ω. If also $v(z) = \operatorname{Im} f(z)$, where f is analytic on $\mathbb{H} \cap \Omega$, then the function

$$g(z) = \begin{cases} f(z), & \text{for } z \in \Omega^+ \\ \overline{f(\overline{z})}, & \text{for } z \in \Omega^- \end{cases}$$

extends to be analytic in Ω.

Note that there is no assumption in Theorem 8.15 about boundary values of the real part of f on $\Omega \cap \mathbb{R}$. The region Ω is not necessarily simply-connected either.

Proof The extended function V is continuous on Ω. To prove the first claim, we need only prove that V has the mean-value property for small circles centered on $\Omega \cap \mathbb{R}$. But, for $x_0 \in \mathbb{R} \cap \Omega$, $V(x_0 + re^{it}) = -V(x_0 + re^{-it})$ so that the mean value over a circle centered at x_0 contained in Ω is zero, the value of V at x_0. Thus V is harmonic in Ω.

Suppose now that $v = \operatorname{Im} f$, where f is analytic on $\mathbb{H} \cap \Omega$. If D is a disk contained in Ω and centered on $\mathbb{R}$, then $V = \operatorname{Im} h$ for some analytic function h on D. It is uniquely defined by requiring that $h = f$ on $\Omega^+ \cap D$. By (8.5) $h(z) = \overline{h(\overline{z})}$ on D, and so $h = g$ on $D \cap \Omega \setminus \mathbb{R}$. Thus h provides the unique analytic extension of g to all of D, and g extends to be analytic on Ω. □

Corollary 8.16 *If the function f in the Schwarz reflection principle is also one-to-one in Ω^+ with $\operatorname{Im} f > 0$ on Ω^+, then its extension g is also one-to-one on Ω.*

Proof By definition, g is one-to-one on Ω^- with $\operatorname{Im} g < 0$ on Ω^-. So if $g(z_1) = g(z_2)$ then $z_1, z_2 \in \mathbb{R}$. Since g is open, it maps a small disk centered at z_j, $j = 1, 2$, onto a neighborhood of $g(z_1) = g(z_2)$. If $z_1 \neq z_2$ then there are two points $\zeta_1, \zeta_2 \in \Omega^+$ near z_1, z_2 (respectively) with $g(\zeta_1) = g(\zeta_2)$, contradicting the assumption that f is one-to-one on Ω^+. □

The function $f(z) = z^2$ is one-to-one on $\mathbb{D}^+$, and real on $\mathbb{D} \cap (-1, 1)$, but does not have a one-to-one extension to $\mathbb{D}$, so the assumption $\operatorname{Im} f > 0$ on Ω^+ cannot be removed.

As an application of Corollary 8.16 we outline how to find a conformal map of $\mathbb{H}$ to the region $\Omega = \{x+iy : y > x^2\}$, which is somewhat different from the approach in Exercise 6.12. First find a map f_1 of $\mathbb{H}$ onto the half-strip $S = \{z = x + iy : 0 < x < b, y > 0\}$ so that $f_1(\infty) = \infty$ (i.e., $\lim_{|z|\to\infty} |f_1(z)| = \infty$), where b is chosen later. The function $f_2(z) = z^2$ is one-to-one on S with image equal to half of the interior region of a parabola. The composed function $f_2 \circ f_1$ maps an interval $I_a = (-\infty, a]$ onto $I_{b^2} = (-\infty, b^2]$ and $[a, +\infty)$ onto the half-parabola. By Corollary 8.16 we can extend this function by reflecting the domain across the interval I_a and the range across I_{b^2} to obtain a conformal map of $\mathbb{C} \setminus [a, \infty)$ onto the interior of a parabola. The slit plane can be mapped onto $\mathbb{H}$ using $\sqrt{z-a}$ and the parabola can be dilated, translated and rotated using a linear map. With the proper choice of b we obtain the desired region.

This same idea can be used to construct a conformal map from $\mathbb{H}$ to the region outside one branch of a hyperbola, which was also done in Exercise 6.11.

It is sometimes useful to be able to reflect across curves more general than line segments.

Definition 8.17 *Let $I = (0, 1)$. An open analytic arc γ contained in the boundary of a region Ω is called a* **one-sided** *arc if there exists a function g which is one-to-one and analytic in a neighborhood N of I with $g(I) = \gamma$ and $g(N \cap \mathbb{H}) \subset \Omega$ and $g(N \setminus \overline{\mathbb{H}}) \subset \Omega^c$. If $g(N\setminus I) \subset \Omega$ then γ is called a* **two-sided** *arc.*

Corollary 8.18 *Suppose Ω is a simply-connected region and suppose $\gamma \subset \partial\Omega$ is a one-sided analytic arc. Let f be a conformal map of Ω onto $\mathbb{D}$. Then f extends to be analytic and one-to-one in a neighborhood of $\Omega \cup \gamma$.*

Proof By Exercise 3.7 the map f is proper so that $|f(z)| \to 1$ as $z \to \partial\Omega$. If $\gamma \neq \partial\Omega$, choose g and N as in Definition 8.17. Then $f \circ g$ is analytic in $N \cap \mathbb{H}$. By shrinking N if necessary, we may suppose N is symmetric about $\mathbb{R}$, $N \cap \mathbb{H}$ is simply-connected and $f \circ g \neq 0$ on N. We can then define $h = i \log f \circ g$ to be analytic in $N \cap \mathbb{H}$ and

$$\lim_{z\in N\to(0,1)} \operatorname{Im} h(z) = \lim_{z\in N\to(0,1)} -\log|f \circ g(z)| = 0.$$

By Corollary 8.16, h extends to be one-to-one and analytic in N. Therefore f extends to be one-to-one and analytic in $\Omega \cup g(N)$. See Exercise 8.2(c) if $\gamma = \partial\Omega$. □

Corollary 8.18 remains valid if we replace $\mathbb{D}$ by $\mathbb{H}$, although ∞ presents a small complication. It is not true that a conformal map f of Ω onto $\mathbb{H}$ satisfies $\operatorname{Im} f \to 0$ as $z \to \gamma$ because ∞ is a boundary point of $\mathbb{H}$ in $\mathbb{C}^*$. Thus we cannot directly apply the Schwarz reflection principle. However, composing f with an LFT τ of $\mathbb{H}$ onto $\mathbb{D}$, then applying Corollary 8.18, we conclude that $\tau \circ f$ extends to be analytic and one-to-one in a neighborhood of $\Omega \cup \gamma$, so that f extends to be analytic and one-to-one in a neighborhood of $\Omega \cup \gamma$ with the possible exception of a simple pole at a single point of γ.

If h is analytic, but not necessarily one-to-one, in a region Ω whose boundary contains a one-sided analytic arc γ, and if $h(z)$ tends to an analytic arc σ as $z \to \gamma$, then h extends to be analytic in a neighborhood of $\Omega \cup \gamma$. See Exercise 8.2(d).

Care must be taken if γ is a two-sided arc in $\partial\Omega$. For example, if $\Omega = \mathbb{C} \setminus I$, where $I = (-\infty, 0]$, then the function $\sqrt{z}$ defined in Ω has an analytic extension across $I^o = (-\infty, 0)$ from the upper half-plane, and across I^o from the lower half-plane, but these extensions do not agree with the original function.

8.4 Conformal Maps to Polygonal Regions

In this section we will use the Schwarz reflection principle to find a conformal map, called a Schwarz–Christoffel formula or integral, from the upper half-plane onto a region bounded by a polygon.

Suppose Ω is a bounded simply-connected region bounded by a simple closed curve consisting of finitely many straight-line segments with vertices $\{v_j\}_1^n$. If we give $\partial\Omega$ a positive (counter-clockwise) orientation, then at v_j the tangent vector turns by an angle $\pi\alpha_j$, where $-1 < \alpha_j < 1$. See Figure 8.6.

By the Schwarz reflection principle, Theorem 8.15, if φ is a conformal map of Ω onto $\mathbb{D}$, then φ extends analytically and one-to-one across (the interior of) each boundary segment. If B_j is a small ball centered at v_j, then the map $\varphi_j(z) = (z - v_j)^{1/(1-\alpha_j)}$ is one-to-one and analytic in $\Omega \cap B_j$, and maps $\partial\Omega \cap B_j$ onto a straight-line segment. The inverse of this map composed with φ then extends to be analytic and one-to-one in a neighborhood of 0, again by the Schwarz reflection principle. Thus φ extends to be a one-to-one and continuous map of $\overline{\Omega}$ onto $\overline{\mathbb{D}}$. Similarly, if $f(z)$ is a conformal map of $\mathbb{H}$ onto Ω then $f(z)$ extends to be one-to-one and continuous on $\overline{\mathbb{H}}$ and analytic on $\mathbb{R}$ except at the **prevertices** $x_j = f^{-1}(v_j)$, $j = 1, \ldots, n$.

We may assume that the vertices are ordered so that $-\infty < x_1 < x_2 < \cdots < x_n < \infty$. If $f(x)$ and $f(x+h)$ belong to the same edge of the polygon $\partial\Omega$ then $f(x+h) - f(x)$ is parallel to $\partial\Omega$. Writing

$$f'(x) = \lim_{h>0\to 0} \frac{f(x+h) - f(x)}{h},$$

we deduce that $f'(x)$ points in the tangential direction to $\partial\Omega$. In other words, for $x_j < x < x_{j+1}$, $\arg f'(x)$ is given by the direction of the line segment from v_j to v_{j+1}. Since $f' \neq 0$ on a simply-connected region containing $\overline{\mathbb{H}} \setminus \{x_j\}_1^n$, we can define $\log f'(z)$ so as to be analytic on $\overline{\mathbb{H}} \setminus \{x_j\}_1^n$, and hence $\arg f'(z)$ is a bounded harmonic function on $\mathbb{H}$ which is continuous on $\overline{\mathbb{H}} \setminus \{x_j\}_1^n$, with a jump discontinuity on $\mathbb{R}$ of $\pi\alpha_j$ at x_j, where $\pi\alpha_j$ is the turning angle at $v_j = f(x_j)$. See Figure 8.6.

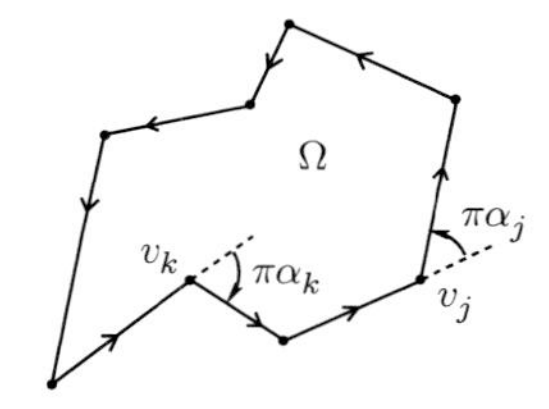

Figure 8.6 Turn angles at the vertices of a polygon.

We can build a bounded harmonic function on $\mathbb{H}$ with exactly these jump discontinuities on $\mathbb{R}$ as follows. The function $\arg(z-a)$, for $a \in \mathbb{R}$, with $0 \leq \arg(z-a) \leq \pi$ on $\mathbb{H} \cup \mathbb{R} \setminus \{a\}$, is harmonic on $\mathbb{H}$ and has a jump discontinuity of $-\pi$ at a. Let c_0 be the constant $c_0 = \arg f'(x)$ for $x > x_n$. Then

$$u(z) = c_0 - \sum_{j=1}^{n} \alpha_j \arg(z - x_j)$$

is a bounded harmonic function on $\mathbb{H}$ which equals $\arg f'$ on $\mathbb{R} \setminus \{x_1, \ldots, x_n\}$. By Lindelöf's maximum principle applied to $u - \arg f'$, we conclude that $u(z) = \arg f'(z) = \operatorname{Im} \log f'(z)$ on $\mathbb{H}$. Thus

$$f'(z) = c \prod_{j=1}^{n} (z - x_j)^{-\alpha_j},$$

for some constant c. The function f can now be found by integration.

We have proved the following theorem:

Theorem 8.19 (Schwarz–Christoffel) *Suppose Ω is a bounded simply-connected region whose positively oriented boundary $\partial\Omega$ is a polygon with vertices $v_1, \ldots, v_n$. Suppose the tangent direction on $\partial\Omega$ increases by $\pi\alpha_j$ at v_j, $-1 < \alpha_j < 1$. Then there exists $x_1 < x_2 < \cdots < x_n$ and constants c_1, c_2 so that*

$$f(z) = c_1 \int_{\gamma_z} \prod_{j=1}^{n} (\zeta - x_j)^{-\alpha_j} d\zeta + c_2$$

is a conformal map of $\mathbb{H}$ onto Ω, where the integral is along any curve γ_z in $\mathbb{H}$ from i to z.

The definition of f in Theorem 8.19 does not depend on the choice of $\gamma_z \in \mathbb{H}$ by Cauchy's theorem, because $\mathbb{H}$ is simply-connected. We remark that this proof of the Schwarz–Christoffel theorem uses the existence of the desired conformal map, so that it cannot be used to replace the geodesic zipper algorithm in the proof of the Riemann mapping theorem in Section 8.2.

One way to understand the powers in the product in Theorem 8.19 is to note that, near x_j, $f(z) - v_j$ behaves like a power $c_j(z - x_j)^{\beta_j}$, where $\pi\beta_j$ is the interior angle at v_j. So f' behaves like $c_j\beta_j(z-x_j)^{\beta_j - 1}$. If the turning angle at v_j is $\pi\alpha_j$, then $\pi\beta_j + \pi\alpha_j = \pi$, so that $\beta_j - 1 = -\alpha_j$. The remarkable part of Theorem 8.19 is that this local behavior gives a formula for f'.

The difficulty in using a Schwarz–Christoffel formula is that it does not tell us how to find the prevertices x_j. As in the proof of Morera's Theorem 4.19, the integral in Theorem 8.19 is always analytic, in fact locally conformal, but in general it does not represent a conformal map. In the formulation above, $-1 < \alpha_j < 1$ and the tangent direction has a total increase of 2π around $\partial\Omega$, so that $\sum_{j=1}^{n} \alpha_j = 2$. Thus the integral is absolutely convergent, even if the integration takes place on $\mathbb{R}$ instead of $\mathbb{H}$. The length of the segment from v_j to v_{j+1} can be expressed as

$$L_k = |c_1| \int_{x_k}^{x_{k+1}} \prod_{j=1}^{n} |x - x_j|^{-\alpha_j} dx, \tag{8.6}$$

because $\arg f'$ is constant on this segment. So the prevertices can be written as a solution of the system (8.6) of (highly non-linear) singular integral equations. For numerical applications, because the derivative is singular at some of the prevertices, Riemann sums are too inaccurate to compute the integrals, so that Gauss–Jacobi quadrature is needed.

The Schwarz–Christoffel theorem on the unit disk has a similar form. See Exercise 8.4. The Schwarz–Christoffel theorem can also be extended to cover unbounded regions whose boundary consists of line segments and half-lines or full lines. Allowing an angle $\pi\alpha_j$ to equal π corresponds to an infinite channel between two parallel half-lines. The case when the angles $\pi\alpha_j$ are allowed to equal $-\pi$ corresponds to a "cut" along a line segment into a polygonal region. See Exercises 8.5 and 8.13.

We end this chapter with one of the more important open problems in complex analysis called the **Brennan conjecture**. Suppose that T is a polygonal tree, a connected finite union of line segments and the half-line $[0, \infty]$, whose complement is a simply-connected region Ω. The conformal map ψ of the upper half-plane $\mathbb{H}$ onto Ω can be normalized so that it is asymptotic to z^2 for $|z|$ large. Thus $\psi'(z)$ is asymptotic to $2z$ near ∞. Indeed, the total turning of the tangent direction following a positive orientation of $\partial\Omega$ is $-\pi$, so that $\sum \alpha_j = -1$, where $\pi\alpha_j$ is the change in the tangent direction at the vertex $v_j = \psi(x_j)$. By the Schwarz–Christoffel theorem,

$$\psi'(z) = \frac{2}{\prod (z - x_j)^{\alpha_j}}.$$

The Brennan conjecture is about the second derivative of ψ at the "tips" of the tree: prove that for every polygonal tree

$$\sum_{k:\alpha_k=-1} \prod_{j\neq k} |x_k - x_j|^{2\alpha_j} \leq 1.$$

This is equivalent to proving that, for any conformal map f of the disk into the plane, $\int |f'|^{-p} dxdy < \infty$ for $0 < p < 2$. The conformal map $f(z) = (1 - z)^2$ defined on $\mathbb{D}$ has a derivative whose reciprocal is not square integrable.

8.5 Exercises

A

8.1 Let f be an analytic function on $|z| < 2$ such that f is real valued on a subarc of $|z| = 1$. Prove that f is a constant.

8.2 (a) Using the Cayley transform, show that reflection $z \to \bar{z}$ across $\mathbb{R}$ corresponds to "reflection" $z \to 1/\bar{z}$ across $\partial\mathbb{D}$.

(b) Suppose g is analytic in $\mathbb{D}$ and $|g(z)| \to 1$ as $z \to I$, where $I \subset \partial\mathbb{D}$ is an open arc. Show that g extends to be meromorphic in $\mathbb{C} \setminus (\partial\mathbb{D} \setminus I)$ using $1/\overline{g(1/\bar{z})}$. The extension will have poles at the reflection of the zeros of g.

(c) Suppose Ω is a simply-connected region and suppose $\partial\Omega$ is a closed analytic curve. Prove that if f is a conformal map of Ω onto $\mathbb{D}$ then f and f^{-1} extend to be one-to-one and analytic in a neighborhood of $\overline{\Omega}$, respectively $\overline{\mathbb{D}}$.

(d) Prove that if g is analytic in a region Ω_1 whose boundary contains a one-sided analytic arc γ, and if $g(z)$ tends to an analytic arc σ as $z \to \gamma$, then g extends to be analytic in a neighborhood of $\Omega \cup \gamma$.

(e) Suppose Ω is a region whose boundary contains a simple arc γ which is locally analytic, in the sense that each point of γ is contained in a one-sided analytic subarc of γ. Prove γ is a one-sided analytic arc. Hint: Find a simply-connected region $\Omega_1 \subset \Omega$ so that $\gamma \subset \partial\Omega_1$.

(f) If $\gamma : [0,1] \to \mathbb{C}$ is a simple arc whose image can be covered by a finite union of (images of) analytic arcs, then γ is (the image of) an analytic arc.

8.3 Find a conformal map of $\mathbb{H}$ onto $\{x + iy : y > x^2\}$ so that $z_0 = i$ is mapped to the focus. See the outline in Section 8.3.

8.4 Derive a formula for a conformal map of the unit disk onto a region bounded by a polygon by a change of variables in the Schwarz–Christoffel formula in Theorem 8.19. Use the fact that the sum of the turning angles is 2π to simplify your expression. This formula is also called a Schwarz–Christoffel formula.

8.5 Show that the Schwarz–Christoffel formula still holds if angles $\pi\alpha_j$ are allowed to equal $-\pi$ (slits) as described in Section 8.4.

8.6 If f is a conformal map of a rectangle of width R and height 1 onto the upper half-plane, then prove it can be reflected across a vertical edge to obtain a map from a rectangle twice as long onto the slit plane. Prove that, by repeated reflections, the map f can be extended to be a meromorphic function on $\mathbb{C}$ such that $f(z+2R) = f(z)$ and $f(z+2i) = f(z)$. In general, an **elliptic function** is a meromorphic function invariant under two linearly independent shifts. See Example 11.5. The inverse of an elliptic function, such as the Schwarz–Christoffel integral for a rectangle map, is an example of an **elliptic integral**. We will investigate elliptic functions more thoroughly in Section 16.3.

B

8.7 Suppose g is a non-constant analytic function on a region Ω and suppose $\{g_n\}$ is a sequence of analytic functions converging uniformly on compact subsets of Ω to g. If K is a compact subset of $g(\Omega)$ then prove there is an $n_0 < \infty$ so that K is a subset of $g_n(\Omega)$ for all $n \ge n_0$. Hint: Apply the proof of Hurwitz's Theorem 8.8 to $g_n - w$ and $g - w$ for all w in a small ball containing $g(z_0)$.

8.8 Find a conformal map of $\mathbb{D}$ onto the interior of the ellipse given by $x^2/4 + y^2/9 = 1$. Hint: Consider the map $\frac{1}{2}(z + \frac{1}{z})$ on the top half of an annulus. Your answer will involve a Schwarz–Christoffel integral. The prevertices can be put in a symmetric position (at the vertices of a rectangle on $\partial\mathbb{D}$) by choosing the value at 0 and the argument of the derivative at 0, so that it involves only one unknown parameter. Give an equation whose solution is this parameter. In general, it is not possible to find a formula for the parameter in terms of elementary functions.

8.9 (a) Given a continuous function $f(e^{it})$, $0 \le t \le \pi$, find a harmonic function u on $\mathbb{D}^+ = \mathbb{D} \cap \{\text{Im} z > 0\}$ such that u is continuous on $\overline{\mathbb{D}^+}$, $u = f$ on $\partial\mathbb{D}^+ \cap \{\text{Im} z > 0\}$ and $\frac{\partial u}{\partial y} = 0$ on $(-1, 1)$. Hint: Find u so that $u(\overline{z}) = u(z)$.

(b) Suppose g is continuous on $\partial\mathbb{D}$ and $\int_0^{2\pi} g d\theta = 0$. Find a harmonic function v on $\mathbb{D}$ with the property that $\frac{\partial v}{\partial \eta} = g$ on $\partial\mathbb{D}$, where η is the unit inward normal to $\partial\mathbb{D}$. This is called a **Neumann problem**. Hint: Consider $f(e^{it}) = \int_0^t g(e^{i\theta})d\theta$, $u = PI(f)$, and use the Cauchy–Riemann equations on $|z| = r$, then let $r \to 1$.

(c) Given bounded continuous functions $f(x)$, $-\infty < x < 0$, and $g(x)$, $0 < x < \infty$, find a harmonic function u on $\mathbb{H}$ such that u is continuous on $\mathbb{H} \cup \mathbb{R} \setminus \{0\}$, $u = f$ on $(-\infty, 0)$ and $\frac{\partial u}{\partial y} = g$ on $(0, \infty)$.

(d) Suppose φ is analytic and one-to-one on a neighborhood of $\overline{\mathbb{H}}$ and $\Omega = \varphi(\mathbb{H})$. Suppose E is a connected subset of $\partial\Omega$. Given continuous functions f on E and g on $\partial\Omega \setminus E$, describe how to find a harmonic function u on Ω so that $u = f$ on E° and $\frac{\partial u}{\partial \eta} = g$ on $(\partial\Omega \setminus E)^\circ$, where $\eta = \eta_\zeta$ is the unit inner normal to $\partial\Omega$ at $\zeta \in \partial\Omega$. This is called a **mixed Dirichlet–Neumann problem**.

8.10 The intersection of two disks is called a lens. Suppose z_i, $i = 1, \ldots, n$, are distinct points in $\mathbb{C}$. Suppose D_i^+ and D_i^- are (open) disks whose intersection is a lens $L_i = D_i^+ \cap D_i^-$ with endpoints z_{i-1} and z_i. If $(D_i^+ \cup D_i^-) \cap L_j = \emptyset$ for $j = 1, \ldots, i-1$, and if ∂D_i^+ is tangent to ∂D_{i-1}^- and ∂D_i^- is tangent to ∂D_{i-1}^+, then we say that $\{L_i\}$ is a tangential lens-chain. Prove that if $\{L_i\}$ is a tangential lens-chain then the geodesic zipper algorithm constructs a conformal map of the half-plane onto a simply-connected region whose boundary is contained in $\cup \overline{L_i}$. This idea can be used to reduce the number of data points z_i in the geodesic zipper algorithm by using long thin lenses where the boundary is flat.

8.11 Prove that the basic map in the geodesic zipper algorithm $\sqrt{z^2+1}$ of $\mathbb{H} \setminus [0, i]$ onto $\mathbb{H}$ can be extended as a conformal map of $\mathbb{C} \setminus [-i, i]$ onto $\mathbb{C} \setminus [-1, 1]$, by writing it in the form $z\sqrt{1 + 1/z^2}$, where the square root is defined on $\mathbb{C} \setminus (-\infty, 0]$. Show that it can also be computed by setting $w = \sqrt{z^2+1}$ (any choice of the square root), then replacing w with $-w$ when $\mathrm{Re}z \cdot \mathrm{Re}w < 0$ or when $\mathrm{Re}z = 0$ and $\mathrm{Im}z \cdot \mathrm{Im}w < 0$. The latter method is faster numerically. Find similar methods for $\sqrt{z^2-1}$.

8.12 Find a conformal map of $\mathbb{D}$ onto a regular n-gon. Use symmetry to find the prevertices explicitly.

8.13 Formulate and prove the Schwarz–Christoffel theorem for an unbounded simply-connected region whose boundary consists of line segments and half-line(s). Hint: The turning angle can be understood by intersecting the region with a large disk.

8.14 Suppose φ is a conformal map of $A_1 = \{z : 1 < |z| < R_1\}$ onto $A_2 = \{z : 1 < |z| < R_2\}$. Prove $R_1 = R_2$ and $\varphi(z) = \lambda z$ or $\varphi(z) = \lambda R_1/z$ for some constant λ with $|\lambda| = 1$. Hint: Reflect.

8.15 Find a conformal map f of the upper half-plane $\mathbb{H}$ onto $\mathbb{H} \setminus I$, where I is the line segment from 0 to $e^{i\pi a}$, $0 < a < 1$, as follows. First show that such a map can be chosen so that $|f| \to \infty$ as $z \in \mathbb{H} \to \infty$, then prove f extends to be continuous on $\mathbb{R}$. Show that f can be chosen so that $f(a) = 0$ and $f(a-1) = 0$. Note that $\arg f$ is bounded and is constant on each of the three intervals comprising $\mathbb{R} \setminus \{a, a-1\}$. Construct a bounded harmonic function in $\mathbb{H}$ which has the same values on these intervals using $\arg(z-a)$ and $\arg(z+1-a)$. Apply Lindelöf's maximum principle then exponentiate to find Cf, for some positive constant C. Using logarithmic differentiation, find the preimage of the tip of the slit I. Use this information to find C.

These maps can be used in place of the basic maps φ_j in the geodesic zipper algorithm; however, the inverses are not elementary functions. The maps are very important for a recent application of conformal mapping and probability to statistical mechanics called SLE.

C

8.16 The geodesic zipper algorithm finds conformal maps of the upper half-plane $\mathbb{H}$ onto two complementary regions $\Omega^{\pm} \subset \mathbb{C}^*$. Suppose Ω^+ is a bounded region and Ω^- is unbounded.

(a) Describe how to find conformal maps f_+ of $\mathbb{D}$ onto Ω^+ and f_- of $\mathbb{C} \setminus \overline{\mathbb{D}}$ onto $\Omega^- \setminus \{\infty\}$.

(b) Check that the map f_+ extends to be continuous and one-to-one on $\overline{\mathbb{D}}$ and that f_- extends to be continuous and one-to-one on $\mathbb{C} \setminus \mathbb{D}$. See Figure 8.5.

(c) The extended maps $f_\pm$ do not agree on $\partial\mathbb{D}$. However, $\varphi = f_+^{-1} \circ f_-$ is an increasing homeomorphism of $\partial\mathbb{D}$ onto $\partial\mathbb{D}$ (called a **welding homeomorphism**). Prove that any increasing homeomorphism of $\partial\mathbb{D}$ onto $\partial\mathbb{D}$ can be uniformly approximated by a welding homeomorphism constructed via the geodesic zipper algorithm. Hint: Two increasing maps of $[0, 2\pi]$ to itself that agree on a fine mesh are uniformly close.

8.17 For $h \geq 1$, let φ_h be a conformal map of the upper half-plane $\mathbb{H}$ onto the rectangle $R_h = \{z : |\mathrm{Re}z| < \pi/2,\ 0 < \mathrm{Im}z < h\}$, with $\varphi_h(0) = 0$ and $\varphi(\infty) = ih$. Prove:

(a) there is a unique $v_h > 1$ so that φ can be chosen to map $v_h, 1/v_h, -1/v_h, -v_h$ onto the vertices of R_h;

(b) $\varphi_h(z) = \varphi_{2h}(J^{-1}(1/(v_h z)))$, where J is the Joukovski map $J(z) = \frac{1}{2}(z + 1/z)$ and J^{-1} is chosen to map the lower half-plane onto $\mathbb{H} \cap \{|z| < 1\}$;

(c) $v_h^2 = J(v_{2h})$.

(d) Derive a formula for φ_h in terms of $\varphi_{2^n h}$.

(e) Let $\psi(z) = i\log(J^{-1}(z)) + \pi/2$, where J^{-1} is chosen to map $\mathbb{H}$ onto $\mathbb{H} \cap \{|z| > 1\}$. Prove that ψ is a conformal map of $\mathbb{H}$ onto the half-strip $\{z : |\mathrm{Re}z| < \pi/2, \mathrm{Im}z > 0\}$.

(f) Replace $\varphi_{2^n h}(z)$ with $\psi(v_{2^n h} z)$ in the formula in (d) to obtain an explicit approximation ψ_n for φ_h.

(g) Estimate the error $\varphi_h - \psi_n$ when $n = 4$ and $h > 1$. The approximation is accurate enough that it is virtually a formula for φ_h.

9 Calculus of Residues

9.1 Contour Integration and Residues

In this chapter we will learn one of the main applications of complex analysis to engineering problems, but we will concentrate on the math, not the engineering. Cauchy's theorem says that if f is analytic in a region Ω and if γ is a closed curve in Ω which is homologous to 0, then $\int_\gamma f(z)dz = 0$. What happens if f has an isolated singularity at $a \in \Omega$, but otherwise is analytic? Expanding f in its Laurent series about a, we have

$$\begin{aligned} f(z) &= \sum_{n=-\infty}^{\infty} b_n(z-a)^n \\ &= \frac{b_{-1}}{z-a} + \frac{d}{dz}\sum_{\substack{n=-\infty \\ n\neq -1}}^{\infty} \frac{b_n}{(n+1)}(z-a)^{n+1}. \end{aligned}$$

If Δ is a disk centered at a with $\overline{\Delta} \subset \Omega$ then, by the fundamental theorem of calculus and Proposition 4.13,

$$\int_{\partial\Delta} f(z)dz = b_{-1}\int_{\partial\Delta} \frac{dz}{z-a} = 2\pi i b_{-1}, \tag{9.1}$$

provided $\partial\Delta$ is oriented in the positive or counter-clockwise direction.

Definition 9.1 *If f is analytic in $\{0 < |z-a| < \delta\}$ for some $\delta > 0$, then the* **residue of f at** *a, written* $\text{Res}_a f$, *is the coefficient of $(z-a)^{-1}$ in the Laurent expansion of f about $z = a$.*

Theorem 9.2 (Residue theorem) *Suppose f is analytic in Ω except for isolated singularities at $a_1, \ldots, a_n$. If γ is a cycle in Ω with $\gamma \sim 0$ and $a_j \notin \gamma, j = 1, \ldots, n$, then*

$$\int_\gamma f(z)dz = 2\pi i \sum_k n(\gamma, a_k)\text{Res}_{a_k} f.$$

Usually the residue theorem is applied to curves γ such that $n(\gamma, a_k) = 0$ or 1, so that the sum on the right is $2\pi i$ times the sum of the residues of f at points enclosed by γ. If f has infinitely many singularities clustering only on $\partial\Omega$ then we can shrink Ω slightly so that it contains only finitely many a_j and still have $\gamma \sim 0$.

Proof Let Δ_k be a disk centered at a_k, $k = 1, 2, \ldots, n$, such that $\overline{\Delta}_m \cap \overline{\Delta}_k = \emptyset$ if $m \neq k$. Orient $\partial \Delta_k$ in the positive or counter-clockwise direction. Then

$$\gamma - \sum_k n(\gamma, a_k)\partial \Delta_k \sim 0$$

in the region $\Omega \setminus \{a_1, \ldots, a_n\}$. By Cauchy's theorem,

$$\int_\gamma f(z)dz - \sum_{k=1}^n n(\gamma, a_k) \int_{\partial \Delta_k} f(z)dz = 0.$$

Then Theorem 9.2 follows from (9.1). □

9.2 Some Examples

In this section we first illustrate several useful techniques for computing residues, then use the residue theorem to compute some integrals. The emphasis is on the techniques instead of proving general results that can be quoted.

Example 9.3 $f(z) = \dfrac{e^{3z}}{(z-2)(z-4)}$ has a simple pole at $z = 2$ and hence

$$\text{Res}_2 f = \lim_{z \to 2}(z-2)f(z) = \frac{e^6}{-2}.$$

The residue at $z = 4$ can be calculated similarly.

Example 9.4 $g(z) = \dfrac{e^{3z}}{(z-2)^2}$.

Expand e^{3z} in a series expansion about $z = 2$:

$$g(z) = \frac{e^6 e^{3(z-2)}}{(z-2)^2} = \frac{e^6}{(z-2)^2} \sum_{n=0}^{\infty} \frac{3^n}{n!}(z-2)^n = \frac{e^6}{(z-2)^2} + \frac{3e^6}{z-2} + \ldots,$$

so that

$$\text{Res}_2 g = 3e^6.$$

In this case $\lim_{z \to 2}(z-2)^2 g(z)$ is not the coefficient of $(z-2)^{-1}$ and $\lim_{z \to 2}(z-2)g(z)$ is infinite. Of course the full series for e^{3z} was not necessary to compute the residue. We can find the appropriate coefficient in the series expansion for e^{3z} by computing the derivative of e^{3z}. More generally, if $G(z)$ is analytic at $z = a$ then

$$\text{Res}_a \frac{G(z)}{(z-a)^n} = \frac{G^{(n-1)}(a)}{(n-1)!}.$$

Example 9.5 Another technique that can be used with simple poles, when the pole is not already written as a factor of the denominator, is illustrated by the example $h(z) = e^{az}/(z^4 + 1)$. Then h has simple poles at the fourth roots of -1. If $\omega^4 = -1$, then

$$\text{Res}_\omega h = \lim_{z\to\omega} \frac{(z-\omega)e^{az}}{z^4+1} = \frac{e^{a\omega}}{\lim_{z\to\omega} \frac{z^4+1}{z-\omega}}.$$

Note that the denominator is the limit of difference quotients for the derivative of z^4+1 at $z=\omega$ and hence

$$\text{Res}_\omega \frac{e^{az}}{z^4+1} = \frac{e^{a\omega}}{4\omega^3} = -\frac{\omega e^{a\omega}}{4}.$$

Example 9.6 Another method using series is illustrated by the example

$$k(z) = \frac{\pi \cot \pi z}{z^2}.$$

To compute the residue of k at $z=0$, note that $\cot \pi z$ has a simple pole at $z=0$ and hence k has a pole of order 3, so that

$$\frac{\pi \cot \pi z}{z^2} = \frac{b_{-3}}{z^3} + \frac{b_{-2}}{z^2} + \frac{b_{-1}}{z} + b_0 + \dots .$$

Then

$$\pi \cos \pi z = \left(\frac{\sin \pi z}{z}\right)(b_{-3} + b_{-2}z + b_{-1}z^2 + b_0 z^3 + \dots).$$

Inserting the series expansions for cos and sin we obtain

$$\pi\left(1 - \frac{\pi^2}{2}z^2 + \dots\right) = \left(\pi - \frac{\pi^3}{6}z^2 + \dots\right)(b_{-3} + b_{-2}z + b_{-1}z^2 + \dots).$$

Equating coefficients,

$$\pi = \pi b_{-3}, \qquad -\frac{\pi^3}{2} = -\frac{\pi^3}{6}b_{-3} + \pi b_{-1},$$

and $\text{Res}_0 k = b_{-1} = -\frac{\pi^2}{3}$.

Now let's apply these ideas and the residue theorem to compute some integrals.

Example 9.7 If γ is the circle centered at 0 with radius 3, then

$$\int_\gamma \frac{e^{3z}}{(z-2)(z-4)} dz = -2\pi i \frac{e^6}{2},$$

by the residue theorem and Example 9.3.

Example 9.8 $\displaystyle\int_{-\infty}^{\infty} \frac{dx}{x^4+1}.$

Construct a contour γ consisting of the interval $[-R, R]$ followed by the semi-circle C_R in $\mathbb{H}$ of radius R, with $R > 1$. See Figure 9.1.

By the residue theorem with $f(z) = 1/(z^4+1)$,

$$\int_{-R}^{R} f(z)dz + \int_{C_R} f(z)dz = 2\pi i(\text{Res}_{z_1} f + \text{Res}_{z_2} f), \tag{9.2}$$

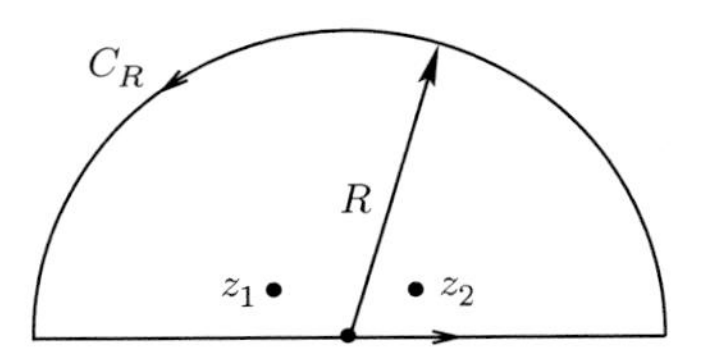

Figure 9.1 Half-disk contour.

where z_1 and z_2 are the roots of $z^4+1=0$ in the upper half-plane $\mathbb{H}$. Note that

$$\left|\int_{C_R} f(z)dz\right| \le \int_0^{2\pi} \frac{Rd\theta}{R^4-1} \to 0$$

as $R \to \infty$. Since the integral $\int_{\mathbb{R}}(x^4+1)^{-1}dx$ is convergent, it equals $\lim_{R\to\infty}\int_{-R}^{R}(x^4+1)^{-1}dx$, so that, by (9.2) and example 9.5,

$$\int_{-\infty}^{\infty} \frac{1}{x^4+1}dx = -\frac{2\pi i}{4}(z_1+z_2) = \frac{\pi}{\sqrt{2}}.$$

The technique in Example 9.8 can be used to compute the integral of any rational function with no poles on $\mathbb{R}$ if the degree of the denominator is at least 2 plus the degree of the numerator. This latter condition is needed for the absolute convergence of the integral.

Example 9.9 $\displaystyle\int_0^{2\pi} \frac{1}{3+\sin\theta}d\theta.$

Set $z=e^{i\theta}$. Then

$$\int_0^{2\pi} \frac{1}{3+\sin\theta}d\theta = \int_{|z|=1} \frac{1}{(3+\frac{1}{2i}(z-1/z))}\frac{dz}{iz} = \int_{|z|=1} \frac{2dz}{z^2+6iz-1}.$$

The roots of $z^2+6iz-1$ occur at $z_1, z_2 = i(-3\pm\sqrt{8})$. Only $i(-3+\sqrt{8})$ lies inside $|z|=1$. By the residue theorem and the method for computing residues in Example 9.3 or Example 9.5,

$$\int_0^{2\pi} \frac{1}{3+\sin\theta}d\theta = 2\pi i \text{Res}_{i(-3+\sqrt{8})}\frac{2}{z^2+6iz-1} = \frac{2\pi}{\sqrt{8}}.$$

The technique in Example 9.9 can be used to compute

$$\int_0^{2\pi} R(\cos\theta, \sin\theta)d\theta,$$

where $R(\cos\theta, \sin\theta)$ is a rational function of $\sin\theta$ and $\cos\theta$, with no poles on the unit circle. An integral on the circle, as in Example 9.9, can be converted to an integral on the line using the Cayley transform $z=(i-w)/(i+w)$ of the upper half-plane onto the disk. It is interesting to note that you obtain the substitution $x=\tan\frac{\theta}{2}$ which you might have learned in calculus.

Example 9.10 $\displaystyle\int_{-\infty}^{\infty} \frac{\cos x}{x^2+1}dx.$

A first guess might be to write $\cos z = (e^{iz}+e^{-iz})/2$, but if $y=\text{Im}z$ then $|\cos z| \sim e^{|y|}/2$ for large $|z|$. This won't allow us to find a closed contour where the part off the real line makes only a small contribution to the integral. Instead, we use $e^{iz}/(z^2+1)$ then take real parts of the resulting integral. Using the same contour as in Example 9.8, we have the estimate

$$\left|\int_{C_R} \frac{e^{iz}}{z^2+1} dz\right| \leq \int_{C_R} \frac{e^{-y}}{R^2-1} |dz| \leq \frac{\pi R}{R^2-1} \to 0$$

as $R \to \infty$, where $y = \operatorname{Im} z > 0$. By the residue theorem and the method in Example 9.5,

$$\int_{-\infty}^{\infty} \frac{e^{ix}}{x^2+1} dx = 2\pi i \sum_{\operatorname{Im} a>0} \operatorname{Res}_a \frac{e^{iz}}{z^2+1} = 2\pi i \frac{e^{i \cdot i}}{2i} = \frac{\pi}{e}.$$

In this particular case, we did not have to take real parts. The integral itself is real because $\sin x/(x^2+1)$ is odd.

9.3 Fourier and Mellin Transforms

The technique in Example 9.10 can be used to compute

$$\int_{-\infty}^{\infty} f(x) e^{i\lambda x} dx \tag{9.3}$$

for $\lambda > 0$, provided f is meromorphic in the closed upper half-plane $\mathbb{H} \cup \mathbb{R}$ with no poles on $\mathbb{R}$ and $|f(z)| \leq K/|z|^{1+\varepsilon}$ for some $\varepsilon > 0$ and all large $|z|$ with $\operatorname{Im} z > 0$. If the integral (9.3) is desired for all real λ, then for negative λ use a contour in the lower half-plane, provided f is meromorphic and $|f(z)| \leq K/|z|^{1+\varepsilon}$ in $\operatorname{Im} z < 0$. As the reader can deduce, if f is meromorphic in $\mathbb{C}$ and satisfies this inequality for all $|z|$ large, then f is rational and the degree of the denominator is at least 2 plus the degree of the numerator. The integral in (9.3), usually with λ replaced by $-2\pi\lambda$, is called the **Fourier transform of** f, as a function of λ.

Example 9.11 $\displaystyle\int_{-\infty}^{\infty} \frac{x \sin \lambda x}{x^2+1} dx.$

This example cannot be done by the method in Example 9.10 because the integrand does not decay fast enough to prove $\int_{C_R} |f(z)||dz| \to 0$, where $f(z) = ze^{i\lambda z}/(z^2+1)$. Indeed, it is not even clear a priori that the integral in Example 9.11 exists.

We may suppose $\lambda > 0$, because sine is odd. Let $\gamma = \gamma_1+\gamma_2+\gamma_3+\gamma_4$, where $\gamma_1 = [-A, B]$, $A, B > 0$, $\gamma_2 = \{B + iy : 0 \leq y \leq A+B\}$, $\gamma_3 = \{x + i(A+B) : B \geq x \geq -A\}$ and $\gamma_4 = \{-A + iy : A+B \geq y \geq 0\}$, orienting γ as indicated in Figure 9.2.

To prove convergence of the integral in Example 9.11, we will let A and B tend to ∞ independently, and use the estimate

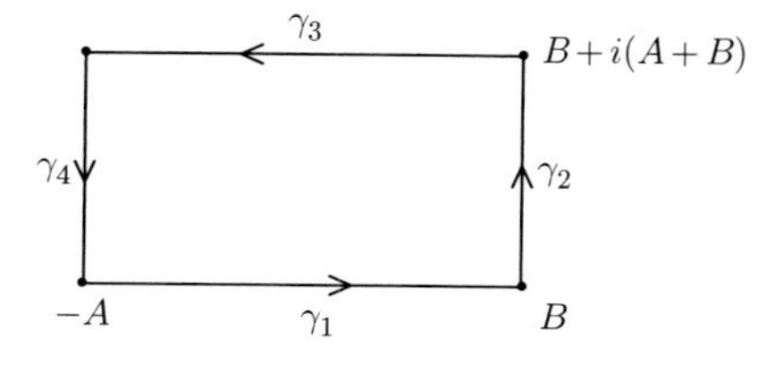

Figure 9.2 Rectangle contour.

$$|z/(z^2+1)| \le |z|/(|z|^2-1) \le 2/|z|$$

when $|z|^2 > 2$. For A and B large,

$$\left|\int_{\gamma_3} \frac{ze^{i\lambda z}}{z^2+1}dz\right| \le \int_{-A}^{B} \frac{2}{A+B}e^{-\lambda(A+B)}dx = \frac{2e^{-\lambda(A+B)}}{A+B}(A+B) \to 0$$

as $A+B\to\infty$. Also

$$\left|\int_{\gamma_2} \frac{ze^{i\lambda z}}{z^2+1}dz\right| \le \int_0^{A+B} \frac{2}{B}e^{-\lambda y}dy \le \frac{2}{B}\frac{(1-e^{-\lambda(A+B)})}{\lambda} \to 0$$

as $B\to\infty$. A similar estimate holds on γ_4 as $A\to\infty$. By the residue theorem,

$$\lim_{A,B\to\infty}\int_{-A}^{B}\frac{xe^{i\lambda x}}{x^2+1}dx = 2\pi i\,\mathrm{Res}_i\frac{ze^{i\lambda z}}{z^2+1} = \frac{2\pi i\cdot ie^{-\lambda}}{2i} = i\pi e^{-\lambda}. \tag{9.4}$$

Indeed, by our estimates, the integrals over γ_2, γ_3 and γ_4 tend to 0 as A and B tend to ∞ so that the limit on the left-hand side of (9.4) exists and (9.4) holds. Example 9.11 follows from (9.4) by taking the imaginary parts.

The technique in Example 9.11 can be used to compute (9.3) with the (weaker) assumption $|f(z)| \le K/|z|$ for large $|z|$.

Example 9.12 $\displaystyle\int_{-\infty}^{\infty}\frac{\sin x}{x}dx.$

The main difference between Examples 9.11 and 9.12 is that the function $f(z) = e^{iz}/z$ has a simple pole on $\mathbb{R}$. The function $\sin x/x$ is integrable near 0 since $\sin x/x \to 1$ as $x\to 0$, but $f(x)$ is not integrable. However, the real part of $f(x)$ is odd so that

$$\lim_{\delta\to 0}\int_{-1}^{-\delta}+\int_{\delta}^{1}\frac{e^{ix}}{x}dx$$

exists.

Definition 9.13 *Suppose f is continuous on $\gamma\setminus\{a\}$, and suppose γ' is continuous at $\gamma^{-1}(a)$. Then the* **Cauchy principal value** *of the integral of f along γ is defined by*

$$PV\int_\gamma f(z)dz = \lim_{\delta\to 0}\int_{\gamma\cap\{|z-a|\ge\delta\}} f(z)dz,$$

provided the limit exists.

Note that we have deleted points in a ball *centered* at a. If the limit exists for all balls containing a, then the usual integral of f exists. For example,

$$PV\int_{-1}^{1}\frac{\cos x}{x}dx = 0,$$

because the integrand is odd, but the integral itself does not exist.

Proposition 9.14 *Suppose f is meromorphic in $\{\mathrm{Im}z \ge 0\}$, such that*

$$|f(z)| \le \frac{K}{|z|}$$

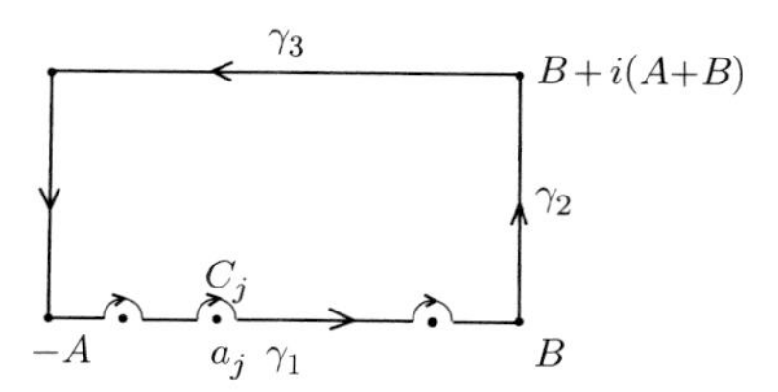

Figure 9.3 Contour avoiding poles on $\mathbb{R}$.

when $\text{Im}z \geq 0$ *and* $|z| > R$. *Suppose also that all poles of* f *on* $\mathbb{R}$ *are simple. If* $\lambda > 0$*, then*

$$PV \int_{-\infty}^{\infty} f(x)e^{i\lambda x}dx = 2\pi i \sum_{\text{Im}a>0} \text{Res}_a e^{i\lambda z}f(z) + 2\pi i \sum_{\text{Im}a=0} \frac{1}{2}\text{Res}_a e^{i\lambda z}f(z). \tag{9.5}$$

Part of the conclusion of Proposition 9.14 is that the integral exists even though the rate of decay at ∞ is possibly slower than our assumptions in (9.3). If $\lambda < 0$, then a similar result holds using the lower half-plane. The integral may not exist if $\lambda = 0$, as the example $f(z) = 1/z$ shows.

Proof Note that f has at most finitely many poles in $\{\text{Im}z \geq 0\}$ because $|f(z)| \leq K/|z|$, so that both sums in the statement of Proposition 9.14 are finite. Construct a contour similar to the rectangle in Figure 9.2, but avoiding the poles on $\mathbb{R}$ using small semi-circles C_j of radius $\delta > 0$ centered at each pole $a_j \in \mathbb{R}$. See Figure 9.3.

The integral of $f(z)e^{i\lambda z}$ along the top and sides of the contour tend to 0 as $A, B \to \infty$, as in Example 9.11. The semi-circle C_j centered at a_j can be parameterized by $z = a_j + \delta e^{i\theta}$, $\pi > \theta > 0$, so that if

$$f(z)e^{i\lambda z} = \frac{b_j}{z - a_j} + g_j(z),$$

where g_j is analytic in a neighborhood of a_j, then

$$\int_{C_j} f(z)e^{i\lambda z}dz = \int_{\pi}^{0} \frac{b_j}{\delta e^{i\theta}}\delta i e^{i\theta}\,d\theta + \int_{C_j} g(z)dz.$$

Because g_j is continuous at a_j and the length of C_j is $\pi\delta$, we have

$$\lim_{\delta\to 0}\int_{C_j} f(z)e^{i\lambda z}dz = -i\pi b_j.$$

By the residue theorem,

$$PV\int_{\mathbb{R}} f(z)e^{i\lambda z}dz - i\pi\sum_j b_j = 2\pi i \sum_{\text{Im}a>0} \text{Res}_a f(z)e^{i\lambda z},$$

and (9.5) holds. □

One way to remember the conclusion of Proposition 9.14 is to think of the real line as cutting the pole at each a_j in half. The integral contributes half of the residue of f at a_j.

For the present example,

$$PV\int_{-\infty}^{\infty}\frac{e^{ix}}{x}dx = \pi i, \tag{9.6}$$

so that, by taking imaginary parts,

$$\int_{-\infty}^{\infty}\frac{\sin x}{x}dx = \pi.$$

Note that $\sin x/x \to 1$ as $x \to 0$ so that $\int \sin x/x dx$ exists as an ordinary (improper) Riemann integral, if we extend the function $\sin x/x$ to equal 1 at $x = 0$. For this reason we can drop the PV in front of the integral.

Example 9.15 $\displaystyle\int_0^{\infty}\frac{x^{\alpha}}{x^2+1}dx,$
where $0 < \alpha < 1$. An integral of the form $\int_0^{\infty} f(x)x^{\beta-1}dx$ is called a **Mellin transform**. By the change of variables $x = e^y$, the Mellin transform of f is the Fourier transform of $f(e^y)$ when β is purely imaginary. Here we work with the original form of the integral and define $z^{\alpha} = e^{\alpha \log z}$ in $\mathbb{C} \setminus [0, +\infty)$, where $0 < \arg z < 2\pi$ and set $f(z) = 1/(z^2 + 1)$. Part of the difficulty here is constructing a closed contour that will give the desired integral. Consider the "keyhole" contour γ consisting of a portion of a large circle C_R of radius R and a portion of a small circle C_δ of radius δ, both circles centered at 0, along with two line segments between C_δ and C_R at heights $\pm\varepsilon$, oriented as indicated in Figure 9.4.

By the residue theorem, for R large and δ small,

$$\begin{aligned}\int_{\gamma} z^{\alpha}f(z)dz &= 2\pi i(\text{Res}_i z^{\alpha}f(z) + \text{Res}_{-i}z^{\alpha}f(z))\\ &= 2\pi i\left(\frac{e^{\alpha \log i}}{2i} + \frac{e^{\alpha \log(-i)}}{-2i}\right) = \pi(e^{i\frac{\pi\alpha}{2}} - e^{i\frac{3\pi\alpha}{2}}).\end{aligned} \tag{9.7}$$

We will first let $\varepsilon \to 0$, then $R \to \infty$ and $\delta \to 0$. Even though the integrals along the horizontal lines are in opposite directions, they do not cancel as $\varepsilon \to 0$. For $\varepsilon > 0$,

$$\lim_{\varepsilon\to 0}(x + i\varepsilon)^{\alpha}f(x + i\varepsilon) = e^{\alpha \log|x|}f(x)$$

and

$$\lim_{\varepsilon\to 0}(x - i\varepsilon)^{\alpha}f(x - i\varepsilon) = e^{\alpha(\log|x|+2\pi i)}f(x),$$

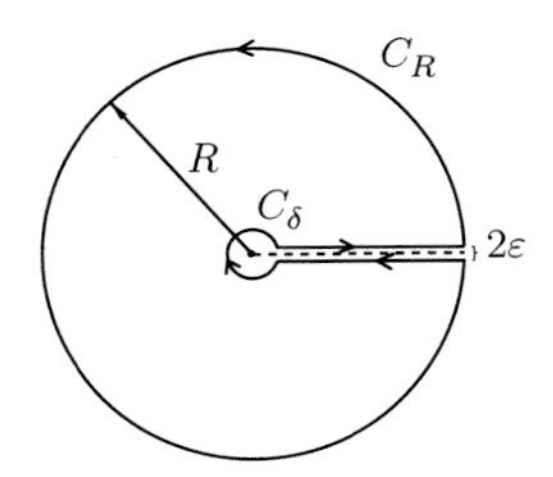

Figure 9.4 Keyhole contour.

because of our definition of $\log z$. Thus the integral over the horizontal line segments tends to

$$\int_{\delta}^{R} (1 - e^{2\pi i\alpha})x^{\alpha} f(x)dx. \tag{9.8}$$

For R large,

$$\left| \int_{C_R} z^{\alpha} f(z)dz \right| \le \int_0^{2\pi} \frac{R^{\alpha}}{R^2 - 1} R d\theta \to 0 \tag{9.9}$$

as $R \to \infty$. Similarly,

$$\left| \int_{C_\delta} z^{\alpha} f(z)dz \right| \le \int_0^{2\pi} \frac{\delta^{\alpha}}{1 - \delta^2} \delta d\theta \to 0 \tag{9.10}$$

as $\delta \to 0$. By (9.7)–(9.10),

$$\int_0^{\infty} \frac{x^{\alpha}}{x^2 + 1} dx = \pi \frac{e^{i\frac{\pi\alpha}{2}} - e^{i\frac{3\pi\alpha}{2}}}{1 - e^{2\pi i\alpha}} = \frac{\pi}{2\cos \alpha\pi/2}.$$

This line of reasoning works for meromorphic f satisfying $|f(z)| \le C|z|^{-2}$ for large $|z|$ and with at worst a simple pole at 0. The function z^{α} can be replaced by other functions which are not continuous across $\mathbb{R}$, such as $\log z$. In this case real parts of the integrals along $[0, \infty)$ will cancel, but the imaginary parts will not. See Exercise 9.11. Mellin transforms are used in applications to signal processing, image filtering, stress analysis and other areas.

In summary, computing integrals using the residue theorem usually involves choosing a meromorphic function and a contour, computing residues, estimating integrals then passing to a limit. This section is the first encounter in this text with estimating integrals, a useful skill. For that reason we have not given general results which could be simply quoted to find special integrals.

9.4 Series via Residues

We can turn this game around. Instead of using sums to compute integrals, we can use integrals to compute sums.

Example 9.16

$$\sum_{n=0}^{\infty} \frac{1}{n^2 + 1}.$$

Set $f(z) = \frac{1}{z^2+1}$ and consider the meromorphic function $f(z)\pi \cot \pi z$. Write

$$\pi \cot \pi z = \pi i \left(\frac{e^{i\pi z} + e^{-i\pi z}}{e^{i\pi z} - e^{-i\pi z}} \right) = \pi i \left(\frac{e^{2\pi iz} + 1}{e^{2\pi iz} - 1} \right). \tag{9.11}$$

Multiplying (9.11) by $z - n$ and letting $z \to n$ shows that $\pi \cot \pi z$ has a simple pole with residue 1 at each integer n. Because the poles are simple, $f(z)\pi \cot \pi z$ has a simple pole with residue $f(n)$ at $z = n$.

Consider the contour integral of $f(z)\pi \cot \pi z$ around the square S_N with vertices $(N + \frac{1}{2})(\pm 1 \pm i)$, where N is a large positive integer. The function $\pi \cot \pi z$ is uniformly bounded on S_N, independent of N. Indeed, the LFT $(\zeta + 1)/(\zeta - 1)$ maps the region $|\zeta - 1| < \delta$ onto a neighborhood of ∞ and is one-to-one, so it is bounded on $|\zeta - 1| > \delta$. The estimate $|e^{2\pi iz} - 1| > 1 - e^{-\pi}$ holds on S_N, as can be seen by considering the horizontal and vertical segments separately, so that $\pi \cot \pi z$ is bounded on S_N. Because $|f(z)| \leq C|z|^{-2}$, we have

$$\int_{S_N} f(z)\pi \cot \pi z dz \to 0.$$

By the residue theorem,

$$0 = \text{Res}_i f(z)\pi \cot \pi z + \text{Res}_{-i} f(z)\pi \cot \pi z + \sum_{-\infty}^{\infty} f(n),$$

and hence

$$\sum_{n=0}^{\infty} \frac{1}{n^2 + 1} = \frac{\pi}{2}\left[\frac{e^{\pi} + e^{-\pi}}{e^{\pi} - e^{-\pi}}\right] + \frac{1}{2}.$$

This technique can be used to compute

$$\sum_{n=-\infty}^{\infty} f(n),$$

provided f is meromorphic with $|f(z)| \leq C|z|^{-2}$ for $|z|$ large. If some of the poles of f occur at integers, then the residue calculation at those poles is slightly more complicated because the poles of $f(z)\pi \cot \pi z$ will not have order 1 at these integers. See Example 9.6. If only the weaker estimate $|f(z)| \leq C|z|^{-1}$ holds, then f has a removable singularity at ∞ and so $g(z) = f(z) + f(-z)$ satisfies $|g(z)| \leq C|z|^{-2}$ for large $|z|$. Applying the techique to g, we can find the symmetric limit

$$\lim_{N\to\infty} \sum_{n=-N}^{N} f(n).$$

9.5 Laplace and Inverse Laplace Transforms

The **Laplace transform** of a function f defined on $(0, +\infty)$ is given by

$$\mathcal{L}(f)(z) = \int_0^{\infty} f(t)e^{-zt}dt$$

provided the integral exists. For example, if $|f(t)| \leq Me^{ct}$ on $(0, +\infty)$ then $\mathcal{L}(f)(z) \equiv \lim_{T\to\infty} \int_0^T f(t)e^{-zt}dt$ exists for $\text{Re}z > c$. The convergence is uniform for $\text{Re}z \geq a > c$ because $|e^{-zt}| \leq e^{-at}$. By Weierstrass's Theorem 4.29, $\mathcal{L}(f)$ is analytic in $\{\text{Re}z > c\}$. Analyticity also follows from Morera's theorem. The Laplace transform is similar to the Fourier transform, except that the domain is rotated and it applies only to functions defined on $(0, \infty)$.

The Laplace transform can be used to convert an ordinary linear differential equation with constant coefficients for a function f into an algebra problem whose solution is the Laplace transform of f. It roughly converts differentiation into multiplication because of the elementary identity:

$$\mathcal{L}(f')(z) = z\mathcal{L}(f)(z) - f(0).$$

The algebra problem for $\mathcal{L}(f)$ is generally easy to solve, but, to find the solution to the original differential equation, we need to find a function whose Laplace transform is the solution to the algebra problem. In other words, given a function F, we would like to find f so that $\mathcal{L}(f) = F$. The function f is called the **inverse Laplace transform** of F. See Exercise 9.4.

One method for computing the inverse Laplace transform f for a rational function F is to use the partial fraction expansion from Section 2.2 and the fact that the Laplace transform of $f(t) = e^{at}t^n/n!$ is $(z-a)^{-(n+1)}$. We can also compute the inverse Laplace transform using residues.

Theorem 9.17 *Suppose F is a rational function and $|F(z)| \to 0$ as $|z| \to \infty$. Set*

$$f(t) = \sum \text{Res}_p \left(F(z)e^{zt}\right),$$

where the sum is taken over all poles p of F. Then $\mathcal{L}f(z) = F(z)$ for $\text{Re}z > \max_p \text{Re}p$, *where the maximum is taken over all poles of F.*

Proof Choose $a \in \mathbb{R}$ so that $F(\zeta)$ is analytic in $\text{Re}\zeta \geq a$ and suppose $\text{Re}z > a$. Let γ_R denote the positively oriented curve consisting of the vertical segment from $a - iR$ to $a + iR$ followed by the semi-circular arc of the circle $C_R(a)$ of radius R centered at a in $\text{Re}\zeta < a$. By the residue theorem, for R sufficiently large,

$$f(t) = \frac{1}{2\pi i}\int_{\gamma_R} F(\zeta)e^{\zeta t}d\zeta.$$

Thus, for $M > 0$,

$$\int_0^M f(t)e^{-zt}dt = \frac{1}{2\pi i}\int_{\gamma_R} F(\zeta)\left[\frac{e^{(\zeta-z)M} - 1}{\zeta - z}\right]d\zeta. \tag{9.12}$$

Because $\text{Re}\zeta \leq a < \text{Re}z$ on γ_R, the integral in (9.12) converges uniformly on compact subsets of $\{\text{Re}z > a\}$ as $M \to \infty$ to $-\int_{\gamma_R} F(\zeta)/(\zeta - z)d\zeta$. This proves the existence of the Laplace transform of f for $\text{Re}z > a$.

Let σ_R denote the positively oriented curve consisting of the vertical segment from $a+iR$ to $a-iR$ followed by the semi-circular arc of the circle $C_R(a)$ in $\text{Re}\zeta > a$. Then $\gamma_R+\sigma_R = C_R(a)$ and, by Cauchy's integral formula for R sufficiently large,

$$\begin{aligned} F(z) - \mathcal{L}(f)(z) &= \frac{1}{2\pi i}\int_{\sigma_R} \frac{F(\zeta)}{\zeta - z}d\zeta + \frac{1}{2\pi i}\int_{\gamma_R} \frac{F(\zeta)}{\zeta - z}d\zeta \\ &= \frac{1}{2\pi i}\int_{C_R(a)} \frac{F(\zeta)}{\zeta - z}d\zeta. \end{aligned} \tag{9.13}$$

Because F tends to 0 at ∞, the integral around $C_R(a)$ in (9.13) tends to 0 as $R \to \infty$, proving Theorem 9.17. □

The ideas behind the proof of Theorem 9.17 can be used to give a formula for the inverse Laplace transform for functions F which are not necessarily rational.

Theorem 9.18 *Suppose F is analytic in* $\text{Re}z \geq c$, $\int_{-\infty}^{\infty} |F(c+iy)|dy = M < \infty$ *and* $|F(z)| \to 0$ *as* $|z| \to \infty$ *with* $\text{Re}z \geq c$. *Then*

$$f(t) = \frac{1}{2\pi}\int_{-\infty}^{\infty} F(c+iy)e^{(c+iy)t}dy$$

exists, $|f(t)| \leq Me^{ct}$, *and* $\mathcal{L}(f)(z) = F(z)$ *for* $\text{Re}z > c$.

Proof We may suppose $c = 0$ because $\mathcal{L}(g)(z) = \mathcal{L}(f)(z+c)$ if $g(t) = e^{-ct}f(t)$. The function $f(t)$ exists because $|e^{iyt}| = 1$ and $M < \infty$. Moreover,

$$|f(t_1) - f(t_2)| \leq 2\int_{|y|>K} |F(iy)|dy + \left|\int_{-K}^{K} F(iy)\left[e^{iyt_1} - e^{iyt_2}\right]dy\right|,$$

so that, by uniform continuity and $M < \infty, f(t)$ is continuous. Then, for $\text{Re}z > 0$,

$$\begin{aligned}\int_0^K f(t)e^{-zt}dt &= \frac{1}{2\pi}\int_{-\infty}^{\infty} F(iy)\int_0^K e^{(iy-z)t}dtdy \\ &= \frac{1}{2\pi}\int_{-\infty}^{\infty} F(iy)\frac{e^{(iy-z)K}-1}{iy-z}dy \qquad (9.14)\\ &\to -\frac{1}{2\pi}\int_{-\infty}^{\infty}\frac{F(iy)}{iy-z}dy,\end{aligned}$$

as $K \to \infty$. Thus $\mathcal{L}(f)(z)$ exists for $\text{Re}z > 0$.

If $|z| < r$ and $\text{Re}z > 0$ then, by Cauchy's integral formula,

$$F(z) = \frac{1}{2\pi i}\int_{-\frac{\pi}{2}}^{\frac{\pi}{2}} \frac{F(re^{it})}{re^{it}-z}ire^{it}dt - \frac{1}{2\pi i}\int_{-r}^{r}\frac{F(iy)}{iy-z}idy.$$

Because $|F(\zeta)| \to 0$ as $|\zeta| \to \infty$ with $\text{Re}\zeta \geq 0$, we conclude from (9.14) that

$$F(z) = \mathcal{L}(f)(z)$$

for $\text{Re}z > 0$. □

The Laplace transform and the residue theorem are used in the next theorem to give a sufficient condition for the existence of an improper integral. The proof is a nice illustration of how to deform a contour to take advantage of analyticity. Theorem 9.19 is used in Exercise 11.15 to prove the prime number theorem.

Theorem 9.19 *Suppose f is piecewise continuous on* $[0, T]$ *for each* $T < \infty$ *and* $|f| \leq M < \infty$ *on* $[0, +\infty)$. *If there exists a function g which is analytic in a neighborhood of* $\{\text{Re}z \geq 0\}$ *such that* $g(z) = \mathcal{L}f(z)$ *for all* $\text{Re}z > 0$, *then, for* $\text{Re}z = 0$, $\mathcal{L}f(z)$ *exists and equals* $g(z)$. *In particular,* $\int_0^{\infty} f(t)dt$ *exists and equals* $g(0)$.

For students familiar with measure theory, the proof below is valid if the assumptions on f are replaced by $f \in L^{\infty}$.

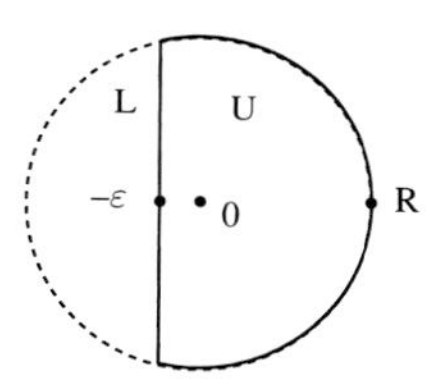

Figure 9.5 Enlarged half-disk.

Proof Set $g_T(z) = \int_0^T f(t)e^{-zt}dt$. Then g_T is entire by Morera's theorem. By a translation, it is enough to show that $\lim_{T\to\infty} g_T(0) = g(0)$.

Fix $R < \infty$ and let $U = \{z : |z| < R \text{ and } \mathrm{Re}z > -\varepsilon\}$, where $\varepsilon = \varepsilon(R) > 0$ is so small that g is analytic in a neighborhood of $\overline{U}$. See Figure 9.5. Then

$$g(0) - g_T(0) = \frac{1}{2\pi i}\int_{\partial U}(g(z) - g_T(z))e^{zT}\left(\frac{z}{R} + \frac{R}{z}\right)\frac{dz}{R} \tag{9.15}$$

by the residue theorem. On the semi-circle $\partial U \cap \{\mathrm{Re}z > 0\}$,

$$|g(z) - g_T(z)| = \left|\int_T^\infty f(t)e^{-zt}dt\right| \leq M\int_T^\infty |e^{-zt}|dt = M\frac{e^{-\mathrm{Re}zT}}{\mathrm{Re}z}, \tag{9.16}$$

and

$$\left|e^{zT}\left(\frac{z}{R} + \frac{R}{z}\right)\right| = e^{\mathrm{Re}zT}\,\frac{2\mathrm{Re}z}{R} \tag{9.17}$$

because $|z| = R$. By (9.16) and (9.17), the contribution to (9.15) from the integral over $\partial U \cap \{\mathrm{Re}z > 0\}$ is bounded by M/R.

For the integral over the rest of ∂U, we treat the integrals involving g and g_T separately. Because g_T is entire, by Cauchy's theorem we can replace the path of integration with the semi-circle $C^- = \{z : |z| = R \text{ and } \mathrm{Re}z \leq 0\}$. Then, as before,

$$|g_T(z)| = \left|\int_0^T f(t)e^{-zt}dt\right| \leq M\frac{e^{-\mathrm{Re}zT}}{-\mathrm{Re}z} \tag{9.18}$$

and

$$\left|e^{zT}\left(\frac{z}{R} + \frac{R}{z}\right)\right| = e^{\mathrm{Re}zT}\left(\frac{-2\mathrm{Re}z}{R}\right). \tag{9.19}$$

By (9.18) and (9.19) we conclude that the integral over C^- is bounded by M/R. To estimate the integral involving g, write the rest of ∂U as the sum of a vertical line segment $L \subset \{\mathrm{Re}z = -\varepsilon\}$ and two small arcs of length $\leq K\varepsilon$. Because R is fixed and g is independent of T, we can bound the integrand on L,

$$\left|g(z)e^{zT}\left(\frac{z}{R} + \frac{R}{z}\right)\frac{1}{R}\right| \leq \frac{K_1}{\varepsilon}e^{-\varepsilon T},$$

where K_1 is a constant depending on R and g. We can bound the integrand on the two small arcs by a constant K_2, again depending on R and g. Choose ε so small that the sum of the integrals over the small arcs is at most M/R, then choose T so large that the integral over the line segment L is at most M/R. Thus there exists $T(R) < \infty$ such that $|g(0) - g_T(0)| \leq 4M/R$ for $T > T(R)$. Letting $R \to \infty$, we conclude $\lim_{T\to\infty} |g(0) - g_T(0)| = 0$, proving Theorem 9.19. □

9.6 Exercises

A

9.1 Find all non-zero residues of the following functions:

(a) $\dfrac{z-1}{(z+1)^2(z-2)}$,

(b) $\dfrac{z^2-2z}{(z+1)^2(z^2+4)}$,

(c) $e^z \csc^2 z$,

(d) ze^{-1/z^2},

(e) $\dfrac{\cot \pi z}{z^6}$.

9.2 Let C be the circle of radius 3 centered at 0, oriented in the positive sense. Find

$$\int_C \frac{e^{\lambda z}}{(z+4)(z-1)^2(z^2+4z+5)} dz.$$

9.3 Suppose that f and g are analytic in a neighborhood of the closed unit disk. Let $\{a_n\}$ denote the zeros of f, including multiplicity, and suppose that $|a_n| < 1$ for all n. Prove

$$\int_0^{2\pi} \frac{e^{i\theta} g(e^{i\theta}) f'(e^{i\theta})}{f(e^{i\theta})} \frac{d\theta}{2\pi} = \sum_n g(a_n).$$

9.4 Solve $y'' - 3y' + 2y = 12e^{4t}$, with $y(0) = 1$ and $y'(0) = 0$, using Laplace transforms and Theorem 9.17.

B

In each of the exercises below, be sure you have defined your functions carefully. Prove all claims about integrals. Draw a picture of any contour you use.

9.5 Find the Fourier transform of $x^3/(x^2+1)^2$. (Note that the integral is not absolutely convergent, so part of the exercise is to prove that the integral converges, i.e., let the limits of integration tend to $\pm\infty$ independently.)

9.6 For $0 < \alpha < 1$, find

$$\int_0^\infty \frac{x^\alpha}{x(x+1)} dx.$$

9.7 Find the inverse Laplace transform of

$$F(z) = \frac{3z^2 + 12z + 8}{(z+2)^2(z+4)(z-1)}$$

in two ways: using partial fractions and using Theorem 9.17.

9.8 Verify

$$\sum_{n=0}^{\infty} \frac{(-1)^n}{(2n+1)^3} = \frac{\pi^3}{32}.$$

9.9 Compute

$$\zeta(6) = \sum_{n=1}^{\infty} \frac{1}{n^6}$$

by calculus of residues. You can use your answer to Exercise 9.1(e). (Remark: This process can be used to compute $\zeta(2n)$. The value of ζ at an odd integer is another story.)

9.10 By means of the calculus of residues, evaluate

$$\int_0^{\infty} \frac{\sqrt{x}\log x}{(1+x^2)} dx.$$

9.11 Find

$$\int_0^{\infty} \frac{x^3+8}{x^5+1} dx$$

using a contour integral of $(\log z)(z^3+8)/(z^5+1)$.

9.12 Find

$$\int_0^1 \frac{1}{\sqrt{x(1-x)}} dx.$$

Put a "dog bone" around the interval $[0, 1]$ and add a large circle. Carefully define the integrand so that it is analytic at ∞, then the integral over the large circle can be found from the series expansion at ∞. Redo the exercise by first making the substitution $x = 1/w$.

10 Normal Families

One of the main goals in this chapter is to give another proof of the Riemann mapping theorem (Theorem 8.13) which is rather elegant and useful for solving other problems. It is based on a notion of convergence of analytic functions, called normal convergence. We also introduce a powerful tool due to Zalcman that characterizes non-normal families in terms of an associated convergent sequence. This result is used to prove classical results of Montel and Picard about sets omitted by analytic and meromorphic functions.

10.1 Normality and Equicontinuity

Definition 10.1 *A collection, or family, $\mathcal{F}$ of continuous functions on a region $\Omega \subset \mathbb{C}$ is said to be* **normal on Ω** *provided every sequence $\{f_n\} \subset \mathcal{F}$ contains a subsequence which converges uniformly on compact subsets of Ω.*

The family $\mathcal{F}_1 = \{f_c(z) = z + c : |c| < 1\}$ is normal in $\mathbb{C}$ but not countable. The family $\mathcal{F}_2 = \{z^n : n = 0, 1, \ldots\}$ is normal in $\mathbb{D}$ but the only limit function, the zero function, is not in $\mathcal{F}_2$. The sequence z^n converges uniformly on each compact subset of $\mathbb{D}$, but does not converge uniformly on $\mathbb{D}$. The family $\mathcal{F}_3 = \{g_n\}$, where $g_n \equiv 1$ if n is even and $g_n \equiv 0$ if n is odd, is normal but the sequence $\{g_n\}$ does not converge.

The first lemma says that normality is local.

Lemma 10.2 *Suppose Ω is a region and suppose $\Omega = \cup_{j=1}^{\infty} \Delta_j$, where $\Delta_j \subset \Omega$ are closed disks. A family of continuous functions $\mathcal{F}$ is normal on Ω if and only if, for each j, every sequence in $\mathcal{F}$ contains a subsequence which converges uniformly on Δ_j.*

Proof The only if part follows by definition. Suppose that, for each j, every sequence in $\mathcal{F}$ contains a subsequence which converges uniformly on Δ_j. Suppose $\{f_n\} \subset \mathcal{F}$. Then there is a subsequence $\{f_n^{(1)}\} \subset \{f_n\}$ such that $\{f_n^{(1)}\}$ converges uniformly on Δ_1. Likewise, there is a sequence $\{f_n^{(2)}\} \subset \{f_n^{(1)}\}$ which converges uniformly on Δ_2, and indeed there is a sequence $\{f_n^{(k)}\} \subset \{f_n^{(k-1)}\}$ which converges uniformly on Δ_k. Then the "diagonal" sequence

$$\{f_k^{(k)}\}$$

converges uniformly on each Δ_j, since it is a subsequence of $\{f_n^{(k)}\}$ for each k. A compact set $K \subset \Omega$ can be covered by finitely many Δ_j, so the diagonal sequence converges uniformly on K. □

The proof of the above lemma shows that we could have used an ostensibly weaker, but equivalent, definition of normality: for each sequence and each compact subset there is a subsequence which converges uniformly on the compact set.

We can define a metric on the space $C(\Omega)$ of continuous functions on a region Ω as follows. Write

$$\Omega = \cup_{j=1}^{\infty} \Delta_j,$$

where each Δ_j is a closed disk with $\Delta_j \subset \Omega$. If $f, g \in C(\Omega)$, set

$$\rho_j(f, g) = \sup_{z \in \Delta_j} \frac{|f(z) - g(z)|}{1 + |f(z) - g(z)|}$$

and

$$\rho(f, g) = \sum_{j=1}^{\infty} 2^{-j} \rho_j(f, g).$$

Then ρ is a metric:

(i) if $\rho(f, g) = 0$, then $f = g$ on each Δ_j and hence $f = g$ on Ω,
(ii) $\rho(f, g) = \rho(g, f)$ for $f, g \in C(\Omega)$,
(iii) $\rho(f, g) \le \rho(f, h) + \rho(h, g)$ for $f, g, h \in C(\Omega)$.

The triangle inequality follows from the observations that $\frac{x}{1+x} = 1 - \frac{1}{1+x}$ is increasing for $x \ge 0$ and, if a and b are non-negative numbers, then

$$\frac{a+b}{1+a+b} \le \frac{a}{1+a} + \frac{b}{1+b}.$$

Note that $\rho_j \le 1$ and $\rho \le 1$.

Proposition 10.3 *A sequence $\{f_n\} \subset C(\Omega)$ converges uniformly on compact subsets of Ω to $f \in C(\Omega)$ if and only if*

$$\lim_n \rho(f_n, f) = 0.$$

In other words, the space $C(\Omega)$ with the topology of uniform convergence on compact subsets is a metric space. Compactness and sequential compactness are the same for metric spaces. A family is normal if and only if its closure is compact in this topology. A sequence which converges uniformly on compact subsets of a region Ω is sometimes said to **converge normally** on Ω or **converge locally uniformly** on Ω. The term "normally convergent" has a different meaning in functional analysis, related to convergence of norms, so it should be avoided. Yet another term that is used is "c.c.," which stands for compact convergence.

Proof If $\rho(f_n, f) \to 0$, then $\rho_j(f_n, f) \to 0$ for each j. This implies f_n converges uniformly to f on Δ_j for each j, because $a/(1+a) < b$ is the same as $a < b/(1-b)$, for $a > 0$ and $0 < b < 1$. Since each compact set $K \subset \Omega$ can be covered by finitely many Δ_j, the sequence f_n converges uniformly on K. Conversely, if f_n converges to f uniformly on each Δ_j, then, given $\varepsilon > 0$, choose n_j so that for $z \in \Delta_j$

$$|f_n(z) - f(z)| < \frac{\varepsilon}{2}, \text{ whenever } n \ge n_j,$$

for $j = 1, 2, 3, \ldots$. Choose m so that

$$\sum_{j=m}^{\infty} 2^{-j} < \frac{\varepsilon}{2}.$$

Then for $n \geq \max\{n_1, \ldots, n_{m-1}\}$

$$\rho(f_n, f) \leq \sum_{j=1}^{m-1} 2^{-j}\frac{\varepsilon}{2} + \sum_{j=m}^{\infty} 2^{-j} < \varepsilon. \qquad \square$$

Normality is related to a strong concept of continuity, called uniform equicontinuity.

Definition 10.4 *A family of functions $\mathcal{F}$ defined on a set $E \subset \mathbb{C}$ is*

(a) **equicontinuous at $w \in E$** *if for each $\varepsilon > 0$ there exist a $\delta > 0$ so that if $z \in E$ and $|z - w| < \delta$, then $|f(z) - f(w)| < \varepsilon$ for all $f \in \mathcal{F}$;*
(b) **equicontinuous on E** *if it is equicontinuous at each $w \in E$;*
(c) **uniformly equicontinuous on E** *if for each $\varepsilon > 0$ there exists a $\delta > 0$ so that if $z, w \in E$ with $|z - w| < \delta$ then $|f(z) - f(w)| < \varepsilon$ for all $f \in \mathcal{F}$.*

For example, if $\mathcal{F}_M$ is the family of analytic functions on $\mathbb{D}$ such that $|f'| \leq M < \infty$ for all $z \in \mathbb{D}$, then $\mathcal{F}_M$ is uniformly equicontinuous on $\mathbb{D}$. Writing f as the integral of its derivative along a line segment from z to w, we obtain $|f(z) - f(w)| \leq M|z - w|$. Thus $|f(z) - f(w)| < \varepsilon$ whenever $|z - w| < \delta = \varepsilon/M$.

Continuity, equicontinuity, uniform continuity and uniform equicontinuity are all related to the following statement:

$$\text{if } |z - w| < \delta \text{ then } |f(z) - f(w)| < \varepsilon.$$

For continuity, δ is allowed to depend on ε, the function f and the point w. For equicontinuity, δ does not depend on f. For uniform continuity, δ depends on ε and the function f, but works for all w. For uniform equicontinuity, δ depends on ε but works for all w and for all $f \in \mathcal{F}$.

A family of functions which is equicontinuous on a compact set K is uniformly equicontinuous on K. When the family has just one function, this is a familiar result about continuous functions; moreover, the proof for equicontinuous families is the same. The next result relates normality and equicontinuity.

Theorem 10.5 (Arzela–Ascoli) *A family $\mathcal{F}$ of continuous functions is normal on a region $\Omega \subset \mathbb{C}$ if and only if*

(i) *$\mathcal{F}$ is equicontinuous on Ω and*
(ii) *there is a $z_0 \in \Omega$ so that the collection $\{f(z_0) : f \in \mathcal{F}\}$ is a bounded subset of $\mathbb{C}$.*

Proof Suppose $\mathcal{F}$ is normal. If $\mathcal{F}$ is not equicontinuous at $z_0 \in \Omega$ then there is an $\varepsilon > 0$, $z_n \in \Omega$ and $f_n \in \mathcal{F}$ such that $|z_n - z_0| < 1/n$ and

$$|f_n(z_n) - f_n(z_0)| \geq \varepsilon \tag{10.1}$$

for $n \geq n_0$. By normality, $\{f_n\}$ contains a subsequence which converges to a continuous function f, uniformly on $|z - z_0| \leq 1/n_0$. Discarding the functions f_n which are not in this subsequence, and then relabeling the sequence and the points z_n, we may suppose (10.1) holds and f_n converges to f uniformly on $|z - z_0| \leq 1/n_0$. Thus

$$\begin{aligned}\varepsilon &\leq |f_n(z_n) - f_n(z_0)| \\ &\leq |f_n(z_n) - f(z_n)| + |f(z_n) - f(z_0)| + |f(z_0) - f_n(z_0)|. \end{aligned} \tag{10.2}$$

For sufficiently large n, the first and third terms on the right-hand side of (10.2) are less than $\varepsilon/3$ by uniform convergence, and the second term is less than $\varepsilon/3$ because the limit function f is continuous on $|z - z_0| \leq 1/n_0$. This contradiction shows that $\mathcal{F}$ is equicontinuous on Ω. If $z_0 \in \Omega$, then $S = \{f(z_0) : f \in \mathcal{F}\}$ is a bounded set, for if $f_n \in \mathcal{F}$ and $|f_n(z_0)| \to \infty$ then, by normality, $\{f_n\}$ contains a subsequence converging at z_0, which is a contradiction.

Conversely, suppose Theorem 10.5 (i) and (ii) hold. We first show that (ii) holds at each $z \in \Omega$. By equicontinuity, each $w \in \Omega$ is contained in the interior of a closed disk $\Delta_w \subset \Omega$ such that

$$|f(z) - f(w)| < 1 \text{ for all } z \in \Delta_w \text{ and all } f \in \mathcal{F}. \tag{10.3}$$

Let $U = \{z_0 \in \Omega : \text{(ii) holds}\}$. If $w \in U$, then by (10.3) $\Delta_w \subset U$, so that U is open. If $w \in \Omega \setminus U$ then, by (10.3) again, $\Delta_w \subset \Omega \setminus U$, so that $\Omega \setminus U$ is open. But Ω is connected and $U \neq \emptyset$ by (ii), so that $U = \Omega$.

Now suppose $\{f_n\} \subset \mathcal{F}$. Let $D = \{z_k\}$ be a countable dense subset of Ω. Since $\{f(z_1) : f \in \mathcal{F}\}$ is a bounded set, we can find a subsequence $\{f_n^{(1)}\}$ such that $\{f_n^{(1)}(z_1)\}$ converges. Likewise we can find a sequence $\{f_n^{(2)}\} \subset \{f_n^{(1)}\}$ such that $\{f_n^{(2)}(z_2)\}$ converges, and indeed there is a sequence $\{f_n^{(k)}\} \subset \{f_n^{(k-1)}\}$ such that $\{f_n^{(k)}(z_k)\}$ converges. Then, as in the proof of Lemma 10.2, the sequence

$$\{f_n^{(n)}\}$$

converges at z_k, for each k. We may relabel this sequence as $\{f_n\}$.

Fix a closed disk Δ with $\Delta \subset \Omega$ and fix $\varepsilon > 0$. By (i) the family $\mathcal{F}$ is uniformly equicontinuous on Δ, so we can find $\delta > 0$ such that if $f \in \mathcal{F}$ and if $z, w \in \Delta$ with $|z - w| < \delta$ then

$$|f(z) - f(w)| < \varepsilon/3. \tag{10.4}$$

Choose a finite set $z_{k(1)}, \ldots, z_{k(N)} \in D \cap \Delta$ so that for each $z \in \Delta$ we have $\min_j\{|z - z_{k(j)}|\} < \delta$. Then find $M < \infty$ so that for $m, p \geq M$ and $1 \leq j \leq N$

$$|f_m(z_{k(j)}) - f_p(z_{k(j)})| < \varepsilon/3. \tag{10.5}$$

Given $z \in \Delta$ and $m, p \geq M$, choose $z_{k(j)}$ so that $|z - z_{k(j)}| < \delta$. Then

$$\begin{aligned}&|f_m(z) - f_p(z)| \\ &\quad \leq |f_m(z) - f_m(z_{k(j)})| + |f_m(z_{k(j)}) - f_p(z_{k(j)})| + |f_p(z_{k(j)}) - f_p(z)|. \end{aligned} \tag{10.6}$$

The first and third terms on the right-hand side of (10.6) are at most $\varepsilon/3$ by (10.4), and the middle term is at most $\varepsilon/3$ by (10.5). Thus, for $m, p > M$,

$$|f_m(z) - f_p(z)| < \varepsilon$$

on Δ. Since $\varepsilon > 0$ is arbitrary, this proves that $\{f_n\}$ is a Cauchy sequence in the uniform topology on Δ. Because $\mathbb{C}$ is complete, the relabeled subsequence $\{f_n\}$ converges on Δ. By Lemma 10.2, $\mathcal{F}$ is normal on Ω. □

In the discussion above, we used the Euclidean distance $|\alpha - \beta|$ to measure the distance between $\alpha = f(z)$ and $\beta = g(z)$. We could similarly consider families of continuous functions with values in a complete metric space. The Arzela–Ascoli theorem holds in this context with the same proof, replacing $|f(z) - f(w)|$ with the metric distance between $f(z)$ and $f(w)$. For example, we could consider continuous functions with values in the Riemann sphere using the chordal distance between any two points on the sphere. In this case, we must allow the north pole $(0, 0, 1)$ as a possible value of a function. Equivalently, we can consider functions with values in the extended plane $\mathbb{C}^* = \mathbb{C} \cup \{\infty\}$ with the chordal metric

$$\chi(\alpha, \beta) = \begin{cases} \frac{2|\alpha-\beta|}{\sqrt{1+|\alpha|^2}\sqrt{1+|\beta|^2}}, & \text{if } \alpha, \beta \in \mathbb{C} \\ \frac{2}{\sqrt{1+|\alpha|^2}}, & \text{if } \beta = \infty. \end{cases}$$

Then, to say that a function f is "continuous" at z_0 with $f(z_0) = \infty$ means that for all $R < \infty$ there is a $\delta > 0$ so that $|f(z)| > R$ for all z with $|z - z_0| < \delta$. A family $\mathcal{F}$ of continuous functions on Ω is normal in the chordal metric if and only if $\mathcal{F}$ is equicontinuous in the chordal metric. The analog of Theorem 10.5(ii) is not needed because $\mathbb{C}^*$ is compact in this topology. See Exercise 10.2. We remind the reader that the topology induced on $\mathbb{C}$ by χ is the same as the usual Euclidean topology, by Corollary 1.5.

For families of analytic functions, the Arzela–Ascoli theorem, together with Cauchy's estimate, Corollary 4.16, gives simple criteria for normality, Theorem 10.7.

Definition 10.6 *A family $\mathcal{F}$ of continuous functions is said to be* **locally bounded** *on Ω if for each $w \in \Omega$ there is a $\delta > 0$ and $M < \infty$ so that if $|z - w| < \delta$ then $|f(z)| \leq M$ for all $f \in \mathcal{F}$.*

An equivalent definition is that $\mathcal{F}$ is locally bounded on Ω if and only if, for each compact $K \subset \Omega$, the set $\{f(z) : f \in \mathcal{F}, z \in K\}$ is bounded.

Theorem 10.7 *The following are equivalent for a family $\mathcal{F}$ of analytic functions on a region Ω:*

(i) *$\mathcal{F}$ is normal on Ω;*
(ii) *$\mathcal{F}$ is locally bounded on Ω;*
(iii) *$\mathcal{F}' = \{f' : f \in \mathcal{F}\}$ is locally bounded on Ω and there is a $z_0 \in \Omega$ so that $\{f(z_0) : f \in \mathcal{F}\}$ is a bounded subset of $\mathbb{C}$.*

Proof Suppose $\mathcal{F}$ is normal. By the proof of Theorem 10.5, for each $w \in \Omega$, $\{f(w) : f \in \mathcal{F}\}$ is bounded. By (10.3), $\mathcal{F}$ is locally bounded.

Now suppose $\mathcal{F}$ is locally bounded on Ω. If $|f| \leq M$ on a closed disk $\overline{B}(z_1, r) \subset \Omega$ centered at z_1 with radius $r > 0$, then, by Cauchy's estimate, $|f'(z)| \leq 4M/r^2$ on $B(z_1, r/2)$. It follows that $\mathcal{F}'$ is locally bounded, and (iii) holds.

Finally, if (iii) holds and $z_1 \in \Omega$, then $|f'(z)| \leq L < \infty$ for z in a disk Δ centered at z_1. Integrating f' along a line segment from z_1 to $z \in \Delta$, we have $|f(z) - f(z_1)| \leq L|z - z_1|$ for all $f \in \mathcal{F}$. So $\mathcal{F}$ is equicontinuous at z_1. By the Arzela–Ascoli theorem, $\mathcal{F}$ is normal. □

We remark that the space of analytic functions on a region $\Omega \subset \mathbb{C}$ is closed in the topology of uniform convergence on compact subsets by Weierstrass's Theorem 4.29.

For families of meromorphic functions, we use the chordal metric χ (see Equation (1.4)) instead of the Euclidean metric. Lemma 10.8 is the analog of Weierstrass's theorem for meromorphic functions.

Lemma 10.8 *If $\{f_n\}$ is a sequence of meromorphic functions which converges uniformly on compact subsets of a region $\Omega \subset \mathbb{C}$ in the chordal metric, then the limit function is either meromorphic on Ω or identically equal to ∞.*

Proof Note that

$$\chi(\alpha, \beta) = \chi(\frac{1}{\alpha}, \frac{1}{\beta}). \tag{10.7}$$

If $\{f_n\}$ is a sequence of meromorphic functions which converges uniformly on compact subsets of Ω in the χ metric, then the limit function f is continuous as a map into the extended plane $\mathbb{C}^*$. If $|f(z_0)| < \infty$ then f is bounded in a neighborhood of z_0, and hence f_n converges to f uniformly in the Euclidean metric in a neighborhood of z_0. By Weierstrass's theorem, f is analytic in a neighborhood of z_0. If $f(z_0) = \infty$, then $1/f_n$ is bounded in a neighborhood of z_0, for n sufficiently large, and hence extends to be analytic by Riemann's removable singularity theorem. By (10.7) and Weierstrass's theorem, $1/f$ is analytic in a neighborhood of z_0. By the identity theorem, if $1/f$ is not identically 0 in a neighborhood of z_0, then the zero of $1/f$ at z_0 is isolated, and hence f has an isolated pole at z_0. The set of non-isolated zeros of $1/f$ is then both open and closed in Ω. Because Ω is connected, either f is identically equal to ∞ or f is analytic except for isolated poles. This proves Lemma 10.8. □

For example, by (10.7)

$$\left\{\frac{1}{z - 1/n}\right\}$$

is a normal family in $\mathbb{C}$ in the χ metric with limit function $1/z$. Be careful because $1/(z - 1/n) - 1/z$ does not converge to 0 in the χ metric.

The equivalence of (i) and (ii) in Theorem 10.7 does not have an analog for meromorphic functions in the chordal metric because all continuous functions are bounded by 2 in the chordal metric. However, there is an analog of the equivalence of (i) and (iii) using the spherical derivative.

Definition 10.9 *If f is meromorphic on a region $\Omega \subset \mathbb{C}$ then*

$$f^{\#}(z) = \lim_{w \to z} \frac{\chi(f(z), f(w))}{|z - w|}$$

is called the **spherical derivative** *of f.*

If z is not a pole of f then

$$f^{\#}(z) = \frac{2|f'(z)|}{1 + |f(z)|^2}.$$

By (10.7), $(1/f)^{\#} = f^{\#}$ so that $f^{\#}(z)$ is finite and continuous at each $z \in \Omega$. It is positive at z if and only if f is one-to-one in a neighborhood of z.

The **spherical distance** $d(p^*, q^*)$ between $p^*, q^* \in \mathbb{S}^2$ is the arc-length of the shortest curve on the sphere containing p^* and q^*. The quantity $f^{\#}$ is called the spherical derivative because

$$f^{\#}(z) = \lim_{w \to z} \frac{d(f(z)^*, f(w)^*)}{|z - w|},$$

where $f(z)^*$ denotes the stereographic projection of $f(z)$.

Theorem 10.10 (Marty) *A family $\mathcal{F}$ of meromorphic functions on a region $\Omega \subset \mathbb{C}$ is normal in the chordal metric if and only if $\mathcal{F}^{\#} = \{f^{\#} : f \in \mathcal{F}\}$ is locally bounded.*

Proof Suppose $\mathcal{F}$ is normal in the chordal metric on Ω and suppose the spherical derivatives are not bounded in any neighborhood of z_0. Then there is a sequence $f_n \in \mathcal{F}$ and $z_n \to z_0$ so that $f_n^{\#}(z_n) \to \infty$. Taking a subsequence, we may suppose that f_n converges uniformly on compact subsets of Ω in the chordal metric to a meromorphic function f, or to ∞, by normality and Lemma 10.8. If $f(z_0) \neq \infty$ then f is bounded in the Euclidean metric in a neighborhood N_{z_0} of z_0. Because f_n converges to f in the chordal metric, f_n must also be bounded in N_{z_0} and thus f_n converges to f in the Euclidean metric on N_{z_0}. By Weierstrass's theorem, f_n' converges to f' uniformly on compact subsets of N_{z_0} and thus $f_n^{\#}$ converges uniformly on compact subsets of N_{z_0} to $f^{\#}$. This contradicts the unboundedness of $f_n^{\#}(z_n)$. If $f(z_0) = \infty$, then we can apply this same argument to $1/f$ and $1/f_n$ using (10.7).

Conversely, suppose the spherical derivatives are bounded by M in a disk Δ centered at z_0. If $z, w \in \Delta$, let $z_j = z + (j/n)(w - z), 0 \leq j \leq n$. Then, for n large,

$$\chi(f(z), f(w)) \leq \sum_{j=1}^{n} \chi(f(z_j), f(z_{j-1})) \approx \sum_{j=1}^{n} f^{\#}(z_j)|z_j - z_{j-1}| \leq M|z - w|.$$

Thus $\mathcal{F}$ is equicontinuous, with the chordal metric, on Δ. By the version of the Arzela–Ascoli theorem with the chordal metric (see Exercise 10.2) $\mathcal{F}$ is normal on Δ, and, by the chordal version of Lemma 10.2, normal on Ω. □

A closer examination of the proof of Marty's theorem shows that if meromorphic functions f_n converge to f in the chordal metric then $f_n^{\#}$ converges to $f^{\#}$. The converse fails as a sequence of constants, or a sequence of rotations of a single function, show.

10.2 Riemann Mapping Theorem Revisited

In this section we give another proof of the Riemann mapping theorem, Theorem 8.13. Much of the proof is the same. We avoid the use of the geodesic algorithm by using normal families instead of Harnack's principle.

Theorem 10.11 (Riemann mapping theorem) *Suppose $\Omega \subset \mathbb{C}$ is simply-connected and $\Omega \neq \mathbb{C}$. Then there exists a one-to-one analytic map f of Ω onto $\mathbb{D} = \{z : |z| < 1\}$. If $z_0 \in \Omega$ then there is a unique such map with $f(z_0) = 0$ and $f'(z_0) > 0$.*

Proof Fix $z_0 \in \Omega$. The idea of the proof is to show that there is a conformal map of Ω into $\mathbb{D}$, vanishing at z_0, with the largest possible derivative at z_0. This forces the image region to be as large as possible and hence to be equal to all of $\mathbb{D}$.

Set

$$\mathcal{F} = \{f : f \text{ is one-to-one, analytic, } |f| < 1 \text{ on } \Omega, f(z_0) = 0, f'(z_0) > 0\}.$$

By the first paragraph of the proof of Theorem 8.11, $\mathcal{F}$ is non-empty. By Theorem 10.7, $\mathcal{F}$ is normal. Let $\{f_n\} \subset \mathcal{F}$ such that

$$\lim_{n\to\infty} f_n'(z_0) = M = \sup\{f'(z_0) : f \in \mathcal{F}\}. \tag{10.8}$$

Replacing f_n with a subsequence, we may suppose that f_n converges uniformly on compact subsets of Ω. By Weierstrass's theorem, the limit function f is analytic and $\{f_n'\}$ converges to f'. Thus $f'(z_0) = M$. In particular, f is not constant, and $M < \infty$. By Hurwitz's theorem (Corollary 8.9), f is one-to-one. Also $f(z_0) = \lim f_n(z_0) = 0$, so that $f \in \mathcal{F}$.

Next we show that f must map Ω onto $\mathbb{D}$. The proof is similar to the proof in Theorem 8.11, though we can avoid an explicit computation. Suppose there is a point $\zeta_0 \in \mathbb{D}$ such that $f \neq \zeta_0$ on Ω. Then

$$g_1(z) = \frac{f(z) - \zeta_0}{1 - \overline{\zeta_0} f(z)} \equiv T_1 \circ f(z)$$

is a non-vanishing function on the simply-connected region Ω and hence it has an analytic square root which is, again, one-to-one. Set $g_2(z) = \sqrt{g_1(z)}$ and

$$g(z) = \frac{g_2(z) - g_2(z_0)}{1 - \overline{g_2(z_0)} g_2(z)} \equiv T_2 \circ g_2(z).$$

Then $\lambda g(z) \in \mathcal{F}$, where $\lambda = |g'(z_0)|/g'(z_0)$. Because T_1 and T_2 are LFTs of $\mathbb{D}$ onto $\mathbb{D}$,

$$\varphi = T_1^{-1} \circ S \circ T_2^{-1},$$

where $S(z) = z^2$, is a two-to-one analytic map (counting multiplicity) of $\mathbb{D}$ onto $\mathbb{D}$ with $\varphi(0) = 0$, and

$$f(z) = \varphi \circ g(z).$$

By Schwarz's lemma (or direct computation), $|\varphi'(0)| < 1$, so that

$$f'(z_0) = |f'(z_0)| = |\varphi'(0)||g'(z_0)| < |g'(z_0)| = \lambda g'(z_0),$$

contradicting the maximality of $f'(z_0)$, and hence f maps $\mathbb{D}$ onto $\mathbb{D}$.

Uniqueness follows from Proposition 6.5. □

We remark that, because the limit function in Theorem 10.11 is unique, every sequence f_n of conformal maps of Ω onto $\mathbb{D}$ with $f_n(z_0) = 0$ and $f_n'(z_0) \to M$ must converge uniformly on compact subsets of Ω to the map f. See Exercise 10.1.

10.3 Zalcman, Montel and Picard

In this section we will prove Zalcman's clever lemma characterizing non-normal families, and use it to prove far-reaching extensions of Liouville's theorem and Riemann's theorem on removable singularities. See Exercise 10.6 for a version for families of analytic functions using the Euclidean metric.

Theorem 10.12 (Zalcman's lemma) *A family $\mathcal{F}$ of meromorphic functions on a region Ω is not normal in the chordal metric if and only if there exists a sequence $\{z_n\}$ converging to $z_\infty \in \Omega$, a sequence of positive numbers ρ_n converging to 0, and a sequence $\{f_n\} \subset \mathcal{F}$ such that*

$$g_n(\zeta) = f_n(z_n + \rho_n\zeta) \tag{10.9}$$

converges uniformly in the chordal metric on compact subsets of $\mathbb{C}$ to a nonconstant function g which is meromorphic in all of $\mathbb{C}$. Moreover, if $\mathcal{F}$ is not normal then $\{z_n\}$ and $\{\rho_n\}$ can be chosen so that

$$g^\#(\zeta) \leq g^\#(0) = 1, \tag{10.10}$$

for all $\zeta \in \mathbb{C}$.

It is remarkable that non-normality can be described in terms of a convergent sequence. If $\{f_n\}$ were convergent then the functions g_n given by (10.9) would converge to a constant on compact subsets of $\mathbb{C}$, since the radii ρ_n tend to 0. Zalcman's lemma says that arbitrarily small disks centered at z_n can be found where f_n is close to a non-trivial meromorphic function, after rescaling.

For example, by Marty's theorem, the family $\mathcal{F} = \{z^n\}$ is normal on $\mathbb{D}$ and on $\mathbb{C} \setminus \overline{\mathbb{D}}$ but not on $\{z : |z| < 2\}$. Uniform convergence fails in every neighborhood of any point on the unit circle. So, set $z_n = 1$ and $\rho_n = 1/n$. Then for $\zeta \in \mathbb{C}$

$$f_n(z_n + \rho_n\zeta) = \left(1 + \frac{\zeta}{n}\right)^n \to e^\zeta$$

as $n \to \infty$, because

$$\frac{n \log(1 + \zeta/n)}{\zeta} = \frac{\log(1 + \zeta/n) - 0}{\zeta/n}$$

is the difference quotient for the derivative of $\log(1 + z)$ at $z = 0$. It is also easy to check that the spherical derivative of e^ζ satisfies (10.10).

Proof One direction of Zalcman's lemma is easy. If $\mathcal{F}$ is normal then any sequence $\{f_n\}$ contains a convergent subsequence, which we can relabel as $\{f_n\}$, and call the limit function f. If $z_n \to z_\infty \in \Omega$ and $\rho_n \to 0$, then

$$g_n(\zeta) = f_n(z_n + \rho_n\zeta) \to f(z_\infty),$$

for each $\zeta \in \mathbb{C}$ as $n \to \infty$ since the family $\{f_n\}$ is uniformly equicontinuous in a neighborhood of z_∞. Thus the limit of g_n is constant.

Conversely, suppose $\mathcal{F}$ is not normal. By Marty's theorem, there exists $w_n \to w_\infty \in \Omega$ and $f_n \in \mathcal{F}$ such that the spherical derivatives satisfy

$$f_n^{\#}(w_n) \to \infty.$$

Without loss of generality, we may suppose $w_\infty = 0$ and $\{|z| \le r\} \subset \Omega$. Then

$$M_n = \max_{|z|\le r}(r - |z|)f_n^{\#}(z) = (r - |z_n|)f_n^{\#}(z_n),$$

for some $|z_n| < r$, since $f_n^{\#}$ is continuous. Note that $M_n \to \infty$, since $w_n \to 0$. Then

$$g_n(\zeta) = f_n(z_n + \zeta/f_n^{\#}(z_n))$$

is defined on $\{|\zeta| \le M_n\}$ because

$$|z_n + \zeta/f_n^{\#}(z_n)| \le |z_n| + M_n/f_n^{\#}(z_n) = |z_n| + r - |z_n| = r.$$

Fix $K < \infty$. If $|\zeta| \le K < M_n$, then

$$\begin{aligned} g_n^{\#}(\zeta) &= \frac{f_n^{\#}(z_n + \zeta/f_n^{\#}(z_n))}{f_n^{\#}(z_n)} \\ &\le \frac{M_n}{r - |z_n + \zeta/f_n^{\#}(z_n)|}\frac{r - |z_n|}{M_n} \\ &\le \frac{r - |z_n|}{r - |z_n| - |\zeta|/f_n^{\#}(z_n)} \\ &= \frac{1}{1 - |\zeta|/M_n} \to 1, \end{aligned} \tag{10.11}$$

as $n \to \infty$. By Marty's Theorem 10.10, and the chordal version of Lemma 10.2, the family $\{g_n\}$ contains a convergent subsequence in the chordal metric. Relabeling the subsequence, (10.9) holds with $\rho_n = 1/f_n^{\#}(z_n)$. By (10.11), the limit function g satisfies the inequality and equality in (10.10) and is meromorphic by Lemma 10.8. Because $g^{\#}(0) = 1$, g is non-constant. Because $\{z_n\}$ is contained in a compact subset of Ω, we can arrange that $z_n \to z_\infty \in \Omega$ by taking subsequences. □

The first consequence of Zalcman's lemma is an improvement of the locally bounded condition for normality.

Theorem 10.13 (Montel) *A family $\mathcal{F}$ of meromorphic functions on a region Ω that omits three distinct fixed values $a, b, c \in \mathbb{C}^*$ is normal in the chordal metric.*

Proof Normality is local by the chordal version of Lemma 10.2, so we may assume $\Omega = \mathbb{D}$. An LFT and its inverse are uniformly continuous in the chordal metric. So we may suppose $a = 0$, $b = 1$ and $c = \infty$ by composing with an appropriate LFT. Without loss of generality, we may assume that $\mathcal{F}$ is the family of all analytic functions on $\mathbb{D}$ which omit the values 0 and 1. Set

$$\mathcal{F}_m = \{f \text{ analytic on } \mathbb{D} : f \ne 0 \text{ and } f \ne e^{2\pi i k 2^{-m}}, k = 1, \ldots, 2^m\}.$$

Then

$$\mathcal{F} = \mathcal{F}_0 \supset \mathcal{F}_1 \supset \mathcal{F}_2 \supset \dots .$$

If $f \in \mathcal{F}_m$ then f is analytic and $f \neq 0$, so that we can define $f^{\frac{1}{2}}$ so as to be analytic. Moreover, $f^{\frac{1}{2}} \in \mathcal{F}_{m+1}$. If $\mathcal{F}$ is not normal then there exists a sequence $\{f_n\} \subset \mathcal{F}$ with no convergent subsequence. Moreover, $\{f_n^{\frac{1}{2}}\}$ is then a sequence in $\mathcal{F}_1$ with no convergent subsequence. By induction, each $\mathcal{F}_m$ is not normal. Thus for each m we can construct a limit function h_m as in Zalcman's lemma. The functions h_m are entire by Exercise 10.3 and non-constant since $h_m^{\#}(0) = 1$. By (10.10) and Marty's theorem $\{h_m\}$ is a normal family. If h is a limit of a subsequence, uniformly on compact subsets of $\mathbb{C}$ in the chordal metric, then h is entire by Exercise 10.3 and non-constant since $h^{\#}(0) = 1$. By Hurwitz's theorem, h omits the 2^m roots of 1 for each m. These points are dense in the unit circle. Since $h(\mathbb{C})$ is connected and open, either $h(\mathbb{C}) \subset \mathbb{D}$ or $h(\mathbb{C}) \subset \mathbb{C} \setminus \mathbb{D}$. Thus, either h or $1/h$ is bounded. By Liouville's theorem, h must be constant, which contradicts $h^{\#}(0) = 1$, and hence $\mathcal{F}$ is normal. □

We remark that the families $\mathcal{F}_m$ in the proof of Montel's theorem are not closed. Indeed, the constant functions 0, 1 and ∞ are in the closure. For example, $\left(\frac{1+z}{2}\right)^n$ omits 0 and 1 in $\mathbb{D}$ and tends to 0 in $\mathbb{D}$ as $n \to \infty$. However, in the proof of Montel's theorem, Zalcman's lemma yielded limit functions which are not constant, and therefore in the family $\mathcal{F}_m$ by Hurwitz's theorem.

The next consequence of Zalcman's lemma can be viewed as an improved version of Riemann's theorem on removable singularities.

Theorem 10.14 (Picard's great theorem) *If f is meromorphic in $\Omega = \{z : 0 < |z - z_0| < \delta\}$, and if f omits three (distinct) values in $\mathbb{C}^*$, then f extends to be meromorphic in $\Omega \cup \{z_0\}$.*

The reader can verify that an equivalent formulation of Picard's great theorem is that an analytic function omits at most one complex number in every neighborhood of an essential singularity. The function $f(z) = e^{1/z}$ does omit the value 0 in every neighborhood of the essential singularity 0, so that Picard's theorem is the strongest possible statement in terms of the range of an analytic function. The weaker statement that a non-constant entire function can omit at most one complex number is usually called **Picard's little theorem**, and can be viewed as an extension of Liouville's theorem.

Proof As before we may assume $a = 0$, $b = 1$, $c = \infty$ and $z_0 = 0$. Let $\varepsilon_n \to 0$. If f omits 0 and 1 in a punctured neighborhood of 0, then, by Montel's theorem, the family $\{f(\varepsilon_n z)\}$ is normal in the chordal metric on compact subsets of $\mathbb{C} \setminus \{0\}$. Relabeling a subsequence, we may suppose $f(\varepsilon_n z)$ converges uniformly on compact subsets of $\mathbb{C} \setminus \{0\}$ to a function g analytic on $\mathbb{C} \setminus \{0\}$ or to $g \equiv \infty$ by Lemma 10.8 and Hurwitz's theorem. If g is analytic, then $|g(z)| \leq M < \infty$ on $|z| = 1$ and hence $|f| \leq M + 1$ on $|z| = \varepsilon_n$ for $n \geq n_0$. But then, by the maximum principle, $|f| \leq M + 1$ on $\varepsilon_{n+1} < |z| < \varepsilon_n$ for $n \geq n_0$. Thus $|f| \leq M + 1$ on $0 < |z| < \varepsilon_{n_0}$, so that, by Riemann's theorem on removable singularities, f extends to be analytic in a neighborhood of 0.

If $g \equiv \infty$, we can apply a similar argument to $1/f(\varepsilon_n z)$ to conclude that $1/f$ extends to be analytic at 0 and hence f is meromorphic in a neighborhood of 0. □

Picard's little theorem follows from Picard's great theorem by considering $f(1/z)$. Picard's little theorem also follows from exactly the same proof by letting $\varepsilon_n \to \infty$ and using Liouville's theorem instead of Riemann's theorem to conclude that f is constant. Test your understanding of the proof by writing out the details. Zalcman's lemma has been used in many situations to render precise the heuristic principle that if $\mathcal{P}$ is a property of meromorphic functions then

$$\{f \text{ meromorphic on } \Omega : f \text{ has } \mathcal{P}\}$$

is normal on Ω if and only if no non-constant meromorphic function on $\mathbb{C}$ has $\mathcal{P}$. For example, $\mathcal{P}$ could be "omits three values."

We end this chapter with a comment about normal families and some open problems. Normal families can be used to prove results such as Theorems 10.15–10.17.

Theorem 10.15 (Koebe) *There exist K, $K > 0$, such that, if f is analytic and one-to-one on $\mathbb{D}$ with $f(0) = 0$ and $f'(0) = 1$, then $f(\mathbb{D}) \supset \{z : |z| < K\}$.*

Theorem 10.16 (Landau) *There exists L, $0 < L < \infty$, such that if f is analytic on $\mathbb{D}$ with $f'(0) = 1$ then $f(\mathbb{D})$ contains a disk of radius L.*

Theorem 10.17 (Bloch) *There exists B, $0 < B < \infty$, such that if f is analytic on $\mathbb{D}$ with $f'(0) = 1$ then there is a region $\Omega \subset \mathbb{D}$ such that f is one-to-one on Ω, and $f(\Omega)$ is a disk of radius B.*

In each theorem, we already know that there is a constant for each f. Normal families can be used to prove that the constants can be taken to be independent of the functions f, but this method of proof does not give any information about what the constants are. See Exercises 10.12, 10.13 and 10.15. The largest K is known to be $1/4$. The conformal map of $\mathbb{D}$ onto $\mathbb{C} \setminus [1/4, \infty)$ is an example where $1/4$ is achieved. The largest L is called **Landau's constant**, and it is known that

$$0.5 < L \leq L_0 = \frac{\Gamma\left(\frac{1}{3}\right)\Gamma\left(\frac{5}{6}\right)}{\Gamma\left(\frac{1}{6}\right)} = 0.544\ldots.$$

See Section 11.4 for the definition of $\Gamma(z)$. It is conjectured that Landau's constant is L_0. The largest B is called **Bloch's constant**, and it is known that

$$\frac{\sqrt{3}}{4} = 0.433\ldots < B \leq B_0 = \frac{1}{\sqrt{1+\sqrt{3}}}\frac{\Gamma\left(\frac{1}{3}\right)\Gamma\left(\frac{11}{12}\right)}{\Gamma\left(\frac{1}{4}\right)} = 0.472\ldots,$$

and it is conjectured that $B = B_0$.

10.4 Exercises

A

10.1 Prove that if a normal family has only one normal limit, then it converges. Hint: View the elements of the family as points in a precompact metric space.

This simple observation is sometimes a useful bootstrap technique to prove convergence of a sequence: find enough properties of a subsequential limit function to prove it is unique. For example, if $\mathcal{F}$ is a normal family of analytic functions on a region Ω, and if B is a disk with $B \subset \overline{B} \subset \Omega$, then any sequence $\{f_n\} \subset \mathcal{F}$ which converges uniformly on B must also converge uniformly on every compact subset of Ω.

10.2 (a) Prove that a family of continuous functions (in the extended sense) on $\Omega \subset \mathbb{C}^*$ is normal in the chordal metric if and only if it is equicontinuous in the chordal metric.

(b) State and prove the Arzela–Ascoli theorem for continuous functions with values in a complete metric space.

10.3 Prove that if $\{f_n\}$ is a sequence of analytic functions which converges uniformly in the chordal metric on compact subsets of a region $\Omega \subset \mathbb{C}$, then the limit function is either analytic or identically ∞. Moreover, if the limit function is analytic then the convergence is uniform in the Euclidean metric on compact subsets of Ω. Hint: Hurwitz.

10.4 Give details for the second proof of Picard's little theorem mentioned in the text.

B

10.5 Suppose Ω is a simply-connected region, $\Omega \neq \mathbb{C}$, and $z_0 \in \Omega$. Let

$$\mathcal{F} = \{f : f \text{ is analytic and } |f(z)| < 1 \text{ on } \Omega\}.$$

Prove that there exists $f_0 \in \mathcal{F}$ such that

$$|f_0'(z_0)| = \sup\{|f'(z_0)| : f \in \mathcal{F}\}.$$

Prove also that f_0 is one-to-one on Ω and $f_0(z_0) = 0$.

10.6 Prove the following analytic version of Zalcman's lemma. Suppose $\mathcal{F}$ is a family of analytic functions on a region Ω whose members vanish at $z_0 \in \Omega$. Then $\mathcal{F}$ is not normal (in the Euclidean metric) if and only if there is a sequence $\{z_n\}$ converging to $z_\infty \in \Omega$, a sequence of positive numbers ρ_n converging to 0, and a sequence $\{f_n\} \subset \mathcal{F}$ such that $g_n(\zeta) = f_n(z_n + \rho_n\zeta) - f_n(z_n)$ converges uniformly on compact subsets of $\mathbb{C}$ to the function $g(\zeta) = c\zeta$, where c is a constant with $|c| = 1$. Hint: Imitate the proof of Zalcman's lemma using f_n' instead of $f_n^{\#}$.

10.7 Let $\{f_n\}$ be a sequence of analytic functions on a region Ω with $|f_n| \le 1$ on Ω. Let K be compact and contained in Ω. Suppose $\{f_n\}$ converges at infinitely many points in K. Then is it true or false that $\{f_n\}$ necessarily converges at every point of Ω?

10.8 Let F_M be the set of functions analytic on the (open) unit disk $\mathbb{D}$ and continuous on the closed unit disk which satisfy

$$\int_0^{2\pi} |f(e^{i\theta})| d\theta \le M < \infty.$$

Show that F_M is a normal family on $\mathbb{D}$ with respect to the Euclidean metric.

10.9 Let B be the set of functions f which are analytic on the unit disk $\mathbb{D}$ and satisfy both $f(0) = 0$ and $f(\mathbb{D}) \cap I = \emptyset$, where I is the interval $[1, 2]$. Prove that B is a normal family (as maps from $\mathbb{D}$ into the complex plane with the Euclidean metric) which contains all of its limit functions.

10.10 Suppose f and g are entire functions with the property that

$$f^n + g^n = 1.$$

If $n = 2$, prove that there is an entire function h so that $f = \cos h$ and $g = \sin h$. Prove that if $n > 2$ then f and g are constant.

10.11 Prove that the set of one-to-one analytic functions on $\mathbb{D}$ with the property that $f(0) = 0$ and $f'(0) = 1$ is normal in the Euclidean metric. Show also that this set contains all its limit points. Hint: The ingredients are Schwarz's lemma applied to f^{-1} to show $f \neq c$ for some $c = c(f) \in \partial\mathbb{D}$, Montel on $\mathbb{D} \setminus \{0\}$, Hurwitz and the maximum principle.

10.12 Prove Koebe's theorem. Hint: See Exercises 8.7 and 10.11.

10.13 Prove Landau's theorem. Hint: See Exercises 8.7 and 10.6.

10.14 This exercise gives an alternative proof of Picard's little theorem, so don't use Picard to answer the questions below.

(a) Prove that a non-constant entire function covers disks of arbitrarily large radii. Hint: See Exercise 10.13.

(b) If g is analytic on a simply-connected region Ω such that g omits the two values 1 and -1, then prove that we can define $h(z) = -\frac{i}{\pi} \log\left(g(z) + \sqrt{g(z)^2 - 1}\right)$ so as to be analytic on Ω. Furthermore, show that h omits the integers.

(c) Repeat the idea in part (b) by constructing an analytic function k with $k(z) = -\frac{i}{\pi} \log\left(h(z) + \sqrt{h(z)^2 - 1}\right)$. Prove that k does not cover arbitrarily large disks.

(d) Deduce Picard's little theorem from (c).

C

10.15 Prove Bloch's theorem. Hint: See Exercise 10.6.

10.16 Give a proof of the Koebe, Landau or Bloch theorem with an explicit constant.

10.17 (A term project.) As a review of many of the concepts covered so far in this text, give all details (from scratch) of the proof in this chapter of the Riemann mapping theorem: trace backward through all of the required results and their proofs.

11 Series and Products

11.1 Mittag-Leffler's Theorem

So far we have learned a little bit about polynomials, rational functions, exponentials and logarithms. Now we will look at functions that arise naturally as limits of these.

Suppose f is meromorphic in a region Ω with a pole at $b \in \Omega$. Then, recalling the Laurent expansion,

$$f(z) = \frac{c_n}{(z-b)^n} + \frac{c_{n-1}}{(z-b)^{n-1}} + \cdots + \frac{c_1}{(z-b)} + a_0 + a_1(z-b) + a_2(z-b)^2 + \ldots,$$

for z near b. The sum of the first n terms

$$S_b(z) = \frac{c_n}{(z-b)^n} + \frac{c_{n-1}}{(z-b)^{n-1}} + \cdots + \frac{c_1}{(z-b)}$$

is called the **singular part of** f **at** b. If f is rational, then, by a partial fraction expansion,

$$f(z) = \sum_{k=1}^{m} S_{b_k}(z) + p(z),$$

where p is a polynomial and $\{b_k\}$ are the poles of f. If f is meromorphic in a region Ω with only finitely many poles $\{b_k\}$ and singular parts S_{b_k}, $k = 1, \ldots, m$, then

$$f(z) = \sum_{k=1}^{m} S_{b_k}(z) + g(z),$$

where g is analytic in Ω. This follows because $f(z) - \sum S_{b_k}(z)$ is analytic at each b_k and therefore in all of Ω. In this section, we will find a similar expansion for meromorphic functions in Ω with infinitely many poles. As before, we say that an infinite sequence $b_k \in \Omega \to \partial\Omega$ as $k \to \infty$ if each compact $K \subset \Omega$ contains only finitely many b_k.

Theorem 11.1 (Mittag-Leffler) *Suppose $b_k \in \Omega \to \partial\Omega$, with $b_k \neq b_j$ if $k \neq j$. Set*

$$S_k(z) = \sum_{j=1}^{n_k} \frac{c_{j,k}}{(z-b_k)^j},$$

where each n_k is a positive integer and $c_{j,k} \in \mathbb{C}$. Then there is a function meromorphic in Ω with singular parts S_k at b_k, $k = 1, 2, \ldots,$ and no other singular parts in Ω.

If Ω is unbounded, we allow $|b_k| \to \infty \in \partial\Omega \subset \mathbb{C}^*$.

Proof Let

$$K_n = \left\{ z \in \Omega : \operatorname{dist}(z, \partial\Omega) \geq \frac{1}{n} \quad \text{and} \quad |z| \leq n \right\}.$$

Then, as in the proof of Corollary 4.28, K_n is a compact subset of Ω such that each bounded component of $\mathbb{C} \setminus K_n$ contains a point of $\partial\Omega$ and $K_n \subset K_{n+1} \subset \cup_j K_j = \Omega$. Because $b_k \to \partial\Omega$, each K_n contains only finitely many b_k. By Runge's Theorem 4.27, we can find a rational function f_n with poles in $\mathbb{C} \setminus \Omega$ so that

$$\left| \left\{ \sum_{b_k \in K_{n+1} \setminus K_n} S_k(z) \right\} - f_n(z) \right| < 2^{-n}$$

for all $z \in K_n$. Then for each $m = 1, 2, \ldots$

$$\sum_{n \geq m} \left(\left\{ \sum_{b_k \in K_{n+1} \setminus K_n} S_k(z) \right\} - f_n(z) \right)$$

converges uniformly on K_m to an analytic function on K_m by the Weierstrass M-test and Weierstrass's Theorem 4.29. Set

$$f(z) = \sum_{b_k \in K_1} S_k(z) + \sum_{n=1}^{\infty} \left(\left\{ \sum_{b_k \in K_{n+1} \setminus K_n} S_k(z) \right\} - f_n(z) \right). \tag{11.1}$$

Then f is a well-defined analytic function on $\Omega \setminus \{b_k\}$ and $f - S_k$ has a removable singularity at b_k for each $k = 1, 2, \ldots$. □

A more constructive proof of Mittag-Leffler's theorem is useful in some circumstances. We will illustrate the technique in some special cases, and leave the more general case to the reader.

Suppose we would like to find a function meromorphic in $\mathbb{C}$ with singular part $a_k/(z - b_k)$ at $\{b_k\}$, where $|b_k| \to \infty$. Then

$$\sum_{k=1}^{\infty} \frac{a_k}{z - b_k}$$

will work, provided the sum converges. When does it converge? If $|z| \leq R < \infty$, write

$$\sum_{k=1}^{\infty} \frac{a_k}{z - b_k} = \sum_{\{k:|b_k| \leq 2R\}} \frac{a_k}{z - b_k} + \sum_{\{k:|b_k| > 2R\}} \frac{a_k}{z - b_k}. \tag{11.2}$$

The first sum has only finitely many terms. For the second sum, $|z| \leq R < |b_k|/2$, so that

$$\left| \frac{1}{z - b_k} \right| \leq \frac{2}{|b_k|}.$$

Thus, if

$$\sum_{k=1}^{\infty} \frac{|a_k|}{|b_k|} < \infty,$$

then the second sum in (11.2) converges uniformly on $\{|z| \le R\}$ to an analytic function, by Weierstrass again. The right-hand side of (11.2) then is meromorphic in $|z| < R$ with singular part $a_k/(z-b_k)$ at b_k, provided $|b_k| < R$. Since R is arbitrary, the sum in (11.2) is meromorphic in $\mathbb{C}$.

What if $|b_k|$ tends to ∞ more slowly? If we examine the proof, we just need the tail of the series on the left-hand side of (11.2) to converge for each R. Mittag-Leffler's idea was to subtract a polynomial from each term so that the result converges. As we have seen before, if $b_k \neq 0$ and $|z| < |b_k|$, then

$$\frac{1}{z-b_k} = \frac{1}{-b_k(1-\frac{z}{b_k})} = \frac{1}{-b_k}\sum_{j=0}^{\infty}\left(\frac{z}{b_k}\right)^j.$$

So it is natural to subtract a few terms of the expansion to make it smaller:

$$\left|\frac{1}{z-b_k} - \left(\frac{1}{-b_k}\right)\sum_{j=0}^{n_k}\left(\frac{z}{b_k}\right)^j\right| = \left|\frac{1}{-b_k}\sum_{j=n_k+1}^{\infty}\left(\frac{z}{b_k}\right)^j\right|$$
$$\le \left|\frac{1}{b_k}\right|\left|\frac{z}{b_k}\right|^{n_k+1}\frac{1}{1-|\frac{z}{b_k}|}, \tag{11.3}$$

provided $|z| < |b_k|$.

For example, the following proposition holds:

Proposition 11.2 *If* $0 < |b_k| \to \infty$ *and if, for some integer* $n \ge 0$,

$$\sum_k \frac{|a_k|}{|b_k|^{n+2}} < \infty,$$

then

$$f(z) = \sum_{k=1}^{\infty}\left(\frac{a_k}{z-b_k} - \left(\frac{a_k}{-b_k}\right)\sum_{j=0}^{n}\left(\frac{z}{b_k}\right)^j\right)$$

is meromorphic in $\mathbb{C}$ *with singular part* $S_{b_k}(z) = \frac{a_k}{z-b_k}$ *at* b_k, $k = 1, 2, \ldots,$ *and no other poles.*

To prove Proposition 11.2, if $|z| < R$, split the sum into two pieces: a finite sum of the terms with $|b_k| \le 2R$ and a convergent sum of the terms with $|b_k| > 2R$. Then use the estimate in (11.3) with $n_k = n$ for all k.

Example 11.3

$$\frac{\pi^2}{\sin^2 \pi z} = \sum_{n=-\infty}^{\infty}\frac{1}{(z-n)^2}. \tag{11.4}$$

Proof The right-hand side of (11.4) converges uniformly on compact subsets of $\mathbb{C}$. The limit is meromorphic with singular parts $S_n(z) = 1/(z-n)^2$ at $z = n$ and with no other poles. To see this, follow the proof of Proposition 11.2. Fix $R < \infty$ and split the sum into two pieces: a finite sum of terms with $|n| < 2R$ and the remaining infinite sum. In the second sum when

$|z| \le R$, the nth term is uniformly bounded by $1/(|n|-R)^2 \le 4/n^2$. Thus the second (infinite) sum converges uniformly and absolutely on $|z| \le R$ to an analytic function by Weierstrass's theorem.

Note that $\sin \pi z/z$ is the difference quotient for $\frac{d}{dz}\sin \pi z$ at $z = 0$, and so $\sin \pi z/\pi z$ approaches 1 as $z \to 0$. The function $\pi z/\sin \pi z$ has a removable singularity at $z = 0$ and is an even function so that

$$\frac{\pi z}{\sin \pi z} = 1 + \mathrm{O}(z^2)$$

near 0. By squaring and dividing by z^2, we conclude that the singular part of $\pi^2/\sin^2 \pi z$ at $z = 0$ is $1/z^2$. If n is an integer then $\sin^2 \pi(z-n) = \sin^2 \pi z$. Thus the singular part of the left-hand side of (11.4) at $z = n$ is the same as the singular part of the right-hand side of (11.4) at $z = n$ for each integer n. Set

$$F(z) = \frac{\pi^2}{\sin^2 \pi z} - \sum_{n=-\infty}^{\infty} \frac{1}{(z-n)^2}.$$

Then F is entire and $F(z+1) = F(z)$.

Write $z = x + iy$ and suppose $0 \le x \le 1$. We claim that

$$|F(z)| \to 0, \tag{11.5}$$

as $|y| \to \infty$. If so, F is bounded in the strip $0 \le x \le 1$, and, since $F(z+1) = F(z)$, the function F is bounded in $\mathbb{C}$. By Liouville's theorem, F is constant and, by (11.5), $F \equiv 0$, proving Example 11.3. To see (11.5), first observe that

$$|\sin \pi z| = \frac{1}{2}|e^{-\pi y + i\pi x} - e^{\pi y - i\pi x}| \to \infty$$

as $|y| \to \infty$. Thus the left-hand side of (11.4) tends to 0 as $|y| \to \infty$, with $0 \le x \le 1$. Likewise, in the strip $0 \le x \le 1$, the right-hand side of (11.4) is dominated by

$$\frac{1}{y^2} + \sum_{n \ne 0} \frac{1}{(|n|-1)^2 + y^2},$$

which also tends to 0 as $|y| \to \infty$, as can be seen by comparing the sum with the integral $\int_1^\infty dx/(x^2+y^2)$. This proves (11.5) and Example 11.3. □

Example 11.4

$$\pi \cot \pi z = \frac{1}{z} + \sum_{n \ne 0} \left(\frac{1}{z-n} + \frac{1}{n} \right). \tag{11.6}$$

Proof First observe that

$$\frac{d}{dz}\pi \cot \pi z = -\frac{\pi^2}{\sin^2 \pi z} = -\sum \frac{1}{(z-n)^2}.$$

But

$$\pi \cot \pi z \ne \sum \frac{1}{z-n},$$

because the latter sum does not converge. The difficulty is that, for $|z| \leq R$, the nth term behaves like $-1/n$ for large n. However,

$$\frac{1}{z-n} + \frac{1}{n} = \frac{z}{(z-n)n} \sim \frac{1}{n^2},$$

for large n and $\sum n^{-2} < \infty$. To prove Example 11.4, suppose $|z| \leq R$ and split the sum on the right-hand side of (11.6) into the sum of the terms with $|n| \leq 2R$ and the sum of the terms with $|n| > 2R$. The first sum is finite, and the second sum has terms satisfying

$$\left|\frac{1}{z-n} + \frac{1}{n}\right| = \left|\frac{z}{(z-n)n}\right| \leq \frac{R}{\frac{n}{2} \cdot n} = \frac{2R}{n^2}.$$

Because $\sum 2R/n^2 < \infty$, the right-hand side of (11.6) is meromorphic in $|z| \leq R$ with poles only at the integers and with prescribed singular parts $1/(z-n)$, for each $R < \infty$. Furthermore, by Weierstrass's Theorem 4.29 and Example 11.3, the right-hand side of (11.6) has derivative

$$-\frac{1}{z^2} - \sum_{n \neq 0} \frac{1}{(z-n)^2} = -\frac{\pi^2}{\sin^2 \pi z} = \frac{d}{dz}\pi \cot \pi z.$$

Thus, the two sides of (11.6) differ by a constant, C. Convergence of the right-hand side of (11.6) is absolute, so that we can add the terms involving n and $-n$ to obtain

$$\frac{1}{z} + \sum_{n=1}^{\infty} \frac{2z}{z^2 - n^2}, \tag{11.7}$$

which is clearly odd. Since $\pi \cot \pi z$ is also odd, we must have $C = 0$, proving (11.6) as well as the equality with (11.7). □

A slightly subtle point is that the convergence of (11.7) is not the same as the convergence of (11.6). Convergence of (11.7) is the same as convergence of symmetric partial sums in (11.6) from $-N$ to N, which is a priori weaker than allowing the upper and lower limits of the partial sums to tend to ∞ independently.

We used the function $\pi \cot \pi z$ in Section 9.4 to compute sums using residues.

Example 11.5 (Weierstrass $\mathcal{P}$ function) Suppose $w_1, w_2 \in \mathbb{C} \setminus \{0\}$ with w_1/w_2 not real. In other words, w_1 and w_2 are not on the same line through the origin. There is no non-constant entire function f satisfying $f(z + w_1) = f(z + w_2) = f(z)$ for all z, by Liouville's theorem. See Exercise 11.6. But there are meromorphic functions with this property. The Weierstrass $\mathcal{P}$ function is defined by

$$\mathcal{P}(z) = \frac{1}{z^2} + \sum_{(m,n)\neq(0,0)} \left(\frac{1}{(z - mw_1 - nw_2)^2} - \frac{1}{(mw_1 + nw_2)^2}\right), \tag{11.8}$$

where the sum is taken over all pairs of integers except $(0, 0)$.

To prove convergence of this sum, we first observe that there is a $\delta > 0$ such that $|mw_1 + nw_2| \geq \delta$ unless $m = n = 0$, for, if $|m_j w_1 + n_j w_2| \to 0$, then

$$\left|\frac{w_1}{w_2} + \frac{n_j}{m_j}\right| \to 0,$$

contradicting the assumption that w_1/w_2 is not real. Thus $\{\zeta_{m,n} = mw_1 + nw_2 : m, n \in \mathbb{Z}\}$, where $\mathbb{Z}$ denotes the integers, forms a lattice of points in $\mathbb{C}$ with no two points closer than δ. If we place a disk of radius $\delta/2$ centered at each point of the lattice, then the disks are disjoint. The area of the annulus $k \le |\zeta| \le k+1$ is $(2k+1)\pi$ so there are at most Ck lattice points in this annulus, for some constant C depending on δ.

For $|z| < R$, we split the sum in (11.8) into a finite sum of terms with $|\zeta_{m,n}| \le 2R$ and the sum of terms with $|\zeta_{m,n}| > 2R$. Note that if $|z| < R$ and $|\zeta| > 2R$, then

$$\left|\frac{1}{(z-\zeta)^2} - \frac{1}{\zeta^2}\right| = \left|\frac{2z\zeta - z^2}{\zeta^2(z-\zeta)^2}\right| \le \frac{R(2|\zeta|+R)}{|\zeta|^2|\zeta/2|^2} \le \frac{10R}{|\zeta|^3}.$$

We conclude that for $K \ge R$ and $|z| < R$

$$\sum_{|\zeta_{m,n}|>2K}\left|\frac{1}{(z-\zeta_{m,n})^2} - \frac{1}{\zeta_{m,n}^2}\right| = \sum_{k=2K}^{\infty}\sum_{k<|\zeta_{m,n}|\le k+1}\left|\frac{1}{(z-\zeta_{m,n})^2} - \frac{1}{\zeta_{m,n}^2}\right|$$

$$\le \sum_{k=2K}^{\infty} Ck\frac{10K}{k^3} < \infty.$$

By Weierstrass's theorem, the Weierstrass $\mathcal{P}$ function is meromorphic in $\mathbb{C}$ with singular part $S(z) = 1/(z - mw_1 - nw_2)^2$ at $mw_1 + nw_2$ and no other poles.

Next we show that $\mathcal{P}(z + w_1) = \mathcal{P}(z)$. By Weierstrass's theorem,

$$\mathcal{P}'(z) = -\frac{2}{z^3} - \sum_{\zeta_{m,n}\neq 0}\frac{2}{(z-\zeta_{m,n})^3}.$$

By the same estimate, this series converges absolutely so that we can rearrange the terms, obtaining $\mathcal{P}'(z + w_1) = \mathcal{P}'(z)$, and hence $\mathcal{P}(z + w_1) - \mathcal{P}(z)$ is a constant. The series for $\mathcal{P}$ is even, so $\mathcal{P}(z + w_1) = \mathcal{P}(z)$ when $z = -w_1/2$, and thus $\mathcal{P}(z + w_1) = \mathcal{P}(z)$ for all z. A similar argument shows that $\mathcal{P}(z + w_2) = \mathcal{P}(z)$.

It is possible to replace the proof in the preceding paragraph with an estimate of the difference of two partial sums. However, when z is replaced by $z + w_1$, the term involving z will be paired with the wrong value at 0. An estimate of the error made by replacing it with the correct value must be made, as well as an estimate of the terms in the difference of the partial sums which do not cancel, since the summation is shifted by w_1.

The Weierstrass $\mathcal{P}$ function is a basic example of an elliptic function. Elliptic functions are important in the application of complex analysis to number theory. See Exercises 11.6 and 8.6, and Section 16.3.

11.2 Weierstrass Products

Just as infinite sums are used to create meromorphic functions with prescribed poles, infinite products are used to create analytic functions with prescribed zeros. If $\{f_j\}$ are analytic and $f_j(z_j) = 0$, then $\prod_{j=1}^n f_j$ vanishes at $z_1, z_2, \ldots, z_n$. If we want a function to vanish at infinitely many z_j, it is natural to try

$$\lim_{n\to\infty}\prod_{j=1}^{n} f_j.$$

However, this limit may not exist. Moreover, it is possible for the product of infinitely many non-zero numbers to converge to zero, which might create additional zeros. To resolve this difficulty, we will first treat the convergence of infinite products of complex numbers.

Proposition 11.6 *Suppose $p_j \in \mathbb{C} \setminus \{0\}$. Then $\prod_{j=1}^n p_j$ converges to a non-zero complex number P as $n \to \infty$ if and only if*

$$\sum_{j=1}^{\infty} \log p_j$$

converges to a complex number S, where $\log p_j$ is defined so that $-\pi < \arg p_j \le \pi$. Moreover, if convergence holds then $P = e^S$ and $\lim p_j = 1$.

Proof Let $S_n = \sum_{j=1}^n \log p_j$ and $P_n = \prod_{j=1}^n p_j$. If $S_n \to S$ then $P_n = e^{S_n} \to e^S \in \mathbb{C} \setminus \{0\}$. Conversely, if $P_n \to P \in \mathbb{C} \setminus \{0\}$, then, by altering p_1 if necessary, we may suppose $P > 0$. Then $\log P_n \to \log P$, where the logarithm is defined so that $-\pi < \arg P_n \le \pi$. Note that

$$|P_{n-1}||p_n - 1| = |P_n - P_{n-1}| \to 0,$$

so that $p_n \to 1$ and hence $\log p_n \to 0$. But

$$\log P_n = \left(\sum_{j=1}^{n} \log p_j \right) + 2\pi i k_n$$

for some integers k_n. Because $\log p_n \to 0$ and

$$\log p_n + 2\pi i (k_n - k_{n-1}) = \log P_n - \log P_{n-1} \to 0,$$

we must have $k_n = k_{n_0}$ for n_0 sufficiently large and $n \ge n_0$. It follows that $\lim_n \sum_{j=1}^n \log p_j$ exists, proving Proposition 11.6. □

One possibility for a definition of absolute convergence of an infinite product $\prod p_j$ would be the convergence of the partial product $\prod_{j=1}^n |p_j|$. However, $\prod_{j=1}^n e^{ij} = e^{i\frac{n(n+1)}{2}}$ does not converge even though the product of the absolute values is 1. The notion of absolute convergence was useful for infinite sums because it implies that the terms in the sum can be rearranged without affecting convergence.

Definition 11.7 *If p_j are non-zero complex numbers then we say $\prod_{j=1}^{\infty} p_j$* **converges absolutely** *if $\sum |\log p_j|$ converges.*

It follows that if p_j are non-zero complex numbers such that $\prod_{j=1}^{\infty} p_j$ converges absolutely then any rearrangement of the terms in the infinite product will not affect $\lim_n \prod_{j=1}^n p_j$. The next lemma is useful for determining absolute convergence.

Lemma 11.8 *If p_j are non-zero complex numbers then $\prod_{j=1}^{\infty} p_j$ converges absolutely if and only if*

$$\sum_{j=1}^{\infty} |p_j - 1|$$

converges.

Proof

$$\lim_{z\to 1}\frac{\log z}{z-1} = \frac{d}{dz}\log z\bigg|_{z=1} = 1,$$

so that if $p_j \to 1$ then, for j sufficiently large,

$$\left|\frac{\log p_j}{p_j - 1} - 1\right| < \frac{1}{2}$$

and thus

$$\frac{1}{2}|p_j - 1| \le |\log p_j| \le \frac{3}{2}|p_j - 1|,$$

and Lemma 11.8 follows. □

Lemma 11.8 says that the product $\prod_{j=1}^{\infty} p_j$ converges absolutely if and only if the sum $\sum_{j=1}^{\infty}(p_j - 1)$ converges absolutely. The analogous statement for (non-absolute) convergence is false. If $p_j = e^{(-1)^j/\sqrt{j}}$ then $\sum \log p_j$ converges by the alternating series test and hence $\prod p_j$ converges, but $\sum_j (p_j - 1)$ diverges. Similarly, if $p_j = 1 + i\frac{(-1)^j}{\sqrt{j}}$ then $\sum(p_j - 1)$ converges but $\sum \log p_j$ diverges, and hence $\prod p_j$ diverges. See Exercise 11.1.

As mentioned at the start of this section, we would like to consider infinite products of analytic functions $\prod f_j(z)$, but we would like to allow the f_j to have zeros. For that reason, we make the following definition.

Definition 11.9 *Suppose $\{f_j\}$ are analytic on a region Ω. We say that $\prod_{j=1}^{\infty} f_j(z)$* **converges** *on Ω if*

$$\lim_{n\to\infty}\prod_{j=0}^{n} f_j(z)$$

converges uniformly on compact subsets of Ω to a function f which is not identically equal to 0.

If $F_m(z) = \prod_{j=0}^{m} f_j(z)$ converges to $f \not\equiv 0$ on Ω then f is analytic on Ω by Weierstrass's Theorem 4.29. Moreover, $F_m^{(k)}$ converges to $F^{(k)}$ uniformly on compact subsets of Ω, for $k = 1, 2, \ldots$. By Hurwitz's Theorem 8.8, the zero set of f is the union of the zero sets, counting multiplicity, of $\{f_n\}$. Also note that f_m converges uniformly on compact subsets of Ω to the constant function 1. For example, by the proof of Lemma 11.8, if $\sum |f_j(z) - 1|$ converges uniformly on compact subsets of Ω then $\prod_{j=1}^{\infty} f_j(z)$ converges on Ω.

The next result, called the **Weierstrass product theorem** is the analog for products of Mittag-Leffler's theorem.

Theorem 11.10 (Weierstrass) *Suppose Ω is a bounded region. If $\{b_j\} \subset \Omega$ with $b_j \to \partial\Omega$, and if n_j are positive integers, then there exists an analytic function f on Ω such that f has a zero of order exactly n_j at b_j, $j = 1, 2, \ldots$, and no other zeros in Ω.*

Theorem 11.10 is also true for unbounded regions Ω. See Exercise 11.7.

Proof Let

$$K_n = \left\{ z \in \Omega : |z - w| \geq \frac{1}{n} \text{ for all } w \in \partial\Omega \right\}.$$

Then, as in the proof of Corollary 4.28, K_n is a compact subset of $\mathbb{C}$ such that each component of $\mathbb{C} \setminus K_n$ contains a point of $\partial\Omega$ and $K_n \subset K_{n+1}$. Choose $a_j \in \partial\Omega$ so that

$$\text{dist}(b_j, \partial\Omega) = |b_j - a_j|.$$

As in the proof of Corollary 4.28, if $b_j \notin K_n$ then the line segment from b_j to a_j does not intersect K_n. Thus we can define $\log((z - b_j)/(z - a_j))$ so as to be analytic in $\mathbb{C} \setminus K_n$.

Each K_n contains at most finitely many b_k because $b_k \to \partial\Omega$. By Runge's Theorem 4.27, we can find a rational function r_n with poles in $\mathbb{C} \setminus \Omega$ so that

$$\left| \left\{ \sum_{b_k \in K_{n+1} \setminus K_n} n_k \log\left(\frac{z - b_k}{z - a_k}\right) \right\} - r_n(z) \right| < 2^{-n}, \tag{11.9}$$

for all $z \in K_n$. Then

$$\sum_{n \geq m} \left(\left\{ \sum_{b_k \in K_{n+1} \setminus K_n} n_k \log\left(\frac{z - b_k}{z - a_k}\right) \right\} - r_n(z) \right)$$

converges uniformly on K_m to an analytic function on K_m by Weierstrass's Theorem 4.29. Set

$$f(z) = \prod_{b_k \in K_1} \left(\frac{z - b_k}{z - a_k}\right)^{n_k} \prod_{n=1}^{\infty} \left(\prod_{b_k \in K_{n+1} \setminus K_n} \left(\frac{z - b_k}{z - a_k}\right)^{n_k} \right) e^{-r_n(z)}.$$

Then f is a well-defined analytic function on Ω with a zero of order n_k at b_k, $k = 1, 2, \ldots$, and no other zeros. □

Here is a more constructive version when $\Omega = \mathbb{C}$.

Theorem 11.11 *Suppose $\{a_n\} \subset \mathbb{C} \setminus \{0\}$ and suppose g is a non-negative integer such that*

$$\sum_{n=1}^{\infty} \frac{1}{|a_n|^{g+1}} < \infty.$$

Then

$$\prod_{n=1}^{\infty} \left(1 - \frac{z}{a_n}\right) \exp\left(\sum_{j=1}^{g} \frac{1}{j} \left(\frac{z}{a_n}\right)^j \right) \tag{11.10}$$

converges and represents an entire function with zeros at $\{a_n\}$ and no other zeros.

The number g in Theorem 11.11 is called the **genus** of the infinite product. In the case when $g = 0$, we interpret the sum in (11.10) as equal to 0. The a_n in Theorem 11.11 do not need to be distinct, so that the theorem can be used to construct functions with zeros of order more than one.

Proof By integrating the series for $1/(1-w)$ we have

$$\log(1-w) = -\sum_{j=1}^{\infty} \frac{1}{j} w^j,$$

where the sum converges and therefore is analytic for $|w| < 1$. If $|z| < R$ and $|a| > 2R$, then

$$\left| \log\left(1 - \frac{z}{a}\right) + \sum_{j=1}^{g} \frac{1}{j} \left(\frac{z}{a}\right)^j \right| \le C \left|\frac{R}{a}\right|^{g+1},$$

where C is a constant. This follows from the fact that $\left(\log(1-w) + \sum_{j=1}^{g} \frac{1}{j} w^j\right)/w^{g+1}$ is analytic on the unit disk and therefore bounded on the disk of radius $1/2$. (The interested reader can prove that $C = 2$ works by estimating the tail of the series.)

If $|z| \le R$ then

$$\sum_{|a_n|>2R} \left| \log\left(1 - \frac{z}{a_n}\right) + \sum_{j=1}^{g} \frac{1}{j} \left(\frac{z}{a_n}\right)^j \right| \le CR^{g+1} \sum_{|a_n|>2R} \frac{1}{|a_n|^{g+1}} < \infty.$$

Thus, by Weierstrass's theorem,

$$\prod_{|a_n|>2R} \left(1 - \frac{z}{a_n}\right) e^{\sum_{j=1}^{g} \frac{1}{j} \left(\frac{z}{a_n}\right)^j}$$

is analytic and non-zero in $\{|z| < R\}$. The finite product with terms $|a_n| \le 2R$ then gives the zeros in $|z| \le R$. □

Example 11.12 The function

$$f(z) = z \prod_{\substack{n=-\infty \\ n \ne 0}}^{\infty} \left(1 - \frac{z}{n}\right) e^{\frac{z}{n}}$$

converges since $\sum n^{-2} < \infty$ ($g = 1$) and vanishes precisely at the integers. Thus we can write

$$\sin \pi z = e^{G(z)} z \prod_{\substack{n=-\infty \\ n \ne 0}}^{\infty} \left(1 - \frac{z}{n}\right) e^{\frac{z}{n}},$$

where G is entire. Indeed, $\sin \pi z$ has a simple zero at each integer and no other zeros, so, if we divide $\sin \pi z$ by z times the product, the resulting function is entire and never equal to 0 in $\mathbb{C}$, and hence has an analytic logarithm. To find G, we take the logarithmic derivative f'/f:

$$\pi \cot \pi z = G'(z) + \frac{1}{z} + \sum_{n \ne 0} \left(\frac{1}{z-n} + \frac{1}{n} \right).$$

But then, by (11.6), we must have $G'(z) \equiv 0$, and hence G is constant. Since

$$\lim_{z \to 0} \frac{\sin \pi z}{z} = \pi,$$

we must have $e^{G(0)} = \pi$, and we obtain

$$\sin \pi z = \pi z \prod_{n \neq 0} \left(1 - \frac{z}{n}\right) e^{\frac{z}{n}}.$$

We now give a few corollaries of Theorem 11.10, the Weierstrass product theorem.

Corollary 11.13 *If Ω is a region then there is a function f analytic on Ω such that f does not extend to be analytic in any larger region.*

Proof Take a sequence $\{a_n\} \subset \Omega \to \partial\Omega$ such that $\partial\Omega \subset \overline{\{a_n\}}$. By the Weierstrass product theorem, we can find f analytic on Ω, with $f(a_n) = 0$ but f not identically zero. If f extends to be analytic in a neighborhood of $b \in \partial\Omega$ then the zeros of the extended function would not be isolated. □

In several complex variables, a similar result is not true. If $B_r = \{(z, w) : |z|^2 + |w|^2 < r^2\}$, then any function which is analytic on $B_2 \setminus B_1$ extends to be analytic on B_2.

Corollary 11.14 *Suppose Ω is a region and $a_n \to \partial\Omega$, with $a_n \neq a_m$ when $n \neq m$, and suppose $\{c_n\}$ are complex numbers. Then there exists f analytic on Ω such that*

$$f(a_n) = c_n, \quad n = 1, 2, \ldots.$$

Results like Corollary 11.14 are usually called interpolation theorems. See Corollary 11.16 for a version of this result when the zeros are not distinct.

Proof By the Weierstrass product theorem, we can find G analytic on Ω with a simple zero at each a_n. Let

$$d_n = \lim_{z \to a_n} \frac{G(z)}{z - a_n} = G'(a_n).$$

Since the zero of G at a_n is simple, $d_n \neq 0$. By Mittag-Leffler's theorem, we can find F meromorphic on Ω with singular part

$$S_n(z) = \frac{c_n/d_n}{z - a_n}$$

at a_n and no other poles in $\mathbb{C}$. Then $f(z) = F(z)G(z)$ is analytic on $\Omega \setminus \{a_n\}$ and

$$\lim_{z \to a_n} F(z)G(z) = \lim_{z \to a_n} (z - a_n)F(z)\frac{G(z)}{z - a_n} = \frac{c_n}{d_n} d_n = c_n.$$

Thus the singularity of f at each a_n is removable and f extends to be analytic on Ω with $f(a_n) = c_n$, $n = 1, 2, \ldots$. □

Corollary 11.15 *If f is meromorphic in Ω then there are functions g and h, analytic on Ω, such that*

$$f = \frac{g}{h}.$$

Proof Let $\{a_n\}$ be the poles of f, where the list is written such that a pole of order k occurs k times in this list. By the Weierstrass product theorem, there is a function h analytic on Ω with zeros $\{a_n\}$ and no other zeros. Then $g = fh$ is analytic on $\Omega \setminus \{a_n\}$ with removable singularity at each a_n. Thus g extends to be analytic in Ω and $f = g/h$. □

The idea behind the proof of Corollary 11.14 can be used to prove the analog of Mittag-Leffler's theorem for Taylor polynomials.

Corollary 11.16 *Suppose p_n is a polynomial of degree d_n for $n = 1, 2, \ldots$, and suppose $a_n \to \partial\Omega$. Then there is a function f analytic on Ω such that, for each $n = 1, 2, \ldots$,*

$$f(z) - p_n(z - a_n)$$

has a zero of order at least $d_n + 1$ at a_n. In other words, the Taylor polynomial for f at a_n of degree d_n is the prescribed function $p_n(z - a_n)$.

Proof By the Weierstrass product theorem, we can find h analytic on Ω with zeros a_n of order $d_n + 1$. Let $S_n(z)$ be the singular part of $p_n(z - a_n)/h(z)$. This means that $S_n(z) - p_n(z - a_n)/h(z)$ is analytic in a neighborhood of a_n. By Mittag-Leffler's theorem, we can find g meromorphic on Ω with singular part $S_n(z)$ at a_n. Then $g(z) - p_n(z - a_n)/h(z)$ extends to be analytic near a_n and so $g(z)h(z) - p_n(z - a_n)$ is analytic near a_n with a zero of order at least $d_n + 1$ at a_n. This proves the corollary with $f = gh$. □

11.3 Blaschke Products

Weierstrass's product theorem shows that the zero set of an analytic function on the unit disk can be any sequence tending to the unit circle. More information on the growth of the analytic function will give us more information on how quickly the zeros approach the circle. This is the content of Jensen's formula.

Theorem 11.17 (Jensen) *Suppose f is meromorphic on $|z| \le R$ with zeros $\{a_k\}$ and poles $\{b_j\}$. Suppose also that 0 is not a zero or a pole of f. Then*

$$\frac{1}{2\pi}\int_{-\pi}^{\pi} \log|f(Re^{it})|dt = \log|f(0)| + \sum_{|a_k|<R} \log\frac{R}{|a_k|} - \sum_{|b_j|<R} \log\frac{R}{|b_j|} \tag{11.11}$$

Proof Replacing $f(z)$ by $f(Rz)$, we may assume $R = 1$. First suppose that f has no poles or zeros on $|z| = 1$. Write

$$f(z) = \frac{\prod_k \frac{z-a_k}{1-\overline{a_k}z}}{\prod_j \frac{z-b_j}{1-\overline{b_j}z}} g(z),$$

where g is analytic on $|z| \le 1$ and has no zeros on $|z| \le 1$. Then

$$|f(0)| = \frac{\prod_j |a_j|}{\prod_k |b_k|}|g(0)|$$

and $\log|f(e^{it})| = \log|g(e^{it})| = \mathrm{Re}\log g(e^{it})$, where $\log g(z)$ is analytic on $|z| \le 1$. Note that if $z = e^{it}$ then $dz/(iz) = dt$. So, by Cauchy's theorem,

$$\frac{1}{2\pi}\int_{-\pi}^{\pi}\log|f(e^{it})|dt = \mathrm{Re}\frac{1}{2\pi}\int_{-\pi}^{\pi}\log g(e^{it})dt$$
$$= \mathrm{Re}\log g(0) = \log|f(0)| - \sum_j \log|a_j| + \sum_k \log|b_k|.$$

Thus (11.11) holds if f has no zeros or poles on $|z| = R$.

To complete the proof of Jensen's theorem, by Corollary 11.15 we may suppose f is analytic on $|z| < R_1$, for some $R_1 > R$. The right side of (11.11) is continuous as a function of R, $0 < R < R_1$, and equality holds in (11.11) except for isolated R. So it suffices to prove that the left side of (11.11) is non-decreasing when f is analytic. The function $v_M = \max(-M, \log|f|)$ is subharmonic, for $M < \infty$. As in the paragraph following Definition 7.2, to avoid a foray into measure theory, we define $\int \log|f(Re^{it})|dt$ to be the decreasing limit of $\int v_M(Re^{it})dt$. If $r < R_1$ then using the Poisson integral formula, find u_M harmonic on $|z| < r$ with $u_M = v_M$ on $|z| = r$. Then $v_M - u_M$ is subharmonic and ≤ 0 in $|z| < r$ by the maximum principle. Thus for $0 < s < r$

$$\int_0^{2\pi} v_M(se^{it})dt \le \int_0^{2\pi} u_M(se^{it})dt = 2\pi u_M(0) = \int_0^{2\pi} v_M(re^{it})dt.$$

Letting $M \to \infty$ shows that the left-hand side of (11.11) is non-decreasing. □

See Exercise 11.12 for a more direct proof of the equality in (11.11) when f has a zero or pole on $|z| = R$.

Corollary 11.18 *Suppose f is analytic in $\mathbb{D}$, $f \not\equiv 0$, with zeros $\{a_n\}$, and suppose*

$$\sup_{r<1}\int_{-\pi}^{\pi}\log|f(re^{it})|dt < \infty.$$

Then

$$\sum_{n=0}^{\infty}(1 - |a_n|) < \infty.$$

Proof Since f has only finitely many zeros at 0, we may divide them out and suppose $f(0) \ne 0$. Then, by Jensen's formula,

$$\sum_{n=0}^{\infty}\log\frac{1}{|a_n|} < \infty.$$

But $\lim_{x\to 1}\frac{\log x}{x-1} = 1$, so that $\sum \log\frac{1}{|a_n|} < \infty$ if and only if $\sum(1 - |a_n|) < \infty$. □

We remark that

$$\int \log|f(re^{it})|dt \le C_p \int |f(re^{it})|^p dt$$

for some constant C_p depending on $p > 0$. So if $\int_{|z|=r}|f|^p dt$ is uniformly bounded, then the hypothesis of Corollary 11.18 holds.

As a partial converse to Corollary 11.18, we have the following theorem.

Theorem 11.19 *If $\{a_n\} \subset \mathbb{D}$ such that $\sum(1 - |a_n|) < \infty$ then*

$$B(z) = \prod_{n=0}^{\infty} \frac{|a_n|}{-a_n} \frac{z - a_n}{1 - \overline{a_n} z}$$

converges uniformly and absolutely on compact subsets of $\mathbb{D}$, where we define the convergence factor $|a_n|/(-a_n)$ to be equal to 1 if $a_n = 0$. The function B is analytic on $\mathbb{D}$, is bounded by 1 and has zero set exactly equal to $\{a_n\}$.

The function in Theorem 11.19 is called a **Blaschke product**.

Proof Without loss of generality, $a_n \neq 0$ for all n. Then for $|z| \leq r < 1$

$$\left| \frac{|a_n|}{-a_n} \left(\frac{z - a_n}{1 - \overline{a_n} z} \right) - 1 \right| = \frac{(1 - |a_n|)\,||a_n| z + a_n|}{|a_n(1 - \overline{a_n} z)|} \leq \frac{2(1 - |a_n|)}{(\inf |a_n|)(1 - r)}.$$

By Weierstrass's M-test and Lemma 11.8, B converges uniformly and absolutely on $|z| \leq r$. Note that the partial products for B are all bounded by 1 and analytic on $\mathbb{D}$. □

Theorem 11.19 gives the following factorization result, which allows us to divide out the zeros of a bounded analytic function without affecting the supremum norm.

Theorem 11.20 *If f is bounded and analytic on $\mathbb{D}$ with zero set $\{a_n\}$ (counting multiplicity), and if B is the Blaschke product with zero set $\{a_n\}$, then*

$$f(z) = B(z) e^{g(z)}$$

for some analytic function g with

$$\sup_{z \in \mathbb{D}} |e^{g(z)}| = \sup_{z \in \mathbb{D}} |f(z)|.$$

Proof Without loss of generality, $\sup_{\mathbb{D}} |f| = 1$. Write

$$B_N(z) = \prod_{n=1}^{N} \frac{|a_n|}{-a_n} \frac{z - a_n}{1 - \overline{a_n} z}.$$

Then, by Corollary 3.11, $|f(z)/B_N(z)| \leq 1$. Choose r_n so that $B \neq 0$ on $|z| = r_n$. Then f/B_N converges uniformly to f/B on $|z| = r_n$ and, by the maximum principle, the convergence is uniform on $|z| \leq r_n$. Now let $r_n \to 1$. Thus $h = f/B$ is bounded by 1 on $\mathbb{D}$ and non-vanishing. This implies $g = \log h$ can be defined as an analytic function on $\mathbb{D}$. By construction, $\sup_{\mathbb{D}} |h| \leq \sup_{\mathbb{D}} |f| = 1$, but also $|B| \leq 1$, so $\sup_{\mathbb{D}} |h| = \sup_{\mathbb{D}} |f|$. □

It is difficult to overestimate the importance of Blaschke products in function theory on $\mathbb{D}$.

11.4 The Gamma and Zeta Functions

In this section we introduce two important functions, Euler's gamma function and the Riemann zeta function.

As we saw in Example 11.12, the simplest entire function with zeros at the integers is $\sin \pi z$. By the same convergence proof,

$$G(z) \equiv \prod_{n=1}^{\infty} \left(1 + \frac{z}{n}\right) e^{-z/n} \tag{11.12}$$

is the simplest entire function with zeros only at the negative integers. Moreover,

$$zG(z)G(-z) = \frac{1}{\pi} \sin \pi z. \tag{11.13}$$

Note that $G(z-1)/(zG(z))$ has no zeros and hence

$$e^{\gamma(z)} zG(z) = G(z-1), \tag{11.14}$$

where $\gamma(z)$ is entire. By Weierstrass's theorem and the uniform convergence of the product (11.12), we can take logarithmic derivatives and

$$\begin{aligned}\gamma'(z) + \frac{1}{z} + \sum_{n=1}^{\infty} \left(\frac{1}{z+n} - \frac{1}{n}\right) &= \sum_{n=1}^{\infty} \left(\frac{1}{z-1+n} - \frac{1}{n}\right) \\ &= \frac{1}{z} - 1 + \sum_{n=1}^{\infty} \left(\frac{1}{z+n} - \frac{1}{n+1}\right).\end{aligned}$$

Thus

$$\gamma'(z) = -1 + \sum_{n=1}^{\infty} \left(\frac{1}{n} - \frac{1}{n+1}\right) = 0$$

and γ is a constant. By (11.12) and (11.14),

$$\begin{aligned}1 = G(0) = e^{\gamma} G(1) &= \lim_{m\to\infty} e^{\gamma} \prod_{n=1}^{m} \left(1 + \frac{1}{n}\right) e^{-1/n} \\ &= \lim_{m\to\infty} e^{\gamma} (m+1) e^{-(1+\frac{1}{2}+\frac{1}{3}+\cdots+\frac{1}{m})}.\end{aligned}$$

Thus

$$\gamma = \lim_{m\to\infty} \left(1 + \frac{1}{2} + \frac{1}{3} + \cdots + \frac{1}{m} - \log(m+1)\right) \approx 0.57722\ldots.$$

The proof above shows that this limit, called **Euler's constant**, exists.

The **gamma function** is defined by

$$\Gamma(z) = \frac{1}{zG(z)e^{\gamma z}} = e^{-\gamma z} \frac{1}{z} \prod_{n=1}^{\infty} \frac{1}{1 + z/n} e^{z/n}.$$

Then Γ is a non-vanishing meromorphic function with poles at the non-positive integers and no other poles. Moreover, by (11.14),

$$\Gamma(z+1) = z\Gamma(z) \tag{11.15}$$

and, by (11.13),

$$\Gamma(z)\Gamma(1-z)=\frac{\pi}{\sin\pi z}. \tag{11.16}$$

It follows from (11.12) and (11.14) that $\Gamma(1)=1$, and, by (11.15) and induction,

$$\Gamma(n)=(n-1)!\,.$$

So $\Gamma(z+1)$ is a meromorphic extension of the factorial function $n!$. See Exercise 11.14, where Γ is used to estimate the growth of $n!$.

The gamma function also has an integral representation, given by the next theorem.

Theorem 11.21 *If* $\mathrm{Re}z>0$ *then*

$$\Gamma(z)=\int_0^\infty t^{z-1}e^{-t}dt. \tag{11.17}$$

Proof The integral converges uniformly and absolutely on compact subsets of $\mathrm{Re}z>0$ because $|t^{z-1}|=|t|^{\mathrm{Re}z-1}$ is integrable near $t=0$ and because $|t|^{\mathrm{Re}z-1}e^{-t}\le e^{-t/2}$ for t large. By Morera's theorem, the right-hand side of (11.17) is analytic for $\mathrm{Re}z>0$.

Set

$$\Gamma_n(z)=\int_0^n t^{z-1}\left(1-\frac{t}{n}\right)^n dt.$$

Then we claim that for $\mathrm{Re}z=x>0$

$$\lim_{n\to\infty}\Gamma_n(x)=\int_0^\infty t^{x-1}e^{-t}dt \tag{11.18}$$

and

$$\lim_{n\to\infty}\frac{1}{\Gamma_n(z)}=\frac{1}{\Gamma(z)} \tag{11.19}$$

uniformly on compact subsets of $\mathbb{C}$. If (11.18) and (11.19) hold then Theorem 11.21 follows from the uniqueness theorem.

Lemma 11.22 *If* $M<\infty$ *then, as* $n\to\infty$,

$$\left(1-\frac{t}{n}\right)^n e^t \text{ increases to } 1$$

uniformly for $0\le t\le M$.

Proof The slope of the secant line to the graph of $\log u$ with one end at $u=1-t/n$ and the other end at $u=1$ decreases to 1 as $n\to\infty$ since the graph of $\log u$ is concave. Thus

$$n\log\left(1-\frac{t}{n}\right)+t=\left(\frac{\log(1-\frac{t}{n})-\log 1}{(1-\frac{t}{n})-1}-1\right)(-t)\to 0$$

uniformly for $0\le t\le M$. Lemma 11.22 now follows since $-t\le 0$ and e^s is an increasing function of s.

□

By Lemma 11.22,

$$\Gamma_n(x) \le \Gamma_{n+1}(x) \le \int_0^\infty t^{x-1}e^{-t}dt,$$

and (11.18) holds for $x > 0$. Substitute $t = ns$ in the definition of $\Gamma_n(z)$ and obtain

$$\Gamma_n(z)n^{-z} = \int_0^1 s^{z-1}(1-s)^n ds \equiv f_n(z). \tag{11.20}$$

Note that $f_0(z) = 1/z$ and $f_{n+1}(z) = f_n(z) - f_n(z+1)$, so that, by induction,

$$f_n(z) = \frac{n!}{z(z+1)\cdots(z+n)}.$$

Thus

$$\frac{1}{\Gamma_n(z)} = n^{-z} z \prod_{k=1}^n \left(1+\frac{z}{k}\right) = \left(z\prod_{k=1}^n \left(1+\frac{z}{k}\right)e^{-z/k}\right) e^{z(\sum_{k=1}^n \frac{1}{k} - \log n)} \to \frac{1}{\Gamma(z)},$$

as $n \to \infty$, uniformly on compact subsets of $\mathbb{C}$, proving (11.19) and Theorem 11.21. □

The **Riemann zeta function** is defined by

$$\zeta(z) = \sum_{n=1}^\infty n^{-z},$$

where $n^{-z} = e^{-z\log n}$. The zeta function is important because of its connection with the prime numbers, as illustrated in Theorem 11.23 below. By the integral test, the series for $\zeta(z)$ converges uniformly and absolutely on compact subsets of $\text{Re}z > 1$. By Weierstrass's theorem, it is analytic in $\text{Re}z > 1$. The zeta function can be extended to $\text{Re}z > 0$ by comparing the sum to the appropriate integral. For $\text{Re}z > 1$,

$$\begin{aligned}\zeta(z) - \frac{1}{z-1} &= \sum_{n=1}^\infty \left(n^{-z} - \int_n^{n+1} t^{-z}dt\right) \\ &= \sum_{n=1}^\infty \int_n^{n+1}\int_n^t zs^{-z-1}ds\,dt.\end{aligned} \tag{11.21}$$

The estimate

$$\int_n^{n+1}\int_n^t |zs^{-z-1}|ds\,dt \le \frac{|z|}{2}n^{-\text{Re}z-1}$$

shows as above that (11.21) converges uniformly and absolutely on compact subsets of $\text{Re}z > 0$ to an analytic function. Thus $\zeta(z)$ extends to be meromorphic in $\text{Re}z > 0$ with a simple pole at $z = 1$ and no other poles. Figure 11.1 shows the image of $\{z = \frac{1}{2} + iy : -40 \le y \le 40\}$ by the function $\zeta(z)$. As y increases, the "loops" are traced counter-clockwise, passing through 0, except for one loop that crosses $(0, 1)$ before going clockwise around 0. The **Riemann hypothesis**, perhaps the most famous problem in mathematics, is that $\zeta(z) \neq 0$ for $\text{Re}z > \frac{1}{2}$. The next theorem proves the absence of zeros in $\text{Re}z > 1$ and gives the connection between $\zeta(z)$ and the prime numbers.

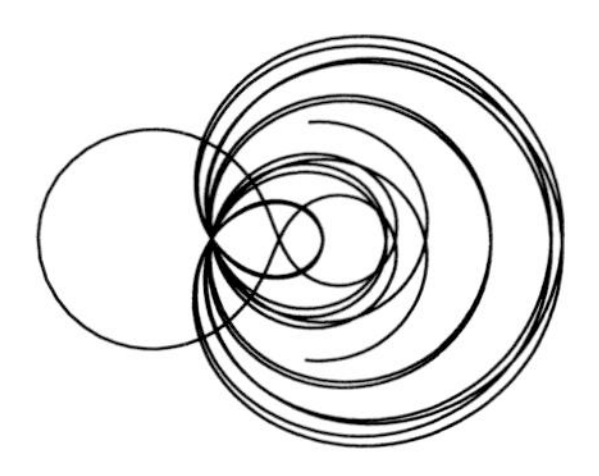

Figure 11.1 $\zeta(\frac{1}{2}+iy)$, $-40 \le y \le 40$.

Theorem 11.23 *Let $\{p_n\}$ be the primes, with $p_1 = 2$, $p_2 = 3, \ldots$. If $\text{Re}z > 1$ then*

$$\frac{1}{\zeta(z)} = \prod_{n=1}^{\infty}(1 - p_n^{-z}). \tag{11.22}$$

In particular, $\zeta(z) \neq 0$ for $\text{Re}z > 1$.

Proof The infinite product in (11.22) converges uniformly and absolutely for $\text{Re}z \geq x_0 > 1$ because

$$|p_n^{-z}| = p_n^{-\text{Re}z} \leq n^{-x_0}.$$

By Lemma 11.8 and the Weierstrass M-test, the right-hand side of (11.22) is analytic in $\text{Re}z > 1$. Because the series defining $\zeta(z)$ is absolutely convergent, we can rearrange the terms to obtain

$$\zeta(z)(1 - 2^{-z}) = \sum n^{-z} - \sum (2n)^{-z} = \sum_{m \text{ odd}} m^{-z}.$$

Similarly,

$$\zeta(z)(1 - 2^{-z})(1 - 3^{-z}) = \sum_{m \text{ odd}} m^{-z} - \sum_{m \text{ odd}} (3m)^{-z} = \sum_{\substack{m \text{ not divisible} \\ \text{by 2 or 3}}} m^{-z}.$$

By induction,

$$\zeta(z)\prod_{n=1}^{k}(1 - p_n^{-z}) = \sum_{m \text{ not div. by } p_1,\ldots,p_k} m^{-z} = 1 + p_{k+1}^{-z} + \ldots.$$

Thus

$$\left|\zeta(z)\prod_{n=1}^{k}(1 - p_n^{-z}) - 1\right| \leq \sum_{n \geq p_{k+1}} |n^{-z}|.$$

Since $p_{k+1} \to \infty$, and since $\sum |n^{-z}|$ converges, the product converges and

$$\zeta(z)\prod_{n=1}^{\infty}(1 - p_n^{-z}) = 1,$$

for $\text{Re}z > 1$. □

As one might expect from (11.20), there is a connection between $\Gamma(z)$ and $\zeta(z)$. By the same substitution used to establish (11.20), it follows from (11.17) that

$$\Gamma(z) = n^z \int_0^\infty s^{z-1} e^{-ns} ds$$

and

$$\sum_{n=1}^\infty n^{-z}\Gamma(z) = \sum_{n=1}^\infty \int_0^\infty s^{z-1} e^{-ns} ds$$
$$= \int_0^\infty s^{z-1} \frac{e^{-s}}{1-e^{-s}} ds.$$

Thus, if $\text{Re}z > 1$ then

$$\zeta(z) = \frac{1}{\Gamma(z)} \int_0^\infty \frac{s^{z-1}}{e^s - 1} ds.$$

The Riemann zeta function is used to prove the celebrated prime number theorem, which describes the rate of growth of the number of primes which are at most x, as $x \to \infty$. The complex analysis ingredients are Theorem 9.19 and Theorem 11.24 See Exercise 11.15.

Theorem 11.24 *The Riemann zeta function $\zeta(z)$ is non-zero for* $\text{Re}z \geq 1$.

Proof We proved in Theorem 11.24 that $\zeta(z) \neq 0$ in $\text{Re}z > 1$. By (11.22) for $\text{Re}z > 1$

$$-\frac{\zeta'(z)}{\zeta(z)} = \sum_{n=1}^\infty \frac{\log p_n}{p_n^z - 1} = \Phi(z) + \sum_{n=1}^\infty \frac{\log p_n}{p_n^z(p_n^z - 1)}, \tag{11.23}$$

where

$$\Phi(z) = \sum_{n=1}^\infty \frac{\log p_n}{p_n^z} = -\frac{d}{dz}\sum_{n=1}^\infty p_n^{-z}.$$

The last sum in (11.23) converges and is analytic for $\text{Re}z > \frac{1}{2}$, so that $\Phi(z)$ extends to be meromorphic in $\text{Re}z > \frac{1}{2}$, with poles only at the zeros of $\zeta(z)$ in $\text{Re}z > \frac{1}{2}$ and at $z = 1$, the pole of ζ. If $\zeta(z) = (z - z_0)^k e^{g(z)}$, where g is analytic near z_0, then

$$-\frac{\zeta'(z)}{\zeta(z)} = \frac{-k}{z - z_0} - g'(z).$$

Thus Φ has a simple pole with integer residue $-k$ at z_0. The residue equals 1 at $z_0 = 1$ by (11.21).

Note that $\zeta(\overline{z}) = \overline{\zeta(z)}$, so that if $\alpha \neq 0$ then

$$\lim_{\varepsilon\downarrow 0} \varepsilon\Phi(1+\varepsilon) = 1, \quad \lim_{\varepsilon\downarrow 0} \varepsilon\Phi(1+\varepsilon \pm i\alpha) = -M, \quad \text{and} \quad \lim_{\varepsilon\downarrow 0} \varepsilon\Phi(1+\varepsilon \pm i2\alpha) = -N,$$

where M and N are non-negative integers. Multiplying the magical inequality

$$\sum_{j=-2}^{2} \binom{4}{2+j} \Phi(1+\varepsilon+ij\alpha) = \sum_{n=1}^\infty \frac{\log p_n}{p_n^{1+\varepsilon}} (p_n^{i\alpha/2} + p_n^{-i\alpha/2})^4 \geq 0$$

by $\varepsilon > 0$, and letting $\varepsilon \to 0$, shows that $6 - 8M - 2N \geq 0$. Since M and N are non-negative integers, we must have that $M = 0$, i.e., $\zeta(1 + i\alpha) \neq 0$. □

11.5 Exercises

A

11.1 (a) Prove that $\prod_{n=1}^{\infty}(1+\frac{i}{n})$ diverges but $\prod_{n=1}^{\infty}|1+\frac{i}{n}|$ converges.
(b) If $p_j = e^{(-1)^j/\sqrt{j}}$ prove that $\sum \log p_j$ converges but $\sum(p_j - 1)$ diverges.
(c) If $p_j = 1 + i(-1)^j/\sqrt{j}$ prove that $\sum(p_j - 1)$ converges but $\sum \log p_j$ diverges.

11.2 Prove directly from the definition that

$$\prod_{n=3}^{\infty} \frac{n^2+\pi}{n^2-1}$$

converges to a non-zero real number. (Convergence can be proved by quoting the right result in the text. Test your understanding of the material by trying to make the estimates yourself.)

11.3 Find an explicit bounded analytic function $f \not\equiv 0$ defined on the unit disk with zeros $\{z_n\}$ such that every point of the unit circle is a cluster point of $\{z_n\}$ and such that $\sum(1-|z_n|) < \infty$.

11.4 If B is the Blaschke product in Theorem 11.19 with zero set $\{a_n\}$ and if $E = \{1/\overline{a_n}\}$, show that B converges uniformly on compact subsets of $\mathbb{C}\setminus\overline{E}$. In particular, if the zeros do not accumulate on an open interval $I \subset \partial\mathbb{D}$ then B is analytic on a neighborhood of I and $|B| = 1$ on I. Hint: The LFT $(z + a_n/|a_n|)/(1/\overline{a_n} - z)$ is bounded on these compact sets.

11.5 Find a meromorphic function with singular parts

$$S_n(z) = \frac{n}{z-n} + \frac{4}{(z-n)^2},$$

for $n = 1, 2, 3, \ldots$. Prove convergence of your sum.

11.6 Suppose w_1, w_2 are non-zero complex numbers such that w_1/w_2 is not real. Prove there are no non-constant entire functions satisfying $f(z+w_1) = f(z+w_2) = f(z)$ for all z. What happens if w_1/w_2 is real?

11.7 (a) Prove Weierstrass's Theorem 11.10 for $\Omega = \mathbb{C}$ by using $z - b_k$ instead of $(z-b_k)/(z-a_k)$. Define $\log(z-b_k)$ by cutting the plane along a line segment from b_k to ∞.
(b) Prove Weierstrass's Theorem 11.10 for unbounded regions Ω. In this case, view $\partial\Omega$ as a subset of $\mathbb{C}^*$ so that we allow subsequences b_j with $|b_j| \to \infty$. Hint: Use the chordal distance instead of the Euclidean distance, and use $z - b_k$ instead of $(z-b_k)/(z-a_k)$ if ∞ is the closest point in $\partial\Omega$ to b_k.

B

11.8 Suppose $\{b_k\} \to \infty$ and $|a_k/b_k| \le M < \infty$ for all $k \ge k_0$. Prove that

$$\sum_k \left(\frac{a_k}{z-b_k} - \left(\frac{a_k}{-b_k}\right) \sum_{j=0}^{k} \left(\frac{z}{b_k}\right)^j \right)$$

is meromorphic in $\mathbb{C}$ with singular part $a_k/(z-b_k)$ at b_k and no other poles in $\mathbb{C}$. The point of this exercise is that the upper limit of the inside sum is equal to k, no matter how slowly $|b_k| \to \infty$. In practice, it is usually best to take as few terms as possible, so in many examples fewer terms are taken.

11.9 Find an explicit entire function g with $g(n \log n) = n^\pi$, $n = 1, 2, 3, \ldots$. Prove that your function works. Hint: Make use of an entire function that vanishes at the integers.

11.10 Find an entire function of least possible genus, with simple zeros at the Gaussian integers:

$$\{m + in : m, n \text{ integers }\},$$

and no other zeros. In other words, find the smallest integer g such that (11.10) converges uniformly and absolutely on compact sets to an analytic function with simple zeros at the Gaussian integers.

11.11 If Ω is a region, then $H(\Omega)$ is the algebra of analytic functions on Ω. One of the first questions you might ask about an algebra is: what are its ideals? Suppose $g_1, g_2 \in H(\Omega)$ with no common zeros. Prove that there are functions $f_1, f_2 \in H(\Omega)$ with

$$f_1(z)g_1(z) + f_2(z)g_2(z) = 1 \tag{11.24}$$

for all $z \in \Omega$. If g_1 and g_2 had a common zero then we could not find f_1 and f_2 satisfying (11.24). This says that the ideal generated by g_1 and g_2 consists of all analytic functions on Ω if and only if g_1 and g_2 have no common zeros. A proof can be based on Corollary 11.16, but there is also a more direct proof using only the Mittag-Leffler Theorem 11.1.

11.12 (a) Show that

$$\int_{-\pi}^{\pi} \log |e^{it} - 1| dt = 0.$$

Hint: Compare the integral with the integral of $\log |re^{it} - 1|$ when $r < 1$, using different estimates for $|t| < \delta$ and $|t| \geq \delta$.

(b) If f is meromorphic in $|z| \leq R$, show that we can assume f has no zero or pole on $|z| = R$ when proving Jensen's theorem because of part (a).

11.13 Let $\Psi(z) = \frac{\Gamma'(z)}{\Gamma(z)}$.

(a) Show $\Psi'(z) + \Psi'(z + \frac{1}{2}) = 4\Psi'(2z)$.

(b) Integrate (a) twice to obtain

$$\Gamma(2z) = e^{az+b}\Gamma(z)\Gamma\left(z + \frac{1}{2}\right)$$

for some constants a and b. (Note: $\log \Gamma(z)$ is only defined locally.)

(c) Find a and b. The resulting formula is called **Legendre's duplication formula**.

11.14 **Stirling's formula** for the gamma function (used in combinatorics, physics and elsewhere).

(a) Suppose $x > \frac{1}{2}$. Substitute $t = (\sqrt{x} + v)^2$ in the integral formula (11.17) for $\Gamma(x)$ to show that

$$\frac{\Gamma(x)e^x\sqrt{x}}{x^x} = 2\int_{-\infty}^{\infty} \varphi_x(v)e^{-v^2} dv,$$

where

$$\varphi_x(v) = \begin{cases} 0, & \text{if } v \le -\sqrt{x} \\ e^{-2vx^{\frac{1}{2}}}\left(1 + \frac{v}{\sqrt{x}}\right)^{2x-1}, & \text{if } v \ge -\sqrt{x}. \end{cases}$$

(b) Prove that

$$\lim_{x\to\infty} \frac{\Gamma(x)e^x\sqrt{x}}{x^x} = 2\int_{-\infty}^{\infty} e^{-2v^2}dv = \sqrt{2\pi}.$$

Hint: Show $\log \varphi_x(v) \le 1$ for each $x > 1/2$. Then show $\log \varphi_x(v) \to -v^2$ as $x \to \infty$, for each v. Use (11.16) to find $\Gamma(\frac{1}{2})$ and then the last equality.

(c) Use part (b) to estimate $n!$ (Stirling's formula).

11.15 (**Prime number theorem**) In this exercise p will always denote a prime number. Let $\pi(x)$ denote the number of primes $\le x$. Set

$$\theta(x) = \sum_{p\le x} \log p \quad \text{and} \quad \Phi(z) = \sum_p \frac{\log p}{p^z}.$$

(a) Prove $\theta(x) \le Cx$. Hint: The binomial coefficient $\binom{2^k}{2^{k-1}}$ is divisible by the product of the primes p satisfying $2^{k-1} < p \le 2^k$, so that, for $2^{k-1} < x \le 2^k$,

$$\theta(x) \le \sum_{j=1}^{k} \theta(2^j) - \theta(2^{j-1}) \le \sum_{j=1}^{k} \log(1+1)^{2^j} \le C2^k \le 2Cx.$$

(b) Prove $\int_1^\infty (\frac{\theta(x)}{x} - 1)\frac{dx}{x} = \int_0^\infty (\theta(e^t)e^{-t} - 1)dt$ converges. Hint:

$$\mathcal{L}\left(\theta(e^t)e^{-t} - 1\right)(z) = \frac{\Phi(z+1)}{z+1} - \frac{1}{z} = -\frac{\zeta'(z+1)}{\zeta(z+1)} - \frac{1}{z} + g(z)$$

is analytic in $\{\text{Re}z > 0\}$. Apply Theorems 11.24 and 9.19.

(c) Prove $\lim_{x\to\infty} \frac{\theta(x)}{x} = 1$. Hint: If $\theta(x) \ge \lambda x$, for $\lambda > 1$ then, because θ is nondecreasing,

$$\int_x^{\lambda x} \left(\frac{\theta(t)}{t} - 1\right)\frac{dt}{t} \ge \lambda - 1 - \log\lambda > 0.$$

If $\theta(x) \le \lambda x$ for $\lambda < 1$, find a similar bound for the integral from λx to x.

(d) Prove that, for $\varepsilon > 0$,

$$\theta(x) \le \pi(x)\log x \le \frac{\theta(x)}{1-\varepsilon} + x^{1-\varepsilon}\log x.$$

Hint: $\theta(x) \ge \theta(x) - \theta(x^{1-\varepsilon})$ and $\pi(x^{1-\varepsilon}) \le x^{1-\varepsilon}$.

(e) Prove $\displaystyle\lim_{x\to\infty} \frac{\pi(x)\log x}{x} = 1.$

C

11.16 (a) Prove that, for $b \in \Omega$,

$$I_b = \{f \in H(\Omega) : f(b) = 0\}$$

is a closed maximal ideal in $H(\Omega)$, the holomorphic functions on Ω, and all closed maximal ideals are of this form. The topology is given by uniform convergence on compact subsets of Ω. Hint: First prove the analog of Exercise 11.11 for n functions instead of 2. Then use that $H(\Omega)$ is closed under uniform convergence on compact sets.

(b) If $z_n \in \Omega$ and $z_n \to \partial\Omega$, then $\{f \in H(\Omega) : \exists m = m(f) \text{ such that } f(z_k) = 0 \text{ for all } k \geq m\}$ is an ideal in $H(\Omega)$ which is dense and not generated by a single function.

(c) Find the principal ideals.

Here is an open problem. Suppose $a_k > 0$, $|z_k| \to \infty$, as $k \to \infty$, and $\sum \frac{a_k}{|z_k|} < \infty$. Then

$$f(z) = \sum_{k=1}^{\infty} \frac{a_k}{z - z_k}$$

is meromorphic in $\mathbb{C}$. Prove $f(w) = 0$ for some w. The three-dimensional analog (also open) is that if $a_k > 0$, $|x_k| \to \infty$ as $k \to \infty$ and $\sum_{k=1}^{\infty} a_k/|x_k|^2 < \infty$, then

$$F(x) = \sum_{k=1}^{\infty} \frac{a_k(x - x_k)}{|x - x_k|^3}$$

has at least one zero $x \in \mathbb{R}^3$. The $\mathbb{R}^3$-valued function F is the gradient of a real-valued function and can be thought of as the gravity force due to the positive masses a_k at $x_k \in \mathbb{R}^3$. The problem asks you to prove that there is always some place in space where you will experience no force. See [5] and [8].

PART III

12 Conformal Maps to Jordan Regions

We define a **Jordan curve** in $\mathbb{C}^*$ to be the homeomorphic image of the unit circle. It may seem obvious, but by no means simple to prove, that a Jordan curve divides $\mathbb{C}^*$ into exactly two regions. We begin the chapter with some examples of strange curves and regions. Then, following Pommerenke [20], we use complex analysis to prove a topological lemma due to Janiszewski, which is quite useful for tackling subtle topological problems in the plane. Janiszewski's lemma is used many times in Section 12.3 to prove the Jordan curve theorem. In Section 12.4, following Garnett and Marshall [10], we prove the useful and important theorem, due to Carathéodory, that a conformal map of the unit disk onto a region bounded by a Jordan curve extends to be a homeomorphism of the closed disk onto the closure of the region.

12.1 Some Badly Behaved Regions

In this section we give some examples of rather badly behaved regions which are useful to keep in mind as we prove the Jordan curve theorem.

Example 12.1 (Jordan curve with positive area) Our first example is a Jordan curve with positive area. First we construct a totally disconnected compact set E of positive area, then we describe how to pass a curve through this set.

The compact set E will be the intersection of compact sets E_j, $j = 0, 1, 2, \ldots$. The initial set E_0 is the closed unit square $E_0 = [0, 1] \times [0, 1]$. Choose $\sigma_1 < 1$ and let $E_1 \subset E_0$ be the union of four squares of side length $\sigma_1/2$ located in each of the corners of E_0. For example, the lower left square is $[0, \sigma_1/2] \times [0, \sigma_1/2]$. The area of E_1 is σ_1^2. Next choose $\sigma_2 < 1$ and let $E_2 \subset E_1$ be the union of 16 squares, one in each of the corners of E_1 with side length $\sigma_1\sigma_2/4$ and total area $(\sigma_1\sigma_2)^2$. Repeating this process we obtain compact sets $E_n \subset E_{n-1}$ consisting of 4^n squares with side length $\prod_{j=1}^n(\sigma_j/2)$ and area $\prod_{j=1}^n \sigma_j^2$. Choose σ_j so that $\prod_{j=1}^\infty \sigma_j > 0$. For instance, we can take $\sigma_j = 1 - 1/(j+1)^2$. Set

$$E = \bigcap_{n=0}^{\infty} E_n.$$

Then E is a compact set of positive area. See Figure 12.1.

The Jordan curve J is constructed in stages also. A **Jordan arc** is the homeomorphic image of the unit interval $[0, 1]$. First we construct a Jordan arc containing the set E, then we close it up to form a Jordan curve. Connect the four squares in E_1 using three line segments

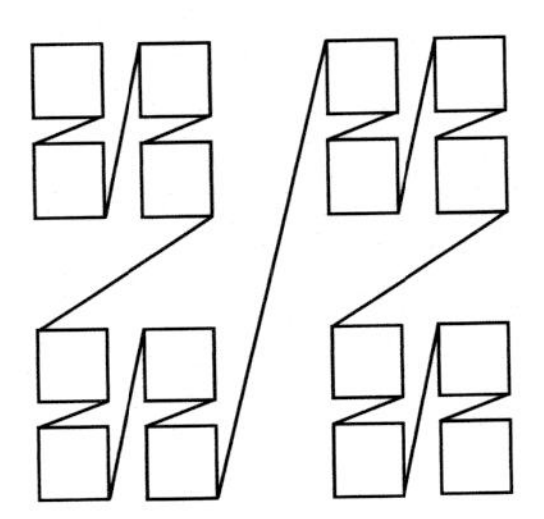

Figure 12.1 E_2 and connecting segments.

which leave the lower right corner of each square and connect to the upper left corner of an adjacent square, as shown in Figure 12.1. There is a natural left-to-right ordering of these three intervals. Inside each square in E_1 we insert three line segments connecting the four squares in E_2 using the same pattern. Repeat this process for each square in each E_n.

Next construct a mapping of a dense subset of the unit interval onto the union of these line segments. Let I_1 consist of three disjoint closed intervals contained in $(0, 1)$ with total length $1/2$. Let I_2 consist of three disjoint closed intervals in each of the four intervals of $[0, 1] \setminus I_1$. Repeat this process so that I_n consists of $3 \cdot 4^{n-1}$ disjoint intervals contained in $[0, 1] \setminus I_{n-1}$ of total length 2^{-n}. Then $\ell(\cup_1^\infty I_n) = \sum_1^\infty 2^{-n} = 1$. Construct a map φ mapping each interval in I_n linearly onto one of the segments connecting a square in E_n, preserving the natural ordering left to right within each square of E_{n-1}, $n = 1, 2, \ldots$. Because the side lengths of the squares in E_n tend to 0 and the lengths of the intervals in $[0, 1] \setminus I_n$ also tend to 0, as $n \to \infty$, the map φ extends to be continuous and one-to-one. It maps the unit interval to a curve which contains the set E. If we then connect the upper left corner of E_0 to the lower right corner of E_0 (in $\mathbb{C} \setminus E_0$) we obtain a (closed) Jordan curve J with positive area.

Example 12.2 (Dense set of spirals) Another example of a Jordan arc J_0 is found by spiraling in to 0 in a clockwise direction then spiraling back out again counter-clockwise with a Jordan arc. For example $\gamma(x) = e^{-(1+i)x}$, $0 \le x < \infty$, followed by $-e^{(1+i)x}$, $-\infty < x \le 0$. See Figure 12.2.

Note that $\arg z$ is neither bounded above nor bounded below on J_0. Starting with the unit circle $\partial\mathbb{D}$, replace a small arc of of $\partial\mathbb{D}$ with a rotated, scaled and translated version of J_0 having the same endpoints. Next choose another subarc of the resulting curve which does not contain the limit point of the spiral and replace it by a rotated and scaled version of J_0 with the same endpoints. If the replaced arc is small enough, the resulting curve J_1 will still be

Figure 12.2 Constructing a dense set of spirals.

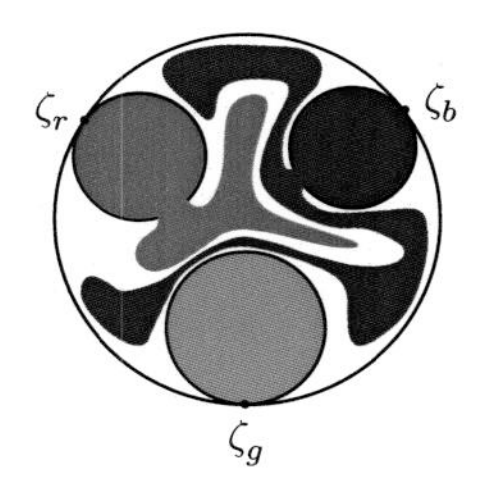

Figure 12.3 Constructing the Lakes of Wada.

a Jordan curve. We can repeat this process countably many times to obtain a closed Jordan curve J with a dense set of spirals. So, there is a dense subset of points $\{a_n\} \subset J$ such that $\arg(z - a_n)$ is neither bounded above nor bounded below for $z \in J$. This curve can even be constructed so that it has finite length. In this example it may be rather difficult to tell whether a given point near the Jordan boundary lies in the "inside" or "outside" region.

Example 12.3 (Lakes of Wada) The third example is three disjoint simply-connected regions with the same boundary. Suppose lakes L_r, L_g, L_b are three simply-connected subsets of $\mathbb{D}$ with disjoint closures and suppose each closure intersects the unit circle in a single point, $\zeta_r, \zeta_g, \zeta_b \in \partial\mathbb{D}$. For example, we can take three disjoint disks in $\mathbb{D}$ which are tangent to the unit circle. Then the dry land $D_0 = \overline{\mathbb{D}} \setminus (L_r \cup L_g \cup L_b)$ is compact, connected and has a simply-connected interior. Increase the size of one of the lakes by allowing water to flow out along an interval on its boundary so that the three lakes remain simply-connected with disjoint closures, each intersecting the unit circle in a single point and such that the remaining dry land $D_1 = \overline{\mathbb{D}} \setminus (L_r \cup L_g \cup L_b) \subset D_0$ is compact, connected and has a simply-connected interior. See Figure 12.3. Repeat this process, cycling through the lakes infinitely many times.

We can also arrange that, after expanding each lake n times, the distance from each point of dry land $z \in D_{3n}$ to each lake is at most $1/n$. The final dry land $D = \cap D_n$ is compact, connected and equal to the boundary of each lake.

The set D in this third example is not a Jordan curve. As we shall see in the proof of the Jordan curve theorem, at most two complementary components of a Jordan curve can have a boundary point in common. A similar construction works for n lakes, giving n disjoint simply-connected regions, all possessing the same boundary.

12.2 Janiszewski's Lemma

The reader may wish to review the definition and elementary properties of the components of an open set $U \subset \mathbb{C}^*$ as given in Exercise 4.2(a). It is not hard to verify also that

(a) if $E \subset U$ is connected, then E is contained in a component of U;
(b) each component U_a of U has boundary ∂U_a contained in $\mathbb{C}^* \setminus U$;
(c) if F is a homeomorphism of $\mathbb{C}^*$ then U_a is a component of U if and only if $F(U_a)$ is a component of $F(U)$.

See Exercise 12.1.

Definition 12.4 *A closed set $E \subset \mathbb{C}^*$* **separates** *points $a, b \notin E$ if a and b belong to distinct components of $\mathbb{C}^* \setminus E$.*

Lemma 12.5 (Janiszewski) *Suppose K_1 and K_2 are compact subsets of $\mathbb{C}$ such that $K_1 \cap K_2$ is connected and $0 \notin K_1 \cup K_2$. If K_1 does not separate 0 and ∞, and if K_2 does not separate 0 and ∞, then $K_1 \cup K_2$ does not separate 0 and ∞.*

The hypothesis that $K_1 \cap K_2$ is connected is essential, for, if K_1 is the closure of the top half of the unit circle and if K_2 is the closure of the bottom half of the unit circle, then $K_1 \cup K_2$ separates 0 and ∞.

Janiszewski's lemma follows from Lemmas 12.6 and 12.7 below.

Lemma 12.6 *If a compact set E separates 0 and ∞ then we cannot define $\log z$ to be analytic in a neighborhood of E.*

In other words, there is no function g which is analytic in a neighborhood of E satisfying $e^{g(z)} = z$, if E separates 0 and ∞.

Proof Suppose g is analytic in an open set $W \supset E$ and $e^{g(z)} = z$, with $0, \infty \in \mathbb{C}^* \setminus W \subset \mathbb{C}^* \setminus E$. Since 0 and ∞ belong to distinct components of $\mathbb{C}^* \setminus W$, by Theorem 5.7 we can find a closed polygonal curve $\sigma \subset W$ so that $n(\sigma, 0) = 1$. But, by the chain rule, $g'(z) = 1/z$, so that $2\pi i = \int_\sigma 1/z\, dz = \int_\sigma g'(z)dz = 0$ by the fundamental theorem of calculus. This contradiction proves Lemma 12.6. □

Lemma 12.7 *Suppose K_1 and K_2 are compact sets such that $K_1 \cap K_2$ is connected and $0, \infty \notin K_1 \cup K_2$. If K_1 does not separate 0 and ∞, and if K_2 does not separate 0 and ∞, then we can define $\log z$ to be analytic in a neighborhood of $K_1 \cup K_2$.*

Proof By hypothesis, we can find simple polygonal curves σ_1 and σ_2 connecting 0 to ∞ with $\sigma_j \cap K_j = \emptyset$, $j = 1, 2$. Then

$$(\sigma_1 \cup \sigma_2) \cap (K_1 \cap K_2) = \emptyset.$$

By item (a) above, the connected set $K_1 \cap K_2$ is contained in one component U of $\mathbb{C}^* \setminus (\sigma_1 \cup \sigma_2)$. See Figure 12.4.

Since $\mathbb{C}^* \setminus \sigma_j$ is simply-connected, $j = 1, 2$, we can find functions f_j analytic on $\mathbb{C}^* \setminus \sigma_j$ such that $e^{f_j(z)} = z$ on $\mathbb{C}^* \setminus \sigma_j$. Then $e^{f_1 - f_2} = 1$ on $\mathbb{C}^* \setminus (\sigma_1 \cup \sigma_2)$. Thus, on each component of $\mathbb{C}^* \setminus (\sigma_1 \cup \sigma_2)$, $f_1 - f_2$ is a constant of the form $2\pi k i$ for some integer k. We may then add a constant to f_2 so that $f_1 = f_2$ on U. Because $K_1 \setminus U$ and $K_2 \setminus U$ are disjoint compact sets, we can find open sets V_j so that

$$K_j \setminus U \subset V_j \subset \overline{V_j} \subset \mathbb{C}^* \setminus \sigma_j,$$

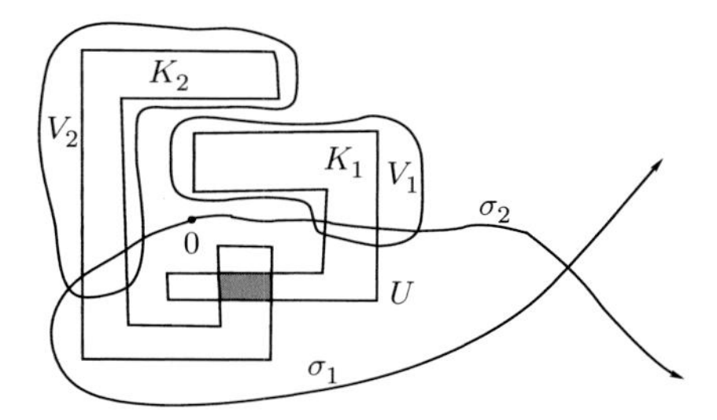

Figure 12.4 Janiszewksi's lemma.

for $j = 1, 2$ and $V_1 \cap V_2 = \emptyset$. Set

$$f(z) = \begin{cases} f_1(z), & \text{for } z \in V_1 \cup U \\ f_2(z), & \text{for } z \in V_2 \cup U. \end{cases}$$

Then f is analytic on $V_1 \cup V_2 \cup U$, an open neighborhood of $K_1 \cup K_2$, and $e^{f(z)} = z$. □

Janiszewski's lemma follows immediately from Lemmas 12.6 and 12.7. Moreover, by item (c), using LFTs, we can replace 0 and ∞ by any two points not in $K_1 \cup K_2$ in the statement of Janiszewski's lemma.

Corollary 12.8 *If J is a Jordan arc in $\mathbb{C}^*$ then $\mathbb{C}^* \setminus J$ is open and connected.*

Proof Since LFTs are homeomorphisms of $\mathbb{C}^*$, it suffices to show that if $0, \infty \notin J$ then J does not separate 0 and ∞. Write $J = \cup_1^n J_k$, where J_k are subarcs of J such that $J_k \cap J_{k-1}$ is a single point. We may choose each J_k so small that, for each k, there is a half-line from 0 to ∞ contained in $\mathbb{C}^* \setminus J_k$ and hence no J_k separate 0 and ∞. But $J_1 \cap J_2$ is a single point, and hence connected, so that, by Janiszewski's lemma, the arc $J_1 \cup J_2$ does not separate 0 and ∞. The intersection of this arc with J_3 again is a single point, and hence $J_1 \cup J_2 \cup J_3$ does not separate 0 and ∞ by Janiszewski's lemma. By induction, J does not separate 0 and ∞. □

12.3 Jordan Curve Theorem

In this section we will prove the Jordan curve theorem.

Theorem 12.9 (Jordan) *If $J \subset \mathbb{C}^*$ is a (closed) Jordan curve then $\mathbb{C}^* \setminus J$ has exactly two components, each of which is simply-connected. Moreover, J is the boundary of each component.*

The proof of Theorem 12.9 is divided into four lemmas.

Lemma 12.10 *If J is a (closed) Jordan curve, and if U is a component of $\mathbb{C}^* \setminus J$, then $\partial U = J$.*

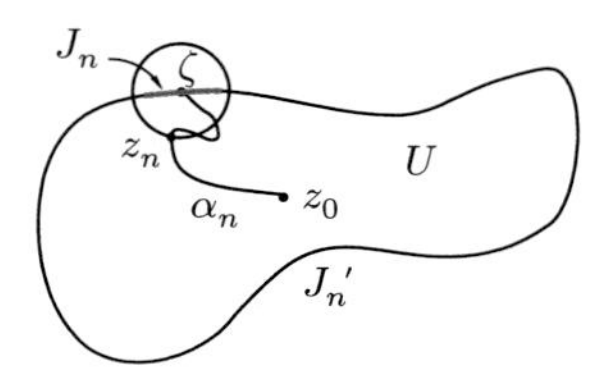

Figure 12.5 Boundary of a component.

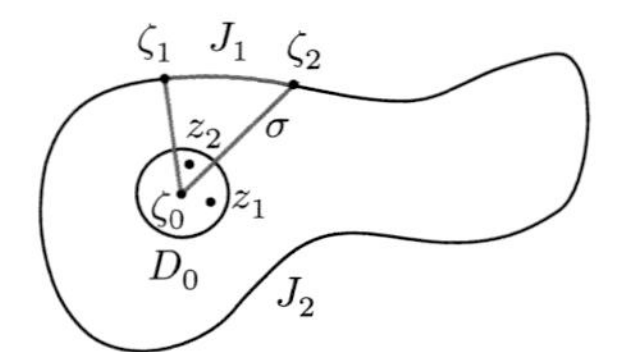

Figure 12.6 A modification of J.

Proof Suppose $\zeta \in J$ and $z_0 \in U$. Because J is the homeomorphic image of the unit circle, given $n < \infty$ we can find Jordan arcs $J_n, J_n' \subset J$ with $J_n \cup J_n' = J$, $\zeta \in J_n$, $\zeta \notin J_n'$ and $J_n \subset D_n$, where $D_n = \{z : |z - \zeta| < 1/n\}$. By Corollary 12.8, J_n' does not separate ζ and z_0. So there is a polygonal curve σ_n from z_0 to ζ such that $\sigma_n \cap J_n' = \emptyset$. See Figure 12.5.

Let z_n be the first intersection of σ_n with ∂D_n. The subarc $\alpha_n \subset \sigma_n$ from z_0 to z_n does not intersect J_n' and does not intersect $J_n \subset D_n$, and hence does not intersect J. Since α_n is connected, we have $\alpha_n \subset U$, and hence $z_n \in U$. But $\lim_n z_n = \zeta \notin U$, so $\zeta \in \partial U$. This shows that $J \subset \partial U$. If $\zeta \in \partial U$, then ζ does not belong to any component of $\mathbb{C}^* \setminus J$ since the components are open. Thus $\zeta \in J$, and we have shown $J = \partial U$. $\square$

Next we will construct a modification σ of the curve J and prove that $\mathbb{C}^* \setminus \sigma$ has exactly two components, and then use this result to prove that $\mathbb{C}^* \setminus J$ has exactly two components.

Take $\zeta_0 \in \mathbb{C}^* \setminus J$. Then there is a straight-line segment $[\zeta_0, \zeta_1]$ with $\zeta_1 \in J$ and $[\zeta_0, \zeta_1) \cap J = \emptyset$. Morever, we can choose another line segment $[\zeta_0, \zeta_2]$ with $\zeta_2 \in J$ and $[\zeta_0, \zeta_2) \cap J = \emptyset$ and $[\zeta_0, \zeta_1] \cap [\zeta_0, \zeta_2] = \{\zeta_0\}$. Otherwise J would be contained in a half-line from ζ_1 to ∞, which is impossible. Write $J = J_1 \cup J_2$, where J_j are Jordan arcs with $J_1 \cap J_2 = \{\zeta_1, \zeta_2\}$. Switching ζ_1 and ζ_2 if necessary,

$$\sigma = J_1 \cup [\zeta_1, \zeta_0] \cup [\zeta_0, \zeta_2]$$

is a (closed) Jordan curve. In other words, σ is the modification of J found by replacing J_2 with the union of two intervals. See Figure 12.6.

Lemma 12.11 *The set $\mathbb{C}^* \setminus \sigma$ has exactly two components.*

Proof If D_0 is an open disk centered at ζ_0 with $J \cap D_0 = \emptyset$, then $D_0 \setminus ([\zeta_1, \zeta_0] \cup [\zeta_0, \zeta_2])$ consists of two connected open circular sectors. By item (a) in Section 12.2, each sector must be contained in a component of $\mathbb{C}^* \setminus \sigma$ and so, by Lemma 12.10, there can be at most two components in $\mathbb{C}^* \setminus \sigma$.

Take z_1 and z_2 in distinct sectors of $D_0 \setminus ([\zeta_1, \zeta_0] \cup [\zeta_0, \zeta_2])$. Then $(\sigma \setminus D_0) \cup \partial D_0$ does not separate z_1 and z_2. Moreover, $\sigma \cap ((\sigma \setminus D_0) \cup \partial D_0) = \sigma \setminus D_0$, which is connected. So if σ also does not separate z_1 and z_2 then, by Janiszewski's lemma, $E = \sigma \cup (\sigma \setminus D_0) \cup \partial D_0$ does not separate z_1 and z_2. But clearly $\partial D_0 \cup [\zeta_1, \zeta_0] \cup [\zeta_0, \zeta_2] \subset E$ does separate. This contradiction proves that z_1 and z_2 are in distinct components of $\mathbb{C}^* \setminus \sigma$, proving Lemma 12.11. □

Lemma 12.12 *The set $\mathbb{C}^* \setminus J$ has at least two components.*

Proof Take $\zeta \in J_1 \setminus \{\zeta_1, \zeta_2\}$ and define

$$\alpha = J_2 \cup [\zeta_2, \zeta_0] \cup [\zeta_0, \zeta_1].$$

Let D_ζ be a disk centered at ζ such that $D_\zeta \cap \alpha = \emptyset$. By Lemma 12.10, $\zeta \in \partial G_1 \cap \partial G_2$, where G_1 and G_2 are the two components of the complement of σ. Take $w_1 \in G_1 \cap D_\zeta$ and $w_2 \in G_2 \cap D_\zeta$. The points w_1 and w_2 are separated by σ, but not by α. Note that $J \cap \alpha = J_2$ is connected so if J also does not separate w_1 and w_2 then, by Janiszewski's lemma, $J \cup \alpha$ does not separate w_1 and w_2. But this is a contradiction since $\sigma \subset J \cup \alpha$. We conclude that J must separate w_1 and w_2 and hence, $\mathbb{C}^* \setminus J$ has at least two components. □

Lemma 12.13 *The set $\mathbb{C}^* \setminus J$ has no more than two components.*

Proof Suppose H_1 and H_2 are the components of $\mathbb{C}^* \setminus J$ containing w_1 and w_2 from the proof of Lemma 12.12. If $\mathbb{C}^* \setminus J$ has another component H_3, then, by Lemma 12.10 again, we can find $w_3 \in H_3 \cap D_\zeta$. But $w_3 \notin \sigma$ and hence $w_3 \in G_1$ or $w_3 \in G_2$. If $w_3 \in G_1$ then w_1 and w_3 are not separated by σ and are not separated by α because $w_1, w_3 \in D_\zeta$ and $D_\zeta \cap \alpha = \emptyset$. But $\sigma \cap \alpha = [\zeta_1, \zeta_0] \cup [\zeta_0, \zeta_2]$ is connected, so that, by Janiszewski's lemma, w_1 and w_3 are not separated by $\sigma \cup \alpha$. But this contradicts the assumption that $J \subset \sigma \cup \alpha$ separates w_1 and w_3. A similar contradiction is obtained if $w_3 \in G_2$, proving Lemma 12.13. □

Proof of Theorem 12.9 Theorem 12.9 now follows from Lemmas 12.12, 12.13, and 12.10, and the observation that the complement in $\mathbb{C}^*$ of one component of $\mathbb{C}^* \setminus J$ is equal to the closure of the other component, and hence is connected. So each component is simply-connected. □

We can use the method of Exercise 5.1 and the Jordan curve theorem to show that two of the displayed points in Figure 5.3 are in the bounded component of the complement of the Jordan curve, and the other point is in the unbounded component. So, there is an arc in the complement which connects two of the points, and there is another arc in the complement connecting the third point to a point far from J. This exercise can be done without assigning a direction to the curve.

12.4 Carathéodory's Theorem

In the course of proving the Schwarz–Christoffel formula in Section 8.4 we proved that a conformal map of the disk onto a region bounded by a simple polygon extends to be a homeomorphism of the closed disk onto the closure of the polygonal region. This is a special case of a more general theorem due to Carathéodory.

A **Jordan region** is a simply-connected region in $\mathbb{C}^*$ whose boundary is a Jordan curve.

Theorem 12.14 (Carathéodory) *If φ is a conformal map of $\mathbb{D}$ onto a Jordan region Ω, then φ extends to be a homeomorphism of $\overline{\mathbb{D}}$ onto $\overline{\Omega}$. In particular, $\varphi(e^{it})$ is a parameterization of $\partial\Omega$.*

For an application of Theorem 12.14, see the first paragraph of Section 8.1.

Proof Using an LFT, we may suppose Ω is bounded. First we show φ has a continuous extension at each $\zeta \in \partial\mathbb{D}$. Let $0 < \delta < 1$ and set

$$\gamma_\delta = \mathbb{D} \cap \{z : |z - \zeta| = \delta\}.$$

The idea of the proof is that the image curve $\varphi(\gamma_\delta)$ cuts off a region U_δ whose closure shrinks to a single point as $\delta \to 0$. See Figure 12.7. It is not hard to show that the area of U_δ decreases to 0, but this is not enough. We use a "length-area principle" to show that the diameter of the boundary of U_δ, and hence the diameter of U_δ, tends to 0.

The open analytic arc $\varphi(\gamma_\delta)$ is the homeomorphic image of the interval $(0, 1)$ with length

$$L(\delta) = \int_{\gamma_\delta} |\varphi'(z)||dz|.$$

By the Cauchy–Schwarz inequality,

$$L^2(\delta) \le \pi\delta \int_{\gamma_\delta} |\varphi'(z)|^2 |dz|,$$

so that, for $r < 1$,

$$\begin{aligned}\int_0^r \frac{L^2(\delta)}{\delta} d\delta &\le \pi \int\int_{\mathbb{D}\cap B(\zeta,r)} |\varphi'(z)|^2 dx\, dy \\ &= \pi \,\text{area}\, (\varphi(\mathbb{D} \cap B(\zeta, r)) < \infty\end{aligned} \tag{12.1}$$

since Ω is bounded.

Therefore there is a decreasing sequence $\delta_n \to 0$ such that $L(\delta_n) \to 0$. When $L(\delta_n) < \infty$, the curve $\varphi(\gamma_{\delta_n})$ has endpoints α_n, β_n, and both of these endpoints must lie on $\partial\Omega$, because φ is proper (see Exercise 3.7). Furthermore,

$$|\alpha_n - \beta_n| \le L(\delta_n) \to 0. \tag{12.2}$$

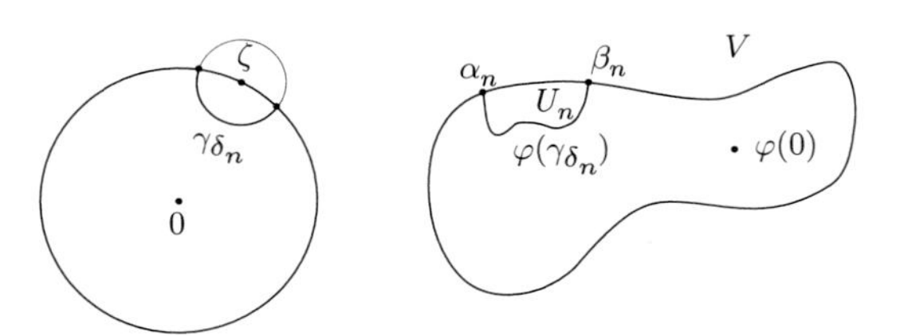

Figure 12.7 Crosscuts γ_{δ_n} and $\varphi(\gamma_{\delta_n})$.

Let γ be a homeomorphism of $\partial\mathbb{D}$ onto $\partial\Omega$. Write $\alpha_n = \gamma(\zeta_n)$ and $\beta_n = \gamma(\psi_n)$. Because γ is uniformly continuous, given $\varepsilon > 0$ there is a $\delta > 0$ so that, if $|\zeta_n - \psi_n| < \delta$, then for ζ in the smaller arc of $\partial\mathbb{D}$ between ζ_n and ψ_n, we have $|\gamma(\zeta) - \gamma(\zeta_n)| < \varepsilon$. But γ^{-1} is also uniformly continuous so there exists $\eta > 0$ such that if $|\alpha_n - \beta_n| < \eta$ then $|\zeta_n - \psi_n| < \delta$. Thus if σ_n is the closed subarc of $\partial\Omega$ of smallest diameter with endpoints α_n and β_n, then, by (12.2),

$$\text{diam}(\sigma_n) \to 0.$$

By the Jordan curve theorem, the curve $\sigma_n \cup \varphi(\gamma_{\delta_n})$ divides the plane into two (connected, open) regions, and one of these regions, say U_n, is bounded. The unbounded component V of the complement of $\partial\Omega$ is connected so if $z \in U_n \cap V$ then there is a polygonal arc γ_z from z to ∞ contained in V. See Exercise 4.2(a). Because $\varphi(\mathbb{D}) \cap V = \emptyset$ and $\sigma_n \subset \partial\Omega$, γ_z does not meet ∂U_n, contradicting the boundedness of U_n. Thus $U_n \cap V = \emptyset$, and hence $U_n \cap \overline{V} = \emptyset$. By the Jordan curve theorem applied to $\partial\Omega$, $U_n \subset \Omega$. Since

$$\text{diam}\,(\partial U_n) = \text{diam}\left(\sigma_n \cup \varphi(\gamma_{\delta_n})\right) \to 0,$$

we conclude that

$$\text{diam}\,(U_n) \to 0. \tag{12.3}$$

Set $D_n = \mathbb{D} \cap \{z : |z - \zeta| < \delta_n\}$. Then $\varphi(D_n)$ and $\varphi(\mathbb{D} \setminus \overline{D}_n)$ are connected sets which do not intersect $\varphi(\gamma_{\delta_n})$. Since φ maps onto Ω, either $\varphi(D_n) = U_n$ or $\varphi(\mathbb{D} \setminus \overline{D}_n) = U_n$. But $\text{diam}\,(\varphi(\mathbb{D} \setminus D_n)) \geq \text{diam}\,(\varphi(B(0, 1/2)) > 0$ and $\text{diam}\,(U_n) \to 0$ by (12.3) so that $\varphi(D_n) = U_n$ for n sufficiently large. Since δ_n is decreasing, $\varphi(D_{n+1}) \subset \varphi(D_n)$ and so $\bigcap \overline{\varphi(D_n)}$ consists of a single point. Thus φ has a continuous extension to $\mathbb{D} \cup \{\zeta\}$.

Let φ also denote the extension $\varphi : \overline{\mathbb{D}} \to \overline{\Omega}$. If $z_n \in \overline{\mathbb{D}}$ converges to $\zeta \in \partial\mathbb{D}$ then we can find $z_n' \in \mathbb{D} \to \zeta$ so that $\varphi(z_n) - \varphi(z_n') \to 0$. By the continuity of φ at ζ, we must have $\varphi(z_n) \to \varphi(\zeta)$, and we conclude that φ is continuous on $\overline{\mathbb{D}}$.

Because $\varphi(\mathbb{D}) = \Omega$, φ maps $\overline{\mathbb{D}}$ onto $\overline{\Omega}$. To show that φ is one-to-one, suppose $\varphi(\zeta_1) = \varphi(\zeta_2)$ but $\zeta_1 \neq \zeta_2$. Because φ is proper, $\varphi(\partial\mathbb{D}) \subset \partial\Omega$ and so we can assume $\zeta_j \in \partial\mathbb{D}, j = 1, 2$. The Jordan curve

$$\{\varphi(r\zeta_1) : 0 \leq r \leq 1\} \;\cup\; \{\varphi(r\zeta_2) : 0 \leq r \leq 1\}$$

bounds a bounded region W. See Figure 12.8.

Arguing exactly as above, replacing U_n, D_n and $\mathbb{D} \setminus \overline{D}_n$ with W and the two components of

$$\mathbb{D} \setminus \left(\{r\zeta_1 : 0 \leq r \leq 1\} \cup \{r\zeta_2 : 0 \leq r \leq 1\}\right),$$

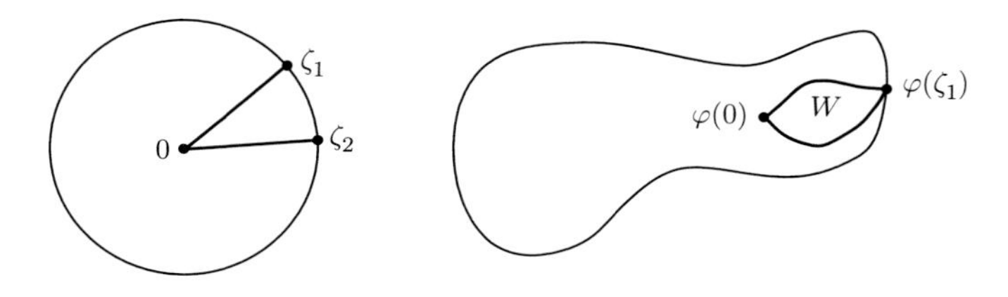

Figure 12.8 Proof that φ is one-to-one.

we conclude that $W \subset \Omega$ and $\varphi^{-1}(W)$ must be one of these two components. Because $\varphi(\partial\mathbb{D}) \subset \partial\Omega$ and φ is proper on $\varphi^{-1}(W)$, we conclude that

$$\varphi(\partial\mathbb{D} \cap \partial\varphi^{-1}(W)) \subset \partial\Omega \cap \partial W = \{\varphi(\zeta_1)\}.$$

Thus φ is constant on an arc of $\partial\mathbb{D}$. It follows that $\varphi - \varphi(\zeta_1) \equiv 0$, by the Schwarz reflection principle and the uniqueness theorem. This contradiction shows that φ must be one-to-one. □

The idea of applying the Cauchy–Schwarz inequality to prove (12.1) is known as the *length-area* principle. It led to the development of extremal length, an important tool in geometric function theory.

Corollary 12.15 *If $h : \partial\mathbb{D} \to \mathbb{C}$ is a homeomorphism then h extends to be a homeomorphism of $\mathbb{C}$ onto $\mathbb{C}$.*

Proof Suppose f and g are conformal maps of the interior and exterior of the disk onto the bounded and unbounded regions in $\mathbb{C} \setminus h(\partial\mathbb{D})$, with $g(\infty) = \infty$. For $|\zeta| = 1$, set

$$F(r\zeta) = \begin{cases} f(rf^{-1}(h(\zeta))), & \text{for } r \leq 1 \\ g(rg^{-1}(h(\zeta))), & \text{for } r \geq 1. \end{cases}$$

□

Corollary 12.15 fails in higher dimensions. Alexander's horned sphere is a homeomorphic image of the ball in $\mathbb{R}^3$. The boundary is the homeomorphic image of the sphere, but the unbounded component of the complement is not "simply-connected" in the sense that there are curves in the complement that cannot be continuously deformed to a point. In particular, the homeomorphism cannot be extended to a homeomorphism of $\mathbb{R}^3$.

The Jordan curve theorem and Carathéodory's theorem yield the following useful result.

Corollary 12.16 *If $J \subset \mathbb{C}$ is a closed Jordan curve then J can be oriented so that $n(J, z) = 1$ for z in the bounded component of the complement of J and $n(J, z) = 0$ for z in the unbounded component of the complement of J.*

Proof Let φ be a conformal map of $\mathbb{D}$ onto the bounded component U of the complement of J. By Carathéodory's theorem, $\varphi_r(z) = \varphi(rz)$ converges uniformly on $\partial\mathbb{D}$ to φ as $r \uparrow 1$. Fix $w_1 \in U$, where U is the bounded component of $\mathbb{C} \setminus J$. Set $\gamma(t) = \varphi(e^{it})$ and $\gamma_r(t) = \varphi(re^{it})$. As in Lemma 4.31, cover $J = \gamma$ by finitely many disks of the form $B_j = \{z : |z - \gamma(t_j)| < \text{dist}(\gamma, w_1)\}$ such that $\gamma([t_{j-1}, t_j]) \subset B_j$ for a finite partition $\{t_j\}$ of $[0, 2\pi]$. Then, for r sufficiently close to 1, $\gamma_r([t_{j-1}, t_j]) \cup \gamma([t_{j-1}, t_j]) \subset B_j$. Then, by adding and subtracting line segments between the endpoints of these arcs, we can write $\gamma - \gamma_r = \sum \beta_j$, where each β_j is a closed curve contained in B_j. Thus for r near 1

$$\int_\gamma \frac{dw}{w - w_1} - \int_{\gamma_r} \frac{dw}{w - w_1} = \sum_j \int_{\beta_j} \frac{dw}{w - w_1} = 0,$$

where the integral $\int_\gamma dw/(w - w_1)$ is defined as in Theorem 4.32. By the argument principle, the number of zeros of $\varphi(z) - w_1$ in $r\mathbb{D} = \{\zeta : |\zeta| < r\}$ is equal to

$$\int_{\partial r\mathbb{D}} \frac{\varphi'(z)}{\varphi(z) - w_1} \frac{dz}{2\pi i} = n(\gamma_r, w_1) = n(\gamma, w_1),$$

where $\partial\mathbb{D}$ has the usual counter-clockwise orientation. Thus $\varphi(e^{it})$, $t \in [0, 2\pi]$, is a parameterization of J so that $n(J, w) = 1$ for each $w \in U$. By property (b) in Section 5.2, $n(J, w) = 0$ for each w in the unbounded component of the complement of J. □

See Exercise 12.10 for the finitely-connected version of Corollary 12.16.

12.5 Exercises

A

12.1 Prove items (a)–(c) in Section 12.2. Hint: If U_a is a component of an open set U then both U_a and $U \setminus U_a$ are open.

12.2 Suppose $G \subset \mathbb{C}^*$ is a connected open set and suppose J is a (closed) Jordan curve, then let H_1 and H_2 be the components of the complement of J. If $\partial G \subset H_1$ then either $G \subset H_1$ or $H_2 \subset G$.

12.3 (a) Prove that if E is connected, if E_a is connected for each index a, and if E_a intersects E for all indices a, then the union of E and all E_a is connected;
(b) Prove that if E is connected then its closure $\overline{E}$ is connected;
(c) Prove that if E is compact and connected, and if U_a is a component of $\mathbb{C}^* \setminus E$, then U_a is simply-connected.
(d) Suppose U is simply-connected and suppose S is connected and $S \cap (\mathbb{C} \setminus U) \neq \emptyset$. Prove that each component of $U \cap (\mathbb{C} \setminus S)$ is simply-connected.

12.4 Prove that if U is a simply-connected region then $\partial U \subset \mathbb{C}^*$ is connected in $\mathbb{C}^*$. Hint: Use the conformal map of $\mathbb{D}$ onto U.

B

12.5 Suppose φ is a conformal map of a region Ω_1 onto a region Ω_2, where both Ω_1 and Ω_2 are bounded by finitely many disjoint Jordan curves. Prove that φ extends to be a homeomorphism of $\overline{\Omega_1}$ onto $\overline{\Omega_2}$.

12.6 A compact set $K \subset \mathbb{C}$ is **locally connected** if for each $\varepsilon > 0$ there exists $\delta > 0$ so that if $z, w \in K$ with $|z - w| < \delta$, then there exists a compact connected set $L \subset K$ with $z, w \in L$ and $\text{diam}(L) < \varepsilon$. Suppose f is a conformal map of $\mathbb{D}$ onto a bounded simply-connected region Ω which extends to be continuous on $\overline{\mathbb{D}}$. Prove that $\partial\Omega$ is locally connected.

12.7 Let Ω be a bounded simply-connected region. Prove that $\partial\Omega$ is locally connected if and only if a conformal map $\varphi : \mathbb{D} \to \Omega$ extends continuously to $\overline{\mathbb{D}}$. See Exercise 12.6. Hint: Apply the proof of Carathéodory's theorem to the map of $\mathbb{D}$ onto Ω.

12.8 (a) Let Ω be a simply-connected region and let $\sigma \subset \Omega$ be a **crosscut**, i.e., an (open) Jordan arc in Ω having endpoints in $\partial\Omega$. Prove that $\Omega \setminus \sigma$ has two components, Ω_1 and Ω_2, each simply-connected.

(b) Let Ω be simply-connected, let $\psi : \Omega \to \mathbb{D}$ be a conformal map onto $\mathbb{D}$, and fix $z \in \Omega$ and $\zeta \in \partial\Omega$. Assume ζ and z can be separated by a sequence of crosscuts $\sigma_n \subset \Omega$ such that length$(\sigma_n) \to 0$. Let U_n be the component of $\Omega \setminus \sigma_n$ such that $z \notin U_n$. Prove that $\psi(U_n) \to \alpha \in \partial\mathbb{D}$.

12.9 Find and read a construction of Alexander's horned sphere and the proof of the claims in the paragraph after the proof of Corollary 12.15. Show also that the notion of simply-connected in $\mathbb{R}^3$ given there is not the same as having a connected complement.

12.10 Suppose Ω is a region bounded by finitely many pairwise disjoint closed Jordan curves $\{J_k\}_1^m \subset \mathbb{C}$. If Ω is unbounded, then we can orient $\partial\Omega$ so that $n(\partial\Omega, z) = 0$ for all $z \in \Omega$ and $n(\partial\Omega, z) = 1$ for all $z \notin \overline{\Omega}$. If Ω is bounded, then we can orient $\partial\Omega$ so that $n(\partial\Omega, z) = 1$ for all $z \in \Omega$ and $n(\partial\Omega, z) = 0$ for all $z \notin \overline{\Omega}$.

12.11 The main spiral from Figure 12.2 terminates at a point z_0. If γ is an arc in the complement terminating at z_0, then either $\arg(\gamma(t) - z_0) \to +\infty$ or $\arg(\gamma(t) - z_0) \to -\infty$. Construct a curve σ which has the property that if γ is an arc in the complement terminating at z_0 then $\limsup \arg(\gamma(t) - z_0) = +\infty$ and $\liminf \arg(\gamma(t) - z_0) = -\infty$. A point with this property is called a **MacMillan twist point**.

13 The Dirichlet Problem

In this chapter we will treat the Dirichlet problem on arbitrary regions. Let $C(\partial\Omega)$ denote the set of continuous functions on the boundary in $\mathbb{C}^*$ of a region Ω. The **Dirichlet problem** on Ω for a function $f \in C(\partial\Omega)$ is to find a harmonic function u on Ω which is continuous on $\overline{\Omega} \subset \mathbb{C}^*$ and equal to f on $\partial\Omega$. If u exists then it is unique by the maximum principle, but it is not always possible to solve the Dirichlet problem. If $f = 0$ on $\partial\mathbb{D}$ and $f(0) = 1$, then $f \in C(\partial\Omega)$ where $\Omega = \mathbb{D} \setminus \{0\}$. But, by Lindelöf's maximum principle, Theorem 7.15, if u is harmonic and bounded on Ω with $u = 0$ on $\partial\mathbb{D}$ then $u(z) = 0$ for all $z \in \Omega$. Thus u extends to be continuous at 0, but $u(0) \neq f(0)$.

There are several approaches to this material. See for example Garnett and Marshall [10]. We will use the elegant Perron process because it is the fundamental method underlying the proof of the uniformization theorem given in Chapter 15.

13.1 Perron Process

Recall that a subharmonic function v on Ω is continuous as a map of Ω into $[-\infty, +\infty)$ and satisfies the mean-value inequality on all sufficiently small circles. See Definition 7.2. There are several key properties of subharmonic functions:

(a) A subharmonic function v on a region Ω satisfies the maximum principle: if there exists $z_0 \in \Omega$ such that $v(z_0) = \sup_{z\in\Omega} v(z)$ then v is constant on Ω.
(b) If v_1, v_2 are subharmonic then $v_1 + v_2$ and $\max(v_1, v_2)$ are subharmonic.
(c) If v is subharmonic on a region Ω and $v > -\infty$ on ∂D, where D is a disk with $\overline{D} \subset \Omega$, then the function v_D which equals v on $\Omega \setminus D$ and equals the Poisson integral of v on D is also subharmonic. In other words, if $D = \{z : |z - c| < r\}$, we can replace v on D by

$$v_D(z) = \int_0^{2\pi} \frac{1 - |(z-c)/r|^2}{|e^{it} - (z-c)/r|^2} v(c + re^{it}) \frac{dt}{2\pi},$$

and still be subharmonic.

See Exercise 13.1. For example, if f is analytic then $|f|^p$, for $p > 0$, $\log|f|$ and $\log(1+|f|^2)$ are subharmonic. See Exercise 13.4.

Definition 13.1 *A family $\mathcal{F}$ of subharmonic functions on a region Ω is called a* **Perron family** *if it satisfies:*

(i) *if $v_1, v_2 \in \mathcal{F}$ then $\max(v_1, v_2) \in \mathcal{F}$,*
(ii) *if $v \in \mathcal{F}$ and D is a disk with $\overline{D} \subset \Omega$, and if $v > -\infty$ on ∂D, then $v_D \in \mathcal{F}$ and*
(iii) *for each $z \in \Omega$ there exists $v \in \mathcal{F}$ such that $v(z) > -\infty$.*

See property (c) of subharmonic functions for the definition of v_D.

Definition 13.2 *If $\mathcal{F}$ is a Perron family on a region Ω then we define*

$$u_{\mathcal{F}}(z) \equiv \sup_{v \in \mathcal{F}} v(z).$$

Theorem 13.3 *If $\mathcal{F}$ is a Perron family on a region Ω then $u_{\mathcal{F}}$ is harmonic on Ω or $u_{\mathcal{F}}(z) = +\infty$ for all $z \in \Omega$.*

The "bootstrap technique" for climbing a cliff is to put your foot above your head, then pull yourself up by your bootstraps or shoelaces. Repeat. The proof of Theorem 13.3 is a classic bootstrap proof.

Proof Fix $z_0 \in \Omega$. Find $v_1, v_2, \cdots \in \mathcal{F}$ such that $\lim_j v_j(z_0) = u_{\mathcal{F}}(z_0)$. Set

$$v_j' = \max(v_1, v_2, \ldots, v_j).$$

By (i) and induction, $v_j' \in \mathcal{F}$, $v_j' \leq v_{j+1}'$ and $\lim v_j'(z_0) = u_{\mathcal{F}}(z_0)$. Suppose D is a disk with $z_0 \in D \subset \overline{D} \subset \Omega$. By (iii), the continuity of subharmonic functions and (i), we may suppose that v_j' is continuous and $v_j' > -\infty$ on ∂D. Let v_j'' equal v_j' on $\Omega \setminus D$ and equal the Poisson integral of v_j' on D. Then, by (ii), $v_j'' \in \mathcal{F}$. Moreover, $v_j' \leq v_j''$ by the maximum principle. The Poisson integral on D of the non-negative function $v_{j+1}' - v_j'$ is non-negative so that $v_j'' \leq v_{j+1}''$. Set $V = \lim_j v_j''$. By Harnack's principle, Theorem 7.20, either $V \equiv +\infty$ in D or V is harmonic in D. Note also that $V(z_0) = u_{\mathcal{F}}(z_0)$ because of the choice of v_j and the maximality of $u_{\mathcal{F}}(z_0)$.

Now take $z_1 \in D$, $w_j \in \mathcal{F}$, with $\lim_j w_j(z_1) = u_{\mathcal{F}}(z_1)$. Set

$$w_j' = \max(v_j'', w_1, w_2, \ldots, w_j)$$

and let $w_j'' \in \mathcal{F}$ equal w_j' on $\Omega \setminus D$ and equal the Poisson integral of w_j' on D. As before, $w_j' \leq w_j'' \leq w_{j+1}''$. Set $W = \lim_j w_j''$. As before, either W is harmonic in D or $W \equiv +\infty$ in D, and $W(z_1) = u_{\mathcal{F}}(z_1)$. Because $v_j'' \leq w_j' \leq w_j''$, we must have that $V \leq W$. But also

$$u_{\mathcal{F}}(z_0) = V(z_0) \leq W(z_0) \leq u_{\mathcal{F}}(z_0)$$

so that $V(z_0) = W(z_0)$. If $u_{\mathcal{F}}(z_0) < \infty$ then $V - W$ is harmonic on D and achieves its maximum value 0 on D at z_0. By the maximum principle, it must equal 0 on D. Because z_1 was arbitrary in D, there are two possibilities, either $u_{\mathcal{F}} \equiv +\infty$ on D or $u_{\mathcal{F}} = V$ on D and hence is harmonic on D. Since z_0 is an arbitrary point in Ω, $\{z : u_{\mathcal{F}}(z) = +\infty\}$ is then both closed and open in Ω. Since Ω is connected, either $u_{\mathcal{F}} \equiv +\infty$ on Ω or $u_{\mathcal{F}}$ is harmonic on Ω. □

Note that the proof of Theorem 13.3 only used local properties of the family $\mathcal{F}$, the mean-value property and the maximum principle on small disks contained in Ω. We will use this observation in Chapter 15.

If $\Omega \subset \mathbb{C}^*$ is a region, and if f is a real-valued function defined on $\partial\Omega$ (a compact subset of $\mathbb{C}^*$) with $|f| \leq M < \infty$ on $\partial\Omega$, set

$$\mathcal{F}_f = \{v \text{ subharmonic on } \Omega : \limsup_{z \in \Omega \to \zeta} v(z) \leq f(\zeta), \text{ for all } \zeta \in \partial\Omega\}. \tag{13.1}$$

Then Definition 1.13(iii) holds since the constant function $-M \in \mathcal{F}_f$. So $\mathcal{F}_f$ is a Perron family. Moreover, each $v \in \mathcal{F}_f$ is bounded by M by the maximum principle. By Theorem 13.3,

$$u_f(z) \equiv \sup_{v \in \mathcal{F}_f} v(z) \tag{13.2}$$

is harmonic in Ω. The function u_f is called the **Perron solution to the Dirichlet problem** on Ω for the function f. It is a natural candidate for a harmonic function in Ω which equals f on $\partial\Omega$. In the next section we will explore to what extent u_f has boundary values equal to f.

13.2 Local Barriers

In this section we develop a criterion for the Perron solution to the Dirichlet problem on Ω for a function f to have boundary value $f(\zeta)$ at $\zeta \in \partial\Omega$. Our account is based on Ransford [21].

Throughout this section, if $\zeta \in \mathbb{C}$ then $B(\zeta, \varepsilon)$ is the open disk centered at ζ with radius $\varepsilon > 0$. If $\zeta = \infty \in \mathbb{C}^*$ then $B(\zeta, \varepsilon)$ is the neighborhood of ζ given by $\{z : |z| > 1/\varepsilon\} \cup \{\infty\} \subset \mathbb{C}^*$. We will refer to $B(\infty, \varepsilon)$ as a disk centered at ∞, because $B(\infty, \varepsilon)$ corresponds to a small disk on the Riemann sphere centered at the north pole.

Properties (i) and (ii) in Definition 13.1 make it somewhat easier to build subharmonic functions than harmonic functions.

Definition 13.4 *If $\Omega \subset \mathbb{C}^*$ is a region and if $\zeta_0 \in \partial\Omega$ then b is called a* **local barrier at ζ_0 for the region Ω** *provided*

(i) *b is defined and subharmonic on $\Omega \cap D$ for some open disk D with $\zeta_0 \in D$,*
(ii) *$b(z) < 0$ for $z \in \Omega \cap D$ and*
(iii) *$\lim_{z \in \Omega \to \zeta_0} b(z) = 0$.*

The function b approaches its supremum on $\Omega \cap D$ as z tends to ζ_0 and it can also approach this value elsewhere on $\partial(\Omega \cap D)$. Note that if b is a local barrier for the region Ω at ζ_0, defined on $\Omega \cap D$, then it is also a local barrier on $\Omega \cap D_1$ for any smaller disk D_1 with $\zeta_0 \in D_1 \subset D$.

Definition 13.5 *If there exists a local barrier at $\zeta_0 \in \partial\Omega$ then ζ_0 is called a* **regular point of $\partial\Omega$**. *Otherwise $\zeta_0 \in \partial\Omega$ is called an* **irregular point of $\partial\Omega$**. *If every $\zeta \in \partial\Omega$ is regular, then Ω is called a* **regular region**.

If ζ_0 is an isolated point of $\partial\Omega$, then there cannot be a local barrier at ζ_0 for Ω by the Lindelöf maximum principle. The next result says that if ζ_0 belongs to a non-trivial component of the boundary, then there is a local barrier at ζ_0.

Theorem 13.6 *Suppose $\Omega \subset \mathbb{C}^*$ is a region and suppose $\zeta_0 \in \partial\Omega$. If $C \subset \partial\Omega$ is closed in $\mathbb{C}^*$ and connected, with $\zeta_0 \in C$ but $C \neq \{\zeta_0\}$, then ζ_0 is regular. Every simply-connected region is regular.*

Proof We claim that each component Ω_1 of $\mathbb{C}^* \setminus C$ is simply-connected. To see this, we may assume $\infty \in C \setminus \{\zeta_0\}$. Let γ be any closed polygonal curve contained in Ω_1. Then $n(\gamma, \zeta) = 0$ for all $\zeta \in C \setminus \{\infty\}$ because C is connected and unbounded, and hence contained in the unbounded component of the complement of γ. If $\beta \notin \Omega_1$ and $n(\gamma, \beta) \neq 0$, let L be a line segment from β to some $\zeta \in C$. Then L must intersect $\partial\Omega_1$ before it intersects γ because $\gamma \subset \Omega_1$. Because $\partial\Omega_1 \subset C$, this intersection gives a point $\zeta_1 \in C$ with $n(\gamma,\zeta_1) = n(\gamma,\beta) \neq 0$. This contradiction proves that every closed polygonal curve in Ω_1 is homologous to 0, and so, by Theorem 5.7, Ω_1 is simply-connected.

If $\zeta_2 \in C \setminus \{\infty\}$ with $\zeta_2 \neq \zeta_0$, then

$$f(z) = \log\left(\frac{z-\zeta_0}{z-\zeta_2}\right)$$

can be defined to be analytic on each component Ω_1 of $\mathbb{C}^* \setminus C$ by Corollary 5.8, and therefore on Ω. Then $b(z) = \text{Re}(1/f(z))$ is a local barrier at ζ_0, defined on $\Omega \cap D$, where $D = \{z : |(z-\zeta_0)/(z-\zeta_2)| < 1/2\}$. The same function works if we only assume Ω is simply-connected. See also Exercise 12.4. □

The local barrier constructed in the proof of Theorem 13.6 can approach 0 along sequences tending to boundary points other than ζ_0. For example, if $C_1 = \{re^{i/(1-r)} : 0 \leq r < 1\}$ is a spiral from the origin to $\partial\mathbb{D}$ then $\mathbb{D} \setminus C_1$ is simply-connected. The local barrier constructed in the proof of Theorem 13.6 with $\zeta_2 = 0$ and $\zeta_0 = 1$ will tend to zero at each point of the unit circle in $|(z-1)/z| < 1/2$ because the imaginary part of the logarithm is unbounded.

Theorem 13.7 *Suppose $\zeta_0 \in \partial\Omega$ is regular. If f is a real-valued function defined on $\partial\Omega$, with $|f| \leq M < \infty$ on $\partial\Omega$, and if f is continuous at ζ_0, then*

$$\lim_{z\in\Omega\to\zeta_0} u_f(z) = f(\zeta_0),$$

where u_f is the Perron solution to the Dirichlet problem for the function f.

See Exercise 13.2 for a partial converse to Theorem 13.7.

Corollary 13.8 *If Ω is a regular region and if $f \in C(\partial\Omega)$ then u_f is harmonic in Ω and extends to be continuous on $\overline{\Omega}$ with $u_f = f$ on $\partial\Omega$.*

The proof of Theorem 13.7 requires two lemmas. If v is subharmonic and not harmonic then $-v$ is not subharmonic. So it does not easily follow that the map $f \to u_f$ is linear. However, the next lemma is not difficult to prove.

Lemma 13.9 *$u_f \leq -u_{-f}$ on Ω.*

Proof If $v \in \mathcal{F}_f$, as defined in (13.1), and if $w \in \mathcal{F}_{-f}$, then

$$\limsup_{z\in\Omega\to\zeta\in\partial\Omega} (v(z) + w(z)) \leq f(\zeta) + (-f(\zeta)) = 0.$$

By the maximum principle, $v + w \leq 0$ on Ω. Taking the supremum over all $v \in \mathcal{F}_f$ and $w \in \mathcal{F}_{-f}$ shows that $u_f + u_{-f} \leq 0$. □

Lemma 13.10 (Bouligand) *Suppose ζ_0 is a regular point of $\partial\Omega$ and suppose b is a local barrier at ζ_0 defined on $\Omega \cap B(\zeta_0,\varepsilon)$. Given $0 < \delta < \varepsilon$, there exists b_δ, which is subharmonic on all of Ω satisfying*

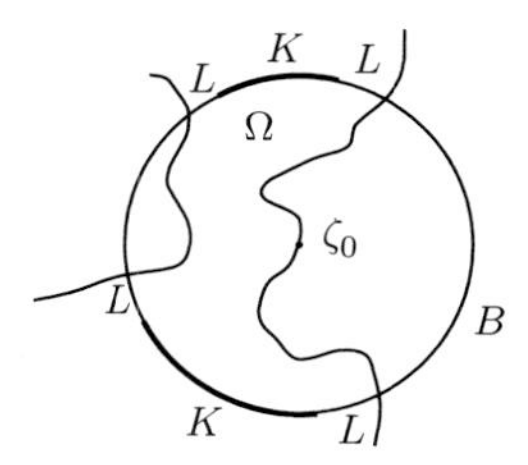

Figure 13.1 Proof of Bouligand's lemma.

(i) $b_\delta < 0$ *on* Ω,
(ii) $b_\delta \equiv -1$ *on* $\Omega \setminus B(\zeta_0, \varepsilon)$,
(iii) $\liminf_{z \in \Omega \to \zeta_0} b_\delta(z) \geq -\delta$.

Condition (iii) is weaker than the corresponding condition (iii) in Definition 13.4, but the function b_δ is defined on all of Ω, not just a neighborhood of ζ_0.

Proof Set $B = B(\zeta_0, r)$ for some $r < \varepsilon$. Then

$$\partial B \cap \Omega = \bigcup_{k=1}^{\infty} I_k,$$

where each I_k is an open arc. Choose $n < \infty$ and compact subarcs $J_k \subset I_k$, $k = 1, \ldots, n$, so that if $K = \cup_{k=1}^{n} J_k$ and $L = (\partial B \cap \Omega) \setminus K$, then the total length of L is less than $\pi\delta r$. See Figure 13.1.

Let $\omega = PI_B(\chi_L)$ be the Poisson integral of the characteristic function of L on B. Then ω is harmonic on B and $0 < \omega < 1$ on B. Because L is open in ∂B, $\omega(z) \to 1$ as $z \in B \to \eta \in L$, by the remark after the proof of Schwarz's Theorem 7.5. Because b is continuous and $b < 0$ on $\Omega \cap B(\zeta_0, \varepsilon)$, and because $K \subset \Omega \cap B(\zeta_0, \varepsilon)$ is compact, there exists $-m < 0$ so that $b \leq -m$ on K. So if $\eta \in K \cup L$ then

$$\limsup_{z \in B \to \eta} \left(\frac{b(z)}{m} - \omega(z) \right) \leq -1. \tag{13.3}$$

Set

$$b_\delta(z) = \begin{cases} \max\left(2\left(\dfrac{b(z)}{m} - \omega(z)\right), -1\right), & \text{on } \Omega \cap B \\ -1, & \text{on } \Omega \setminus B. \end{cases}$$

By (13.3), $b_\delta \equiv -1$ in a neighborhood of $K \cup L$ in Ω. Because $2(\frac{b}{m} - \omega)$ is subharmonic on $B \cap \Omega$, (13.3) and property (b) in Section 13.1 show that b_δ is subharmonic in Ω. Moreover, $b_\delta < 0$ on Ω. Finally, note that $\omega(\zeta_0) = \int_L |dz|/(2\pi r) = |L|/(2\pi r) < \delta/2$. Thus $\liminf_{z \in \Omega \to \zeta_0} b_\delta(z) = -2\omega(\zeta_0) \geq -2\delta/2 = -\delta$. □

Proof of Theorem 13.7 Choose $\varepsilon > 0$ so that $|f(\zeta) - f(\zeta_0)| < \delta$ for $\zeta \in \overline{B} \cap \partial\Omega$, where $B = B(\zeta_0, \varepsilon)$. Let b_δ be the function produced in Bouligand's Lemma 13.10. The idea is to build a subharmonic function which is approximately $f(\zeta_0)$ near ζ_0, but smaller than f away from ζ_0, using b_δ, which is small near ζ_0 and large negative away from ζ_0. Set

$$v(z) = f(\zeta_0) - \delta + (M + f(\zeta_0))b_\delta(z).$$

Then v is subharmonic on Ω and for $\zeta \in \partial\Omega \cap \overline{B}$

$$\limsup_{z \in \Omega \to \zeta} v(z) \le f(\zeta_0) - \delta \le f(\zeta),$$

because $M + f(\zeta_0) \ge 0$ and $b_\delta < 0$. Also if $\zeta \in \partial\Omega \setminus \overline{B}$ then

$$\limsup_{z \in \Omega \to \zeta} v(z) = f(\zeta_0) - \delta - (M + f(\zeta_0)) \le f(\zeta),$$

because $b_\delta = -1$ on $\Omega \setminus B$. Thus $v \in \mathcal{F}_f$ by (13.1). By Lemma 13.10(iii),

$$\liminf_{z \in \Omega \to \zeta_0} u_f(z) \ge \liminf_{z \in \Omega \to \zeta_0} v(z) \ge f(\zeta_0) - \delta + (M + f(\zeta_0))(-\delta).$$

Since $\delta > 0$ is arbitrary, $\liminf_{z \in \Omega \to \zeta_0} u_f(z) \ge f(\zeta_0)$. Replacing f with $-f$, we also have that $\liminf_{z \to \zeta_0} u_{-f}(z) \ge -f(\zeta_0)$. By Lemma 13.9,

$$\limsup_{z \in \Omega \to \zeta_0} u_f(z) \le \limsup_{z \in \Omega \to \zeta_0} -u_{-f}(z) = -\liminf_{z \in \Omega \to \zeta_0} u_{-f}(z) \le f(\zeta_0).$$

This proves Theorem 13.7. □

13.3 Riemann Mapping Theorem Again

One application of Theorems 13.6 and 13.7 is a proof of the Riemann mapping theorem that avoids the use of normal families as well as the geodesic algorithm.

Theorem 13.11 (Riemann mapping theorem) *If $\Omega \subset \mathbb{C}$ is a simply-connected region not equal to all of $\mathbb{C}$, then there exists a conformal map φ of Ω onto $\mathbb{D}$.*

As mentioned before, we adopt the common phrase "conformal map" to mean a one-to-one analytic function. See Proposition 6.5 for the uniqueness portion of the Riemann mapping theorem.

Proof Repeating the first paragraph of the proof of Theorem 8.11, using LFTs and a square root, we may suppose Ω is bounded and $0 \in \Omega$. Then $f(\zeta) = \log|\zeta| \in C(\partial\Omega)$. Let u_f be the Perron solution to the Dirichlet problem for the boundary function f. By Theorems 13.6 and 13.7, u_f is harmonic on Ω and extends to be continuous on $\overline{\Omega}$ and equal to f on $\partial\Omega$. By Theorem 7.10, there is an analytic function g on Ω such that $\mathrm{Re} g = u_f$. Set

$$\varphi(z) = ze^{-g(z)}.$$

Then φ is analytic in Ω, $\varphi(0) = 0$ and $\varphi(z) \ne 0$ if $z \ne 0$. Moreover, $|\varphi(z)| = e^{\log|z| - u_f(z)} \to 1$ as $z \in \Omega \to \partial\Omega$. Thus φ maps Ω into $\mathbb{D}$ by the maximum principle.

Fix $\varepsilon > 0$ and let $K_\varepsilon = \varphi^{-1}(|w| \le 1 - \varepsilon)$. Because $|\varphi| \to 1$ as $z \to \partial\Omega$, the set K_ε is a compact subset of Ω. As in the proof of Runge's Theorem 4.23, we can construct a closed curve $\gamma \subset \Omega$ which winds once around each point of K_ε. The winding number $n(\varphi(\gamma), w)$ is constant in each component of $\mathbb{C} \setminus \varphi(\gamma)$ and $|\varphi| > 1 - \varepsilon$ on γ so that if $|w| < 1 - \varepsilon$ then

$n(\varphi(\gamma), w) = n(\varphi(\gamma), 0)$. By the argument principle, the number of zeros of $\varphi - w$ must equal the number of zeros of φ. But by construction $\varphi = 0$ only at one point, namely 0. Letting $\varepsilon \to 0$, we conclude that each value in $|w| < 1$ is attained exactly once. Thus φ maps Ω one-to-one and onto $\mathbb{D}$. □

The function $u_f(z) - \log|z| = -\log|\varphi(z)|$ constructed in the proof of Theorem 13.11 is called **Green's function**. See Exercise 13.5.

Another application of Theorems 13.6 and 13.7 is the doubly-connected version of the Riemann mapping theorem. A region Ω is **doubly-connected** if $\mathbb{C}^* \setminus \Omega = E_1 \cup E_2$, where E_1 and E_2 are disjoint, connected and closed in $\mathbb{C}^*$.

Theorem 13.12 *If Ω is doubly-connected then there is a conformal map f of Ω onto an annulus $A = \{z : r_1 < |z| < r_2\}$, for some $0 \le r_1 < r_2 \le \infty$.*

See Exercise 8.14 for a corresponding uniqueness statement.

Proof Write $\mathbb{C}^* \setminus \Omega = E_1 \cup E_2$, where E_1 and E_2 are disjoint, connected and closed in $\mathbb{C}^*$. If E_1 and E_2 each consist of one point, then an LFT will map Ω onto A with $r_1 = 0$ and $r_2 = \infty$. If E_1 contains more than one point and E_2 consists of a single point, then, by the Riemann mapping theorem, there exists a conformal map φ of the complement of E_1 in $\mathbb{C}^*$ onto the unit disk with $\varphi(E_2) = 0$. Thus $\varphi(\Omega) = A$ with $r_1 = 0$ and $r_2 = 1$.

If neither E_1 nor E_2 consist of a single point, then, by the Riemann mapping theorem, we can find a conformal map g_1 of $\mathbb{C}^* \setminus E_1$ onto $\mathbb{D}$. We can then find a conformal map g_2 of $\mathbb{C}^* \setminus g_1(E_2)$ onto $\mathbb{D}$. The image $g_2(\partial\mathbb{D})$ is then an analytic curve in $\mathbb{D}$. So we may suppose that $\partial\Omega = \partial\mathbb{D} \cup E_1$, where $E_1 \subset \mathbb{D}$ is an analytic Jordan curve.

By Theorems 13.6 and 13.7, we can find a harmonic function ω on Ω which is continuous on $\overline{\Omega}$ and equal to 1 on $\partial\mathbb{D}$ and equal to 0 on E_1. By the Schwarz reflection principle applied to $\omega - 1$ and ω after a conformal map, ω extends to be harmonic in a neighborhood of $\overline{\Omega}$. Let L be a line segment in Ω connecting E_1 to $\partial\mathbb{D}$. Then $\Omega \setminus L$ is simply connected so that $\omega = \text{Re}g$ for some g analytic on $\Omega \setminus L$. Moreover, $g'(z) = \omega_x - i\omega_y$ is analytic on $\overline{\Omega}$. By Cauchy's theorem

$$\int_\gamma g'(z)dz = \int_{\partial\mathbb{D}} g'(z)dz \equiv ic.$$

for all curves $\gamma \subset \Omega$ with $\gamma \sim \partial\mathbb{D}$. We claim $c > 0$. If so, then $\text{Im}2\pi g/c$ increases by 2π as z traces γ and thus $f(z) = e^{2\pi g/c}$ is continuous in Ω and analytic on $\Omega \setminus L$. By Morera's theorem, f is analytic on Ω. Moreover f'/f is analytic on $\overline{\Omega}$ and hence so is f. Set $A = \{z : 1 < |z| < e^{2\pi/c}\}$. Then $f(\partial\Omega) \subset \partial A$ because $|f| = e^{2\pi\omega/c}$. If $\partial\Omega$ is given the usual positive orientation with respect to Ω, then the cycle $f(\partial\Omega)$ winds exactly once around each $z \in A$ and does not wind around any point $\mathbb{C} \setminus \overline{A}$. By the argument principle, f is a conformal map of Ω onto A.

To see that $c > 0$, note that

$$ic = i\int_{\partial\mathbb{D}} (\omega_x\, dy - \omega_y\, dx) = i\int_{\partial\mathbb{D}} \omega_r\, d\theta.$$

But $0 < \omega < 1$ on Ω and $\omega = 1$ on $\partial\mathbb{D}$ so $\omega_r \geq 0$ on $\partial\mathbb{D}$. If $\int_{\partial\mathbb{D}} \omega_r \, d\theta = 0$, then $\omega_x - i\omega_y \equiv 0$ on $\partial\mathbb{D}$ because ω_θ also $= 0$. By the uniqueness theorem, $\omega_x - i\omega_y \equiv 0$ in Ω which is impossible. This proves $c > 0$ and Theorem 13.12 follows. □

It is known that if $\Omega \subset \mathbb{C}$ is a region whose complement consists of finitely many disjoint compact sets then there is a conformal map of Ω onto a region whose boundary consists of finitely many circles and points. It is an interesting open problem, known as the **Koebe conjecture**, to prove that every region in the plane is conformally equivalent to a region bounded by circles and points.

13.4 Exercises

A

13.1 Prove items (a)–(c) at the beginning of Section 13.1.

13.2 Prove the following partial converse to Theorem 13.6. Suppose $\Omega \subset \mathbb{C}^*$ is a region with $\zeta_0 \in \partial\Omega$, $\zeta_0 \neq \infty$. Set $f(\zeta) = \max\{-|\zeta - \zeta_0|, -1\}$ for $\zeta \in \partial\Omega$. Then $f \in C(\partial\Omega)$. Let u_f be the Perron solution to the Dirichlet problem on Ω with boundary function f. If $\lim_{z\in\Omega\to\zeta_0} u_f(z) = 0$ and if $u_f \not\equiv 0$ then u_f is a local barrier for Ω at ζ_0.

B

13.3 Prove the following alternative way to solve the Dirichlet problem on an annulus $A = \{z : 0 < r < |z| < R < \infty\}$. If f is bounded on ∂A, set $g(z) = f(e^z)$ for z on the boundary of a vertical strip. Let u_1 be the solution to the Dirichlet problem in the simply-connected strip with boundary values g. This can be accomplished by using a conformal map and the Poisson integral formula. Prove $u(z) = u_1(\log z)$ is well defined and harmonic on A. Show that if f is continuous at $\zeta \in \partial A$ then u extends to be continuous and equal to f at ζ.

13.4 Prove that if f is analytic on Ω with $f(\Omega) \subset \Omega_1$ and if v is subharmonic on Ω_1 then $v \circ f$ is subharmonic on Ω. Note that subharmonic functions are not necessarily differentiable so we cannot use the Laplacian to prove this. Hint: First assume f is one-to-one and solve a Dirichlet problem. Then try it for $f(z) = z^m$ on $|z| < r$.

13.5 Suppose $\Omega \subset \mathbb{C}$ is a bounded region and suppose $z_0 \in \Omega$. Then $f(\zeta) = \log|\zeta - z_0| \in C(\partial\Omega)$. Let u_f be the Perron solution to the Dirichlet problem on Ω with boundary function f. If $u_f \not\equiv +\infty$ then $g(z, z_0) = u_f(z) - \log|z - z_0|$ is called **Green's function** on Ω with pole at z_0. The point z_0 is not really a pole in the usual sense, but by (b) below $\lim_{z\to z_0} g(z, z_0) = +\infty$, so the graph of $g(z, z_0)$ is tangent to a vertical line in $\mathbb{R}^3$. If $u_f \equiv +\infty$ we say that Green's function does not exist. If Green's function exists, prove the following:

(a) $g(z, z_0)$ is harmonic in $\Omega \setminus \{z_0\}$;
(b) $g(z, z_0) + \log|z - z_0|$ is bounded in a neighborhood of z_0;
(c) $g(z, z_0) > 0$ for $z \in \Omega$;
(d) $\zeta_0 \in \partial\Omega$ is a regular point if and only if $\lim_{z\to\zeta_0} g(z, z_0) = 0$;

(e) if Ω is regular, and if h satisfies (a), (b), mutatis mutandis, and $\lim_{z\to\partial\Omega} h(z) = 0$, then $h(z) = g(z, z_0)$;

(f) $g_x - ig_y$ is a meromorphic function on Ω with a simple pole with residue $= 1$ at z_0 and no other poles;

(g) if we remove the hypothesis that Ω is bounded then f may no longer be bounded on $\partial\Omega$. As in the proof of the Riemann mapping theorem, if $\partial\Omega$ contains a connected set which is not a single point then we can conformally map Ω onto a bounded region Ω_1 using LFTs and a square root. We can then use Green's function on Ω_1 to construct Green's function on Ω.

Green's function is probably the most important function in the study of harmonic functions on a region, if it exists.

13.6 (a) Suppose Ω is a region such that $\partial\Omega$ consists of finitely many analytic curves. Let $g(z) = g(z, z_0)$ denote Green's function, as in Exercise 13.5. Prove that if f is analytic on Ω and continuous on the closure of Ω then

$$f(z_0) = -\frac{1}{2\pi i}\int_{\partial\Omega} f(\zeta)(g_x - ig_y)(\zeta)d\zeta.$$

(b) Prove that if $f \in C(\partial\Omega)$ and if Ω is bounded by finitely many analytic curves then define

$$u(z) = \int_{\partial\Omega} f(\zeta)\frac{\partial g(\zeta, z_0)}{\partial\eta}\frac{|d\zeta|}{2\pi},$$

where $\eta(\zeta)$ is the unit inner normal to $\partial\Omega$ at ζ. Then u is harmonic in Ω, continuous in $\overline{\Omega}$ and $u = f$ on $\partial\Omega$. Hints: Let u_f be the Perron solution to the Dirichlet problem with boundary function f. Extend g to be harmonic in a neighborhood of $\partial\Omega$ and apply Exercise 7.5(a) to g and u_f on the region $\{z : g(z, z_0) - \varepsilon > 0 \text{ and } |z - z_0| > \varepsilon\}$ then let $\varepsilon \to 0$.

(c) Prove that if Ω is a region bounded by continuously differentiable Jordan curves such that $|\nabla g|$ is continuous on $\overline{\Omega}$ then the conclusion in (a) still holds.

(d) Show that if $\Omega = \mathbb{D}$ then $\frac{1}{2\pi}\frac{\partial g(\zeta, z_0)}{\partial\eta}$ is the Poisson kernel in $\mathbb{D}$ for z_0. For this reason the normal derivative of g is sometimes called the **Poisson kernel on Ω**.

C

13.7 Prove that if Green's function $g(z, z_0)$ exists on Ω then $g(z, z_1)$ exists, for $z_0, z_1 \in \Omega$. Moreover, $g(z_0, z_1) = g(z_1, z_0)$. Hint: See Exercise 7.5(a).

14 Riemann Surfaces

14.1 Analytic Continuation and Monodromy

The root test tells us that each convergent power series $\sum_0^\infty a_n(z-b)^n$ has an open disk centered at b as its natural region of definition. The series converges in the disk and diverges outside the closed disk. We also found that the series is analytic on its disk of convergence, in the sense that, given any other point c in the disk, the power series can be rearranged to be a series converging in a disk centered at c. However, it is possible that the largest disk of convergence centered at c also contains points outside the disk centered at b. This allows us to extend our original series as an analytic function on a larger region. For example, the series $\sum_0^\infty z^n$ converges in the unit disk, but the natural region of definition as an analytic function is $\mathbb{C} \setminus \{1\}$, because it agrees with the function $1/(1-z)$ on the unit disk. Though the process would be laborious, this function can be "discovered" by continually rearranging power series on more and more disks. This is the idea behind analytic continuation.

Suppose U_1 and U_2 are regions such that $U_1 \cap U_2$ is connected. If f_j is analytic on U_j, $j = 1, 2$, and if $f_1 = f_2$ on an infinite set with a cluster point in $U_1 \cap U_2$, then by the uniqueness theorem $f_1 = f_2$ on $U_1 \cap U_2$. Thus

$$f_3(z) = \begin{cases} f_1(z), & \text{if } z \in U_1 \\ f_2(z), & \text{if } z \in U_2 \end{cases} \tag{14.1}$$

is a well-defined analytic function on $U_1 \cup U_2$. In this way, f_2 provides an analytic extension of f_1 on U_1 to f_3 on $U_1 \cup U_2$.

This process can be repeated.

Definition 14.1 *Suppose f_j is analytic on U_j, $j = 0, \ldots, n$, such that $f_j = f_{j+1}$ on $U_j \cap U_{j+1}$, $j = 0, \ldots, n-1$. Then f_n is called a* **direct analytic continuation** *of f_0 to U_n.*

By (14.1) we obtain analytic functions on $U_j \cup U_{j+1}$, but not necessarily an analytic function on all of $\cup_j U_j$. For example, we can define $\log z$ on a sequence of disks whose union contains the unit circle, giving analytic continuations around the unit circle, but there cannot be an analytic function in a neighborhood of the unit circle which equals the definition of $\log z$ on each disk. The difficulty is that the continuation around the circle increases or decreases by $2\pi i$, depending on the direction. The continuation of f_0 on U_0 to f_n on U_n depends not only on the final region U_n, but also on how the intermediary regions U_j are chosen. One way to treat this difficulty is to define analytic continuation along a curve.

Definition 14.2 *If $\gamma : [0,1] \to \mathbb{C}$ is a curve and if f_0 is analytic in a neighborhood of $\gamma(0)$, then an* **analytic continuation of f_0 along γ** *is a finite sequence $f_1, \ldots, f_n$ of functions where $0 = t_0 < t_1 < \cdots < t_{n+1} = 1$ is a partition of $[0,1]$ and f_j is defined and analytic in a neighborhood of $\gamma([t_j, t_{j+1}])$, $j = 0, \ldots, n$, such that $f_j = f_{j+1}$ in a neighborhood of $\gamma(t_{j+1})$, $j = 0, \ldots, n-1$.*

Lemma 4.31 and Theorem 4.32 are sometimes useful for constructing analytic continuations. The function $\log(z)$ can be analytically continued along all curves γ in $\mathbb{C} \setminus \{0\}$ using

$$\int_\gamma \frac{1}{z} dz \tag{14.2}$$

by subdividing γ into arcs, as in Theorem 4.32. But, as mentioned above, the analytic continuation does not agree with an analytic function defined on all of $\mathbb{C} \setminus \{0\}$.

Another way that an analytic continuation arises is the following. If u is harmonic in a region Ω, and if $D \subset \Omega$ is a disk, then we can find an analytic function f on D so that $\mathrm{Re} f = u$. For instance, we can use the Herglotz integral and a linear change of variables to define f as in Corollary 7.7. Moreover, any two analytic functions which have the same real part on a region differ by only a purely imaginary constant because the difference is not open. Fix a point $b \in \Omega$ and an analytic function f_0 defined on a disk D_0 containing b with $\mathrm{Re} f = u$. If γ is a curve in Ω with $\gamma(0) = b$, then by Lemma 4.31 we can cover γ with a finite union of disks D_j so that $\gamma([t_j, t_{j+1}]) \subset D_j$. If f_j is analytic with $\mathrm{Re} f_j = u$ on D_j, then, by adding an appropriate imaginary constant to each f_j, $j \geq 2$, we will create an analytic continuation of f along γ. Thus f can be continued along all curves in Ω. It is not always true, though, that $u = \mathrm{Re} f$ for some analytic function f on all of Ω.

A central question for this chapter is as follows. Suppose we can find an analytic continuation of f_0 along all curves in a region Ω. When can we find a single analytic function on all of Ω which agrees with each of these continuations?

Here are some observations:

(i) An analytic continuation along $\gamma : [0,1] \to \mathbb{C}$ determines a well-defined function on $[0,1]$, but it does not necessarily yield a well-defined function on (the image) $\gamma \subset \Omega$, as indicated by the logarithm example.

(ii) If $f_1, \ldots, f_n$ is an analytic continuation of f_0 then $f_{n-1}, \ldots, f_0$ is an analytic continuation of f_n.

(iii) If $f_1, \ldots, f_n$ is an analytic continuation of f_0 for the partition $0 = t_0 < t_1 \cdots < t_{n+1} = 1$, then we can refine the partition by choosing s with $t_j < s < t_{j+1}$ and using the function f_j on $\gamma([t_j, s])$ and on $\gamma([s, t_{j+1}])$. So if $g_1, \ldots, g_m$ is another analytic continuation along γ of f_0, then we can choose a common refinement so that the two sequences of analytic functions are defined on the same partition $0 = u_0 < u_1 < \cdots < u_{k+1} = 1$. But $f_1 = f_0 = g_1$ in a neighborhood of $\gamma(u_1)$, so by the uniqueness theorem $f_1 = g_1$ on a neighborhood of $\gamma([u_1, u_2])$, and by induction $g_j = f_j$ on a neighborhood of $\gamma([u_j, u_{j+1}])$. In this sense, analytic continuation along a curve is unique.

(iv) Suppose f is an analytic function on a region Ω. If $f_1, \dots, f_n$ is an analytic continuation of f_0 along a curve $\gamma \subset \Omega$, where f_0 is the restriction of f to a neighborhood of $\gamma(0)$, then by the uniqueness theorem and induction $f_j(\gamma(t)) = f(\gamma(t))$ for $t_j \le t \le t_{j+1}$, $0 \le j \le n$.

(v) Suppose that $f_1, f_2, \dots, f_n$ is an analytic continuation of f_0 along γ with partition $0 = t_0 < \cdots < t_{n+1} = 1$. We can choose $\varepsilon > 0$ so that if σ is another curve such that $|\sigma(t) - \gamma(t)| < \varepsilon$, for all $0 \le t \le 1$, then $f_1, f_2, \dots, f_n$ is an analytic continuation of f_0 along σ. Indeed, if $\varepsilon > 0$ is sufficiently small then f_j is defined and analytic in a neighborhood of $\sigma([t_j, t_{j+1}])$ and $f_j = f_{j+1}$ in a neighborhood of $\sigma(t_{j+1})$, $j = 0, \dots, n$.

To extend (v) to curves that are not uniformly close we introduce a concept similar to, but different than, homology. Suppose γ_0 and γ_1 are curves in a region Ω that begin at b and end at c. We say γ_0 is **homotopic in** Ω to γ_1 if there exists a collection of curves $\gamma_s : [0, 1] \to \Omega$, $0 < s < 1$, so that $\gamma_s(t)$, as a function of (t, s), is uniformly continuous on the closed unit square $[0, 1] \times [0, 1]$, with $\gamma_s(0) = b$ and $\gamma_s(1) = c$, $0 \le s \le 1$. If γ_0 is homotopic to γ_1, we write $\gamma_0 \approx \gamma_1$. See Figure 14.1.

The function $H(t, s) \equiv \gamma_s(t)$ is called a **homotopy in** Ω from γ_0 to γ_1. Note that $\gamma_1 \approx \gamma_0$ using γ_{1-s}, so we say simply that γ_0 and γ_1 are homotopic in Ω. It is not hard to prove transitivity: if $\gamma_1 \approx \gamma_2$ and $\gamma_2 \approx \gamma_3$ then $\gamma_1 \approx \gamma_3$. The reader is encouraged to do Exercise 14.3 at this point.

The next lemma says that sufficiently close curves are homotopic and homologous.

Lemma 14.3 *If $\gamma : [0, 1] \to \Omega$ is a curve in a region Ω then there is an $\varepsilon > 0$, depending on the region Ω and the curve γ, such that, if $\sigma : [0, 1] \to \Omega$ is a curve with $|\gamma(t) - \sigma(t)| < \varepsilon$ for all $t \in [0, 1]$, and $\sigma(0) = \gamma(0)$ and $\sigma(1) = \gamma(1)$, then $\sigma \approx \gamma$ and $\gamma - \sigma \sim 0$.*

Proof Choose disks $B_j \subset \Omega$ and a partition $0 = t_0 < t_1 \cdots < t_n < t_{n+1} = 1$ so that $\gamma_j = \gamma([t_j, t_{j+1}]) \subset B_j$. If $\varepsilon > 0$ is sufficiently small then $\sigma_j = \sigma([t_j, t_{j+1}]) \subset B_j$, for each j. Then

$$\gamma_s(t) = (1 - s)\gamma(t) + s\sigma(t)$$

is a homotopy in Ω from γ to σ.

Let $L_j \subset B_j \cap B_{j-1}$ denote the line segment from $\gamma(t_j)$ to $\sigma(t_j)$, and set $L_0 = \{\gamma(0)\}$ and $L_{n+1} = \{\gamma(1)\}$. Then $\alpha_j \equiv \gamma_j + L_{j+1} - \sigma_j - L_j$ is a closed curve contained in $B_j \subset \Omega$ and hence is homologous to 0 in Ω. Thus $\gamma - \sigma = \sum_{j=0}^{n} \alpha_j$ is also homologous to 0. □

Corollary 14.4 *If $\gamma_0 \approx \gamma_1$ in a region Ω then $\gamma_0 - \gamma_1 \sim 0$ in Ω.*

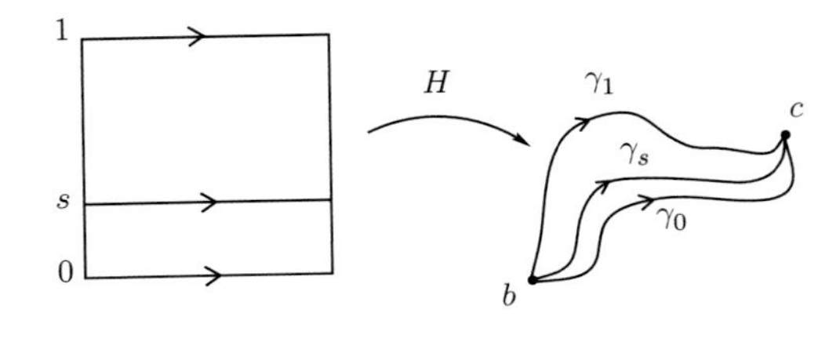

Figure 14.1 Homotopy of γ_0 to γ_1.

Proof If $\gamma_s(t)$ is a homotopy of γ_0 to γ_1 then we can cover $[0, 1]$ with finitely many open intervals J_k so that if $r, s \in J_k$ then $\gamma_r - \gamma_s \sim 0$ by Lemma 14.3. Thus $\gamma_0 - \gamma_1 \sim 0$ by transitivity. □

The converse to Corollary 14.4 fails. See Figure 5.4 and Exercise 14.5. (The reader may wish to defer doing Exercise 14.5 until after reading Example 14.13.) However, homotopy does give us another characterization of simple connectivity.

Theorem 14.5 *A region $\Omega \subset \mathbb{C}$ is simply-connected if and only if every closed curve contained in Ω is homotopic to a constant curve.*

Proof If γ is homotopic to a constant curve then $\gamma \sim 0$ by Corollary 14.4. So if all curves in Ω are homotopic to constant curves then, by Theorem 5.7, Ω is simply-connected. Conversely, if Ω is simply-connected and if γ is a closed curve in $\Omega \neq \mathbb{C}$ beginning and ending at z_0, and if f is a conformal map of Ω onto $\mathbb{D}$ with $f(z_0) = 0$, then $f(\gamma)$ is a closed curve in $\mathbb{D}$ beginning and ending at 0. But then $\gamma_s(t) = f^{-1}(sf(\gamma(t)))$ is a homotopy of the constant curve z_0 to γ_0. If $\Omega = \mathbb{C}$, we can use $z - z_0$ instead of f. □

If a closed curve γ is homotopic to a constant curve, then the constant must be $\gamma(0)$. Some books use the notation $\gamma \approx 0$.

Theorem 14.6 *Suppose $\gamma_s(t)$, $0 \leq s, t \leq 1$, is a homotopy from γ_0 to γ_1 in a region Ω. Suppose f_0 is analytic in a neighborhood of $b = \gamma_0(0) = \gamma_1(0)$ and suppose f_0 can be analytically continued along each γ_s. Then the analytic continuation of f_0 along γ_0 agrees with the analytic continuation of f_0 along γ_1 in a neighborhood of $c = \gamma_0(1) = \gamma_1(1)$.*

Proof By (iii), the analytic continuation of f_0 along each γ_s is unique, $0 \leq s \leq 1$. By (v), for each $s \in [0, 1]$, the analytic continuation of f_0 along γ_s agrees with the analytic continuation of f_0 along γ_u in a neighborhood U_s of c if $|u - s| < \varepsilon$ for some $\varepsilon = \varepsilon(s)$. By compactness, we can cover $[0, 1]$ with finitely many such open intervals $(s_j - \varepsilon_j, s_j + \varepsilon_j)$, for $1 \leq j \leq m$. Then the analytic continuations of f_0 along each γ_s agree on $\cap_{j=1}^m U_{s_j}$. □

Corollary 14.7 (monodromy theorem) *Suppose Ω is simply-connected and suppose f_0 is defined and analytic in a neighborhood of $b \in \Omega$. If f_0 can be analytically continued along all curves in Ω beginning at b then there is an analytic function f on Ω so that $f = f_0$ in a neighborhood of b.*

Proof If $c \in \Omega$ and if γ_0 is a curve in Ω from b to c, let f_n be the analytic continuation of f_0 along γ_0 to a neighborhood of c, and define $f(c) = f_n(c)$. If γ_0 and γ_1 are curves in Ω beginning at b and ending at c then $\gamma_0 \approx \gamma_1$ by Theorem 14.5 and Exercise 14.3(e). So, by Theorem 14.6 the definition of $f(c)$ does not depend on the choice of the curve γ_0. By (v), $f = f_n$ in a neighborhood of c, so that f is analytic at c. Thus f is defined and analytic in Ω and $f = f_0$ in a neighborhood of 0. □

The monodromy theorem can be used to give another proof of the second statement in Theorem 7.10, that a harmonic function u on a simply-connected region Ω is the real part of an analytic function. If f is analytic on a ball $B \subset \Omega$ with $\text{Re}f = u$ on B, then, as noted after Definition 14.2, f can be continued along all curves in Ω. By the monodromy theorem, because Ω is simply-connected, there is an analytic function f on all of Ω with $\text{Re}f = u$.

The monodromy theorem is typically used to find a global inverse of an analytic function with non-vanishing derivative. Suppose f is analytic in Ω with $f' \neq 0$ on Ω. If $c \in \Omega$, then there is a function g analytic in a neighborhood of $f(c)$ so that $g(f(z)) = z$ in a neighborhood of c. If g can be analytically continued along all curves in $f(\Omega)$, and if $f(\Omega)$ is simply-connected, then by the monodromy theorem there is a function G which is analytic on $f(\Omega)$ satisfying $G(f(z)) = z$ for $z \in \Omega$. See Exercises 14.6 and 14.7.

One final remark: analytic continuation really only depends upon the continuity of the functions and the uniqueness theorem on disks, so that the monodromy theorem holds for much more general classes of functions. For example, if two harmonic functions agree on a small disk in a region, then they agree on the entire region. So if we replace "analytic" with "harmonic" or "meromorphic" in our definition of continuation along a curve and in the statement of the monodromy theorem, the theorem remains true.

14.2 Riemann Surfaces and Universal Covers

Analytic continuation along curves arose in the process of trying to understand the inverse of an analytic function. Riemann had a more geometric resolution which is closer to the idea of direct analytic continuation. Instead of considering the function e^z, for example, we can consider the function $f(z) = (\text{Re}\, e^z, \text{Im}\, e^z, \text{Im}\, z) = (e^x \cos y, e^x \sin y, y)$ which maps $\mathbb{C}$ into $\mathbb{R}^3$. Each horizontal line in $\mathbb{C}$ is mapped to a half-line parallel to the $(x, y, 0)$ plane, whose height increases with $\text{Im}z$. The image $R = f(\mathbb{C})$ is an infinite "parking-lot" ramp that spirals around the line $L = \{(0, 0, t) : t \in \mathbb{R}\}$. (The line L is not in the image.) See Figure 14.2.

The function f is then one-to-one on $\mathbb{C}$ and its inverse maps the ramp R onto the plane. We can think of a direct analytic continuation of $\log z$ as determining regions on the ramp, which are mapped into the plane by this inverse map. The image of one "sheet" or "floor" of the ramp

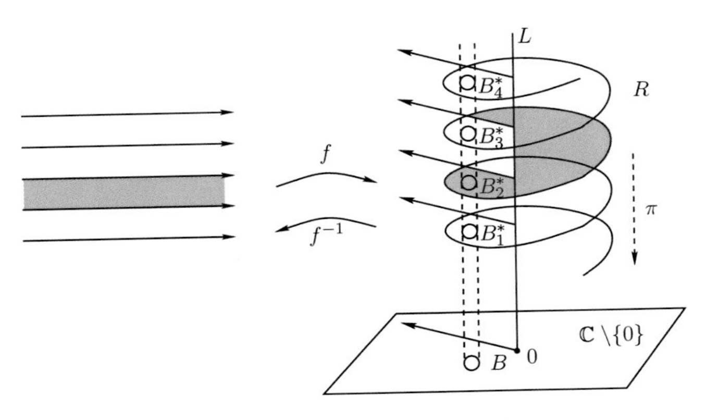

Figure 14.2 Parking-lot ramp surface.

is a horizontal strip of height 2π. A similar analysis for the function z^2 on $\mathbb{C} \setminus \{0\}$ does not quite work, but it is possible to use a map into $\mathbb{R}^4$ instead. Nevertheless, we informally think of many analytic functions with non-zero derivative as maps of regions into $\mathbb{R}^3$ where the image is somehow gradually lifted in the third dimension so that the map becomes one-to-one, and thus we can picture its inverse.

Definition 14.8 *A* **Riemann surface** *is a connected Hausdorff space W, together with a collection of open subsets $U_\alpha \subset W$ and functions $z_\alpha : U_\alpha \to \mathbb{C}$ such that*

(i) $W = \cup U_\alpha$,
(ii) *z_α is a homeomorphism of U_α onto the unit disk $\mathbb{D}$ and*
(iii) *if $U_\alpha \cap U_\beta \neq \emptyset$ then $z_\beta \circ z_\alpha^{-1}$ is analytic on $z_\alpha(U_\alpha \cap U_\beta)$.*

The functions z_α are called **coordinate functions or maps**, and the sets U_α are called **coordinate charts or disks**. We can extend our collection of coordinate functions and disks to form a base for the topology on W by setting $U_{\alpha,r} = z_\alpha^{-1}(r\mathbb{D})$ and $z_{\alpha,r} = \frac{1}{r} z_\alpha$ on $U_{\alpha,r}$, for $r < 1$. We can also compose each z_α with a Möbius transformation of the disk onto itself so that we can assume that for each coordinate disk U_α and each $p_0 \in U_\alpha$ there is a coordinate function z_α with $z_\alpha(p_0) = 0$. For the purposes of the following discussion, we assume that our collection $\{z_\alpha, U_\alpha\}$ includes all such maps. Note that a Riemann surface W is pathwise connected since the set of points that can be connected to p_0 is both open and closed for each $p_0 \in W$.

These concepts have been extended to more general surfaces called differentiable manifolds, where the functions $z_\beta \circ z_\alpha^{-1}$ are required only to be differentiable. However, in that generality, it is necessary also to assume that the surface is second countable. One consequence of the uniformization theorem, which we will prove in Chapter 15, is that all Riemann surfaces are second countable.

More informally, a Riemann surface is a set which locally looks like the plane. Many of the concepts introduced in this book depend only on the local behavior of functions and so can be defined on Riemann surfaces. A function $f : W \to \mathbb{C}$ is called analytic if, for every coordinate function z_α, the function $f \circ z_\alpha^{-1}$ is analytic on $\mathbb{D}$. Harmonic, subharmonic and meromorphic functions on W are defined in a similar way. This means that any theorem about regions in the plane whose proof depends solely on the local behavior of functions is also valid for Riemann surfaces. The key fact needed is that the result or concept should be invariant under the **transition maps** $z_\beta \circ z_\alpha^{-1}$, where z_α and z_β are coordinate maps. By (ii) and (iii), the transition maps are one-to-one and analytic. For example, the characterizations of zeros and poles of analytic functions is the same. The maximum principle can be stated in the following form: if an analytic or harmonic function achieves its maximum on a Riemann surface, then it is constant. Lemma 4.31 holds for Riemann surfaces where disks are replaced by coordinate charts, and can be used to create analytic continuations. Differentiation, however, presents a problem, since, if z_α is a coordinate map, the derivative of $f \circ z_\alpha^{-1}$ will depend on the choice of z_α. However, if both f and g are analytic on a Riemann surface then $f' \circ z_\alpha^{-1}(z)/g' \circ z_\alpha^{-1}(z)$ does not depend on the choice of z_α by the chain rule. Similarly, the property of (real) differentiability does not depend on the choice of coordinate map (though the individual derivatives do). Integration of continuous functions has a similar problem. We can transport an integral over a curve contained in a coordinate chart to the disk, but the result will depend on the choice of the coordinate function.

A Riemann surface does not necessarily lie in some Euclidean space, so we cannot speak of the complement of a Riemann surface or of its boundary. This requires us to use an alternate definition of some concepts such as "simply-connected." A Riemann surface W is called **simply-connected** if every closed curve in W is homotopic to a constant curve. Equivalently, any two curves in W that begin at the same point and end at the same point are homotopic (see Exercise 14.3). This definition agrees with our earlier definition of a simply-connected region in the plane by Theorem 14.5. The monodromy theorem, Corollary 14.7, holds for simply-connected Riemann surfaces because the proof only depends on the local properties of analytic functions and the fact that any two curves with the same initial and terminal points are homotopic. Consequently, every harmonic function on a simply-connected Riemann surface has the form $u = \text{Re}f$ for some analytic function f defined on W. The Perron process also works on a Riemann surface.

Example 14.9 (Riemann sphere) If π is the stereographic projection of $\mathbb{C}$ into the unit sphere $\mathbb{S}^2$ as given in Section 1.3, then coordinate charts on $\mathbb{S}^2$ are given by $\pi(B)$, where B is a disk in $\mathbb{C}$. The corresponding coordinate map is π^{-1} followed by a linear map of B onto $\mathbb{D}$. A coordinate chart containing the north pole is given by $\pi(|z| > r)$ together with the north pole. The corresponding coordinate map is given by π^{-1} followed by r/z, declaring the value of this map at the north pole to be 0.

Example 14.10 (Torus) A **torus** T can be constructed from a rectangle $R = \{(x, y) : 0 < x < a, 0 < y < b\}$ by identifying points on opposite sides of the boundary. We can define an equivalence relation on $\mathbb{C}$ by $z \sim w$ if and only if $z - w = ka + inb$ for some integers k, n. The quotient map $\pi : \mathbb{C} \to \mathbb{C}/\sim$ associated with this equivalence relation identifies the opposite sides of R. We can define charts on the quotient space by $\pi(B)$, where B is a Euclidean ball. The corresponding coordinate maps are given by π^{-1} followed by a linear map of B onto $\mathbb{D}$. It is an exercise to check that the quotient space is a Riemann surface. The torus can be visualized by first forming a cylinder by bending the rectangle in $\mathbb{R}^3$ until the top and bottom edges meet so that the vertical sides of R become circles. The cylinder is then bent until the two circles meet so that the line corresponding to the top and bottom edges of R becomes a circle. It looks like a donut.

In terms of the rectangle R, a small ball centered at a point in the interior of R corresponds to a basic open set in T. The image by the map π of a small disk centered at ic, with $0 < c < b$, is the same as the image of two half-disks in R, one centered at ic and the other centered at $a + ic$, together with the vertical line segment on the boundary of each (which is identified to a single line segment in T). A similar description holds for the open horizontal lines in R. The image of a disk centered at a corner of R is also the image of four quarter-circles in R. These sets in T form the charts on the torus constructed from R, and the chart maps are linear on each portion in R.

Example 14.11 (Riemann surface of an analytic function) If f is analytic on a region $\Omega \subset \mathbb{C}$ with $f' \neq 0$ on Ω, then we can construct a Riemann surface Ω_f associated with f by declaring charts to be the images $f(B(z, r))$ of disks $B(z, r)$ on which f is one-to-one. The associated chart maps are f^{-1} composed with a linear map of $B(z, r)$ onto $\mathbb{D}$. More formally, we write $\Omega = \cup_{j=1}^{\infty} B_j$, where f is one-to-one on each B_j. Set $U = \coprod_{j=1}^{\infty} f(B_j)$, the disjoint union of the sets $f(B_j)$. We then identify $w \in f(B_i)$ and $w \in f(B_j)$ if and only if $w = f(z)$ for some $z \in B_i \cap B_j$. In other words, we identify the copies of $f(B_i \cap B_j)$ in the two images $f(B_i)$ and $f(B_j)$. The corresponding quotient space Ω_f is a Riemann surface. The function f can be viewed as a one-to-one map of Ω onto Ω_f, and f^{-1} becomes a well-defined function on Ω_f. See Exercise 14.4 for the extension of this construction to functions whose derivative vanishes at points of Ω.

The idea behind analytic continuation can be used to create what is called a "covering surface" of a region contained in the plane. Returning to the parking-lot ramp example, Figure 14.2, if B is a disk in $\mathbb{C} \setminus \{0\}$ then the vertical cylinder $\underline{B} \times \mathbb{R}$ cuts out countably many topological disks B_j^* from the ramp. Each B_j^* is mapped onto B by the vertical projection $\pi : B_j^* \to B$. If B has center c and radius r, then $z_{B_j^*} = (\pi - c)/r$ maps B_j^* homeomorphically onto $\mathbb{D}$. Then $\{(z_{B_j^*}, B_j^*) : -\infty < j < \infty\}$ are coordinate maps and charts that make R into a Riemann surface. Let $p = (1, 0) \in \mathbb{C}$ and $p^* = (1, 0, 0) \in R$, so that $\pi(p^*) = p$. Then we can find the center of B_j^*, $c_j^* \in \pi^{-1}(c)$, by following a path $\gamma^* \subset R$ starting at p^*. The vertical projection $\gamma = \pi(\gamma^*)$ is a curve in $\mathbb{C} \setminus \{0\}$ from p to c. Note that if σ^* is another curve from p^* to the center of B_j^* then $\gamma^*(\sigma^*)^{-1}$ is a closed curve on the ramp. See Exercise 14.3 for the definition of the product of two curves (the first **followed by** the second). The surface R is homeomorphic to $\mathbb{C}$ by the map $f(z)$ so that any closed curve in R is homotopic to a constant curve. Indeed, a homotopy in $\mathbb{C}$ transplants to a homotopy in R using the homeomorphism f. Thus $\gamma^* \approx \sigma^*$. Applying the projection map π to the homotopy, we obtain $\gamma \approx \sigma$, where $\sigma = \pi(\sigma^*)$. In fact, a curve α is homotopic to γ in $\mathbb{C} \setminus \{0\}$ if and only if we can find a curve $\alpha^* \subset R$ which begins and ends at the same places as γ^* with $\pi(\alpha^*) = \alpha$. So we can make an identification between the equivalence classes of curves and points of the ramp R.

We can use this idea to create a Riemann surface that "covers" a region $\Omega \subset \mathbb{C}$. Fix a point $b \in \Omega$. Let $[\gamma]$ be the equivalence class under homotopy of a curve $\gamma \subset \Omega$. Let Ω^* be the collection of equivalence classes

$$\Omega^* = \{[\gamma] : \gamma \text{ is a curve in } \Omega \text{ with } \gamma(0) = b\}.$$

Define a projection map $\pi : \Omega^* \to \Omega$ by $\pi([\gamma^*]) = \gamma(1)$. We can make Ω^* into a Riemann surface by describing coordinate maps and charts. If $c \in \Omega$, let $B = B(c, r)$ be a disk in Ω centered at c with radius r. Let γ be a curve in Ω from b to c. For any point $d \in B$, let σ_d be a curve in B from c to d. Let

$$B^* = \{[\gamma\sigma_d] : d \in B\} \tag{14.3}$$

be the equivalence classes of all curves $\gamma\sigma_d$, for all $d \in B$. Since all curves in B beginning at c and ending at d are homotopic, $[\gamma\sigma_d]$ does not depend on the choice of σ_d. By the definition of π, $\pi([\gamma\sigma_d]) = d$ so that π is a one-to-one map of B^* onto B. Note that if $\gamma_1 \approx \gamma$, then we obtain the same set of equivalence classes B^* using γ_1 instead of γ. See Exercise 14.3(c). Thus $B^* \subset \pi^{-1}(B)$ is uniquely determined by the equivalence class of γ, namely a point in B^*, and the disk B. Set $z_{B^*} = (\pi - c)/r$. Then we can give a topology on Ω^* by declaring each set B^* to be open, for all disks $B(c, r) \subset \Omega$ and all equivalence classes $[\gamma]$ of curves $\gamma \subset \Omega$

from b to c. Indeed, the "disks" B^* form a basis for this topology. Equivalently, we give Ω^* the topology required to make each z_{B^*} a homeomorphism.

Theorem 14.12 *Suppose $\Omega \subset \mathbb{C}$ is a region. Then*

(i) *the surface Ω^* is a simply-connected Riemann surface with coordinate functions z_{B^*} and charts B^*;*
(ii) *if B_1^* and B_2^* are coordinate charts with $\pi(B_1^*) = \pi(B_2^*)$ then either $B_1^* \cap B_2^* = \emptyset$ or $B_1^* = B_2^*$;*
(iii) *if $\gamma \subset \Omega$ is a curve beginning at b and if $b^* \in \Omega^*$ with $\pi(b^*) = b$, then there is a unique curve $\gamma^* \subset \Omega^*$, called a* **lift** *of γ, beginning at b^* with $\pi(\gamma^*) = \gamma$;*
(iv) *a curve $\gamma^* \subset \Omega^*$ is closed if and only if $\gamma = \pi(\gamma^*)$ is homotopic to a constant curve in Ω.*

Because Theorem 14.12(ii) holds, we say that π **evenly covers** Ω. It says that each $z \in \Omega$ is contained in a disk B such that $\pi^{-1}(B)$ is the disjoint union of sets homeomorphic to B by the map π. An analytic function f with $f' \neq 0$ on a region W does not necessarily evenly cover $f(W)$.

Proof By construction, Definitions 14.8(i) and (ii) hold. The transition functions are just linear maps and hence analytic. The set Ω^* is pathwise connected, for if $[\gamma] \in \Omega^*$, where $\gamma : [0,1] \to \Omega$, let $\gamma^x(t) = \gamma(xt)$, for $0 \leq x, t \leq 1$. Then $\gamma^x : [0,1] \to \Omega$, and it is not hard to verify that $\gamma^*(x) \equiv [\gamma^x]$ defines a continuous curve in Ω^* from $[\{b\}]$ to $[\gamma]$, where $[\{b\}]$ is the equivalence class containing the constant curve $\{b\}$. To see that Ω^* is Hausdorff, suppose $p^* = [\gamma_p], q^* = [\gamma_q] \in \Omega^*$, where γ_p is a curve from b to $p = \pi(p^*) \in \Omega$ and γ_q is a curve in Ω from b to $q = \pi(q^*) \in \Omega$. Suppose $p^* \neq q^*$. If $p \neq q$, choose disjoint disks B_p and B_q centered at p and q, and form the "disks" B_p^* and B_q^* as in (14.3). Then $B_p^* \cap B_q^* = \emptyset$ since $\pi(B_p^*) \cap \pi(B_q^*) = \emptyset$. If $p = q$, then let $B \subset \Omega$ be a disk centered at p. Form the sets B_p^* and B_q^* using the curves γ_p and γ_q as in (14.3). Since $p^* \neq q^*$, the curves γ_p and γ_q are not homotopic, though they begin at b and end at $p = q$. If $r^* \in B_p^* \cap B_q^*$ then $r^* = [\gamma_p\sigma_d]$ for some curve $\sigma_d \subset B$, and similarly $r^* = [\gamma_q\sigma_e]$, where $\sigma_e \subset B$. But this implies $\gamma_p\sigma_d \approx \gamma_q\sigma_e$. But σ_d and σ_e are contained in the simply-connected region B; they begin at $p = q$ and end at the same point in B so that $\sigma_d^{-1} \approx \sigma_e^{-1}$. By Exercise 14.3,

$$\gamma_p \approx \gamma_p\sigma_d\sigma_d^{-1} \approx \gamma_q\sigma_e\sigma_e^{-1} \approx \gamma_q.$$

This contradicts $p^* \neq q^*$, and so $B_p^* \cap B_q^* = \emptyset$. Thus Ω^* is Hausdorff and we have proved Ω^* is a Riemann surface. This also proves (ii).

Now suppose $\gamma : [0,1] \to \Omega$ is a curve beginning at b, and suppose $b^* = [\sigma] \in \Omega^*$ with $\gamma(0) = b = \pi(b^*) = \sigma(1)$. If γ^x is the portion of γ from $\gamma(0)$ to $\gamma(x)$ then $\gamma^*(x) \equiv [\sigma\gamma^x]$, $0 \leq x \leq 1$, defines a lift of γ to Ω^*. Let B^* be a coordinate disk containing b^*. Because π is a homeomorphism of B^* onto B, if $\gamma^x \subset B$ then there is only one lift of γ^x from Ω to Ω^* beginning at b^*. Cover γ by finitely many balls B_j. If γ^x has a unique lift $(\gamma^x)^*$ beginning at b^*, and if $\alpha_j = \gamma([x,y]) \subset B_j$, then α_j has a unique lift beginning at $(\gamma^x)^*(1)$. Repeating this lifting finitely many times proves that the lift of γ to the curve γ^* is the unique lift beginning at b^*, proving (iii).

Suppose $\gamma^* \subset \Omega^*$ is a closed curve. Then $\gamma = \pi(\gamma^*)$ is a closed curve in Ω. If $[\sigma] = \gamma^*(0)$, then, as above, $\gamma^*(x) = [\sigma\gamma^x]$. Thus $\gamma^*(1) = [\sigma\gamma]$. Since $\gamma^*(0) = \gamma^*(1)$, we have that $[\sigma] = [\sigma\gamma]$. It follows that γ is homotopic to the constant curve $\{\sigma(1) = \gamma(0)\}$.

Conversely, if $\gamma_s(t)$ is a homotopy of the constant curve $\{\gamma(0)\}$ to γ then $[\sigma\gamma_s] = [\sigma]$ for all $s \in [0, 1]$. The lift γ_s^* of $\sigma\gamma_s$ to a curve in Ω^* beginning at $[\sigma]$ ends at $[\sigma\gamma_s]$ and thus is a closed curve. This proves (iv), but also it shows that if γ^* is a closed curve in Ω^* then γ_s^* gives a homotopy of the constant curve $\{\gamma^*(0)\}$ to γ^*. Thus Ω^* is simply-connected, completing the proof of Theorem 14.12. □

We remark that the same construction works if we replace Ω with any Riemann surface W, since it depends only on the local structure of the surface. In this case, disks in Ω are simply replaced by coordinate charts. The constructed surface W^* is called a **universal covering surface** of W and π is called a **universal covering map**. If z_α is a coordinate function on coordinate chart $U_\alpha \subset W$ then $z_\alpha \circ \pi$ is a corresponding coordinate function on U_α^*.

The parking-lot ramp example, Figure 14.2, can be thought of as pasting a copy of the upper half-plane to the lower half-plane along the positive reals. Then another copy of the lower half-plane is pasted to the previous copy of the upper half-plane along the negative reals. This process of pasting half-planes along half-lines is repeated as much as possible. We can make a similar construction for the twice-punctured plane.

Example 14.13 (Covering surface of $\mathbb{C} \setminus \{0, 1\}$) Suppose $\Omega = \mathbb{C} \setminus \{0, 1\}$. Let I_1, I_2, I_3 be the three open intervals in $\mathbb{R} \setminus \{0, 1\}$. We can create the universal cover and covering map for Ω as follows. Let $\mathbb{H}_u = \{z : \text{Im}z > 0\}$ be the upper half-plane and let $\mathbb{H}_l = \{z : \text{Im}z < 0\}$ be the lower half-plane. Let $A = \mathbb{H}_u \cup_j I_j$ and let $B = \mathbb{H}_l \cup_j I_j$. Attach one copy of A to one copy of B by identifying the interval I_1 on A with the interval I_1 on B. The resulting set has unidentified intervals on A and B. For each such interval on A (respectively B) attach a new copy of B (respectively A) along the corresponding interval. Repeat this process with all unidentified intervals on all copies of A and B resulting in new unidentified intervals. Repeat the process indefinitely. The maximal such surface S^* has all edges I_j on every copy of A (respectively B) identified with a corresponding I_j on a copy of B (respectively A). Let π be the projection which maps a point on a copy of A or B to the corresponding point in Ω. If $D = \{z : |z - c| < r\}$ is a disk in Ω then each component D^* of $\pi^{-1}(D)$ lies in the union of a copy of A and a copy of B. The corresponding chart map associated with a chart D^* is $(\pi - c)/r$. It is not hard to verify that S^* is then a Riemann surface.

If γ is a curve in $\mathbb{C} \setminus \{0, 1\}$, then we can "lift" γ to S^* by passing from one half-sheet, a copy of A or B, to another as γ crosses an interval $I_j \in \mathbb{R} \setminus \{0, 1\}$. If $\gamma \subset \Omega$ is a curve beginning, say, at a point $\gamma(0) \in \Omega$ then its homotopy class is determined by which half-sheet it ends up on (and the point $\gamma(1)$). The curve γ is homotopic to a constant curve if and only if its lift to S^* is closed. This gives a one-to-one map of the universal covering surface of $\mathbb{C} \setminus \{0, 1\}$ onto S^*. We will construct another version of this surface in Section 15.1.

This same construction can be done for $\mathbb{C} \setminus E$, where E is a compact subset of $\mathbb{R}$, where we identify half-planes along the open intervals in $\mathbb{R}$ complementary to E. If $E = \{0\}$ then the corresponding S^* is homeomorphic to the parking-lot ramp constructed earlier.

14.3 Deck Transformations

There are natural maps of a universal covering surface to itself, called deck transformations, which can be used to recover a Riemann surface W from its universal cover W^*. Let π be a universal covering map of W^* onto W. Fix a point $b \in W$ and write W^* as the collection of equivalence classes of curves beginning at b. If σ is a closed curve beginning and ending at b, we can define a map $M_{[\sigma]} : W^* \to W^*$ by

$$M_{[\sigma]}([\gamma]) = [\sigma\gamma].$$

Note that this map is one-to-one and onto and does not depend on the choice of the curve in $[\sigma]$. If B^* is a coordinate disk centered at $[\gamma]$ then $M_{[\sigma]}(B^*)$ is a coordinate disk centered at $[\sigma\gamma]$ with the same projection as B^*. So the map $M_{[\sigma]}$ is a homemorphism of W^* onto W^*. These maps $M_{[\sigma]}$ are called the **deck transformations**. The deck transformations form a group $\mathbb{G}$ under composition. If $b^* = [\{b\}]$ is the equivalence class of the constant curve $\{b\}$, then M_{b^*} is the identity map in this group $\mathbb{G}$.

Lemma 14.14 *If $p^*, q^* \in W^*$, then $\pi(p^*) = \pi(q^*)$ if and only if there is a deck transformation M with $M(p^*) = q^*$.*

Proof Recall that $\pi([\gamma]) = \gamma(1)$. If $p^* = [\gamma], q^* = [\alpha] \in W^*$ have the same projection $\gamma(1) = \alpha(1)$, then, setting $\sigma = \alpha\gamma^{-1}$, we have that σ is a closed curve beginning at b and $\sigma\gamma \approx \alpha$ so that $M_{[\sigma]}([\gamma]) = [\alpha]$. Conversely, if $\sigma \subset W$ is a closed curve beginning and ending at b then $\pi([\sigma\gamma]) = \gamma(1) = \pi([\gamma])$, so that $M_{[\sigma]}([\gamma])$ and $[\gamma]$ have the same projection. □

The group $\mathbb{G}$ gives an equivalence relation on W^*, where $p^* \sim q^*$ if and only if there is a deck transformation M with $M(p^*) = q^*$. The quotient space $W^*/\mathbb{G}$ with the quotient topology is a Riemann surface. The coordinate charts on $W^*/\mathbb{G}$ are just the images by the quotient map of the coordinate charts on W^*. Lemma 14.14 says that the map π induces a one-to-one map of $W^*/\mathbb{G}$ onto W. As a group, $\mathbb{G}$ is isomorphic to the group of equivalence classes under homotopy of all closed curves in W beginning at b, which is called the **fundamental group of W at b**. See Exercise 14.3. For this reason, the set of deck transformations is sometimes also called the fundamental group of W.

Definition 14.15 *If W_1 and W_2 are Riemann surfaces and if f is a map of W_1 into W_2, then we say that f is* **analytic** *provided*

$$w_\beta \circ f \circ z_\alpha^{-1}$$

is analytic for each coordinate function z_α on W_1 and w_β on W_2 wherever it is defined.

For example, if W_b^* and W_c^* are the universal covering surfaces of W based on curves beginning at $b \in W$ and $c \in W$, respectively, then let $\sigma \subset W$ be a curve from c to b. Define $N_{[\sigma]} : W_b^* \to W_c^*$ by $N_{[\sigma]}([\gamma]) = [\sigma\gamma]$. Then $N_{[\sigma]}$ is a one-to-one analytic map of W_b^* onto W_c^*.

If f is meromorphic on region $W \subset \mathbb{C}$, then f can also be viewed as a map into the extended plane $\mathbb{C}^*$, or via stereographic projection into the Riemann sphere $\mathbb{S}^2$. Such maps f

are then analytic in the sense of Definition 14.15. So some care must be taken when speaking of analytic functions on a Riemann surface such that both the domain and range surfaces are understood. We shall make the tacit assumption, unless stated otherwise, that an analytic function maps into the plane.

Corollary 14.16 *The deck transformations form a group $\mathbb{G}$ of one-to-one analytic maps of W^* onto W^* with the property that each $p^* \in W^*$ has a neighborhood B^* so that $M(B^*) \cap B^* = \emptyset$ for all deck transformations M not equal to the identity map. The projection map $\pi : W^* \to W$ induces a one-to-one analytic map of $W^*/\mathbb{G}$ onto W.*

We say that a group with the property in Corollary 14.16 is **properly discontinuous**. The proof that the deck transformations are properly discontinuous follows from Theorem 14.12. See also Exercise 14.8.

14.4 Exercises

A

14.1 Create a direct analytic continuation from an analytic continuation along a curve in $\mathbb{C}$. It is sometimes harder to work with direct analytic continuations because the intersection of two regions can have many components. This can be avoided by considering only disks, which are convex and have convex intersections, but in the abstract setting of Riemann surfaces we do not have this Euclidean geometry.

14.2 (a) Prove that the radius of convergence of power series expansions for an analytic function in a region Ω is Lipschitz continuous: if $r(z)$ is the radius of convergence of the series at z then $|r(z_1) - r(z_2)| \leq |z_1 - z_2|$, provided $r(z_j)$ is finite, $j = 1, 2$.

(b) Prove that f_0 has an analytic continuation along γ if and only if there are convergent power series

$$f_t(z) = \sum_{n=0}^{\infty} a_n(t)(z - \gamma(t))^n$$

such that $f_t(\gamma(s)) = a_0(s)$ for s sufficiently close to t.

(c) Prove that the coefficients $a_n(t)$ in part (ii) are continuous.

14.3 All curves in this exercise are contained in a region Ω and all homotopies are assumed to be within Ω.

(a) Prove that if $\gamma_1 \approx \gamma_2$ and $\gamma_2 \approx \gamma_3$ then $\gamma_1 \approx \gamma_3$.

(b) Prove that if σ is a reparameterization of γ (in the same direction) then $\sigma \approx \gamma$.

(c) If σ begins where γ ends, then we define the **product** (γ **followed by** σ) by

$$\gamma\sigma(t) = \begin{cases} \gamma(2t), & \text{for } t \in [0, \frac{1}{2}] \\ \sigma(2t-1), & \text{for } t \in [\frac{1}{2}, 1]. \end{cases}$$

Prove that if $\gamma_1 \approx \gamma_2$ and if $\sigma_1 \approx \sigma_2$ then $\gamma_1\sigma_1 \approx \gamma_2\sigma_2$. Here we assume that σ_j begins where γ_j ends, $j = 1, 2$.

(d) Define γ^{-1} by $\gamma^{-1}(t) = \gamma(1-t)$. Prove that $\gamma\gamma^{-1} \approx \{\gamma(0)\}$, where $\{\gamma(0)\}$ is the constant curve equal to $\gamma(0)$ for all $t \in [0,1]$. Similarly, $\gamma^{-1}\gamma \approx \{\gamma(1)\}$.

(e) Prove $\gamma_0 \approx \gamma_1$ if and only if $\gamma_0\gamma_1^{-1} \approx \{b\}$, where γ_0, γ_1 begin at b and end at c, and where $\{b\}$ is the constant curve at b. Note that in one case the homotopy uses curves with endpoints b and c, and in the other case the homotopy uses curves that begin and end at b. Hint: $\gamma_1^{-1}\gamma_1 \approx \{c\}$ by (d).

Homotopy defines equivalence classes of curves beginning at a fixed point $b \in \Omega$. The product in (c) is not always defined, so we pick a point $b \in \Omega$ and consider all closed curves that begin and end at b. Then homotopy defines equivalence classes of curves beginning and ending at b.

(f) Prove that the equivalence classes of curves which begin and end at $b \in \Omega$ form a group under multiplication with the identity equal to the constant curve at b. This group, written as $\pi_1(\Omega, b)$, is called the **fundamental group of** Ω.

(g) Prove that $\pi_1(\Omega, b)$ is isomorphic to $\pi_1(\Omega, c)$, for $b, c \in \Omega$.

14.4 If f is analytic on Ω then Example 14.11 gives a Riemann surface associated with f on $\Omega \setminus \{z : f'(z) = 0\}$. If $f(z) = z^n$ on $\Omega = \mathbb{C} \setminus \{x \geq 0\}$ then we can visualize the surface in $\mathbb{R}^3$ using the map $z \to (\text{Re}z^n, \text{Im}z^n, \arg z)$ similar to the ramp created for e^z. But we run into difficulties if $\Omega = \mathbb{C} \setminus \{0\}$ because the top edge of this ramp at height $2\pi n$ must be identified with the bottom edge at height 0. We obtain a Riemann surface by the construction in Example 14.11, but we cannot complete the ramp in $\mathbb{R}^3$. Give coordinate maps and charts for the Riemann surface associated with the map z^n on $\mathbb{C} \setminus \{0\}$ explicitly. We can extend this surface Ω_{z^n} to include one more point, namely $\{0\}$, by declaring the image of a ball centered at 0 of radius r by the map z^n on Ω_{z^n} together with 0 to be a coordinate disk. The associated coordinate map is then $z^{1/n}$. Show that this is a Riemann surface. The added point is called a **branch point**. Give a similar construction for zeros and poles of a meromorphic function on a region Ω. Hint: If $f'(z_0) = 0$, write $f(z) - f(z_0) = [(z - z_0)g(z)]^k$, where $g(z_0) \neq 0$.

B

14.5 Prove that the curve illustrated in Figure 5.4 (p. 68) is homologous to 0 but is not homotopic to a constant curve. One elementary approach to this exercise involves mapping an equilateral triangle to the upper half-plane and then applying the Schwarz reflection principle enough times. Another way would be to use Example 14.13, after filling in all of the details. The first approach does not create a universal covering surface for $\mathbb{C} \setminus \{0, 1\}$.

14.6 Suppose p is a polynomial and D is a disk with the property that if $p(z) \in D$ then $z \in D$ and $p'(z) \neq 0$. Prove that there is an analytic function f defined on D so that $p(f(z)) = z$ for all $z \in D$. This exercise is used in the field of complex dynamics. Be careful that your reasoning cannot be used to contradict Exercise 14.7(e).

14.7 Set

$$h(z) = \int_0^z e^{w^2}\, dw,$$

where the integration is along any curve in $\mathbb{C}$ from 0 to z.

(a) Prove that h is well defined and entire.
(b) Find h' and note that it is an even function which is never 0.
(c) Prove that h maps $\mathbb{C}$ onto $\mathbb{C}$,
(d) Prove that for every point $b \in \mathbb{C}$ and each c with $h(c) = b$ there is a neighborhood U of b and an analytic function g on U so that $g(b) = c$ and $h(g(z)) = z$ for $z \in U$.
(e) Show that every curve $\gamma : [0, 1] \to \mathbb{C}$ can be covered by finitely many neighborhoods as in (d), but there is no analytic function G on $\mathbb{C}$ such that $h(G(z)) = z$ for all $z \in \mathbb{C}$.
(f) Find where one of the hypotheses of the monodromy theorem fails. Hint: Look at the preimage of $i[0, c]$ where $c = \int_0^\infty e^{-x^2} dx$.

14.8 Suppose $\mathbb{G}$ is a group of properly discontinuous LFTs mapping $\mathbb{D}$ onto $\mathbb{D}$. Define an equivalence relation on $\mathbb{D}$ by $z \sim w$ if and only if there exists $g \in \mathbb{G}$ with $g(z) = w$. Give coordinate charts and maps on the quotient space $\mathbb{D}/\mathbb{G}$ making it into a Riemann surface. Show that $\mathbb{D}$ is the universal covering surface of $\mathbb{D}/\mathbb{G}$ and that the quotient map is the universal covering map. Here we identify two spaces if there is a one-to-one analytic onto map between them. Do the same for a properly discontinuous group of linear maps on $\mathbb{C}$. This exercise will be used in Chapter 15.

14.9 Prove that the analytic functions (in the sense of Definition 14.15) mapping the Riemann sphere into itself correspond to the rational functions via stereographic projection. Prove that the number of zeros equals the number of poles. The one-to-one analytic functions are the LFTs.

14.10 Prove that if f is continuous and non-vanishing on a simply-connected Riemann surface W (no assumptions about analyticity), then there is a continuous function g so that $f(z) = e^{g(z)}$ on W. Hint: If $|f(z) - f(z_0)| < |f(z_0)|$ on a disk B then we can find such a function g on B. Consider all possible functions g of this form on all coordinate charts in W. For example, it follows from this exercise that a continuous non-vanishing function on $\mathbb{D}$ has a continuous logarithm.

14.11 Suppose W is a region in $\mathbb{C}^*$ such that $\mathbb{C}^* \setminus W$ consists of n compact connected sets. Find $n - 1$ deck transformations such that the smallest group containing these transformations is the set of deck transformations for W on W^*.

14.12 Suppose f is an analytic function on a Riemann surface V with values in another Riemann surface W. If W^* is the universal cover of W and $\pi : W^* \to W$ is the universal covering map, prove there is an analytic function $g : V \to W^*$ so that $f = \pi \circ g$.

15 The Uniformization Theorem

In this chapter we prove a generalization of the Riemann mapping theorem called the uniformization theorem which says that the only universal covering surfaces are the disk, plane and sphere, up to conformal equivalence. This will give us a more concrete realization of all Riemann surfaces using groups of LFTs on these three spaces. The proof illustrates the power of the Perron method and the maximum principle. As is standard, the hyperbolic case, $\mathbb{D}$, is proved by constructing Green's function. The non-hyperbolic cases are treated in a very similar manner by constructing the *dipole* Green's function.

15.1 The Modular Function

We begin the chapter with an example for motivation. The universal covering surface of $\mathbb{C}\setminus\{0,1\}$ in Example 14.13 has another realization using the Schwarz reflection principle. The conjugate analytic map $R(z) = c + r^2/(\overline{z}-\overline{c})$ fixes points on the circle $C = \{z : |z-c|^2 = r^2\}$ and maps the disk D bounded by C to $\mathbb{C}^* \setminus \overline{D}$ and vice versa. The map R is called **reflection** about C. If φ is a conformal map of $\mathbb{H}$ onto D, then $\varphi^{-1} \circ R \circ \varphi(z) = \overline{z}$, the usual reflection about $\mathbb{R}$. The map $\overline{R(z)}$ is an LFT, so it preserves angles between curves, and maps (generalized) circles to (generalized) circles. Because $z \to \overline{z}$ preserves the magnitude of angles between curves, but reverses the direction, and maps circles to circles, we conclude that $R(z)$ also preserves the magnitude of angles between curves, but reverses the direction, and maps circles to circles.

Proposition 15.1 *Suppose C is a circle orthogonal to the unit circle and suppose D_1 and D_2 are the two regions in $\mathbb{D} \setminus C$. Then reflection R about the circle C maps D_1 onto D_2 and D_2 onto D_1. If E is any circle orthogonal to $\partial\mathbb{D}$ then $R(E)$ is a circle orthogonal to $\partial\mathbb{D}$. The map $\tau(z) = \overline{R(z)}$ is an LFT of $\mathbb{D}$ onto $\mathbb{D}$, given by reflection about C followed by reflection about $\mathbb{R}$.*

Proof Reflection about C is the identity on C and preserves orthogonality of circles. So it maps the unit circle to a circle which is orthogonal to C and has the same two points of intersection with C. There is only one such circle, so that reflection about C maps the unit circle to itself, and it must interchange the portion of $\partial\mathbb{D}$ outside C with the portion inside C. The statement about the regions D_1 and D_2 then follows because the map is the identity on C. In particular, $R(\mathbb{D}) = \mathbb{D}$. By the formula for R given before the statement of Proposition 15.1, τ is an LFT. □

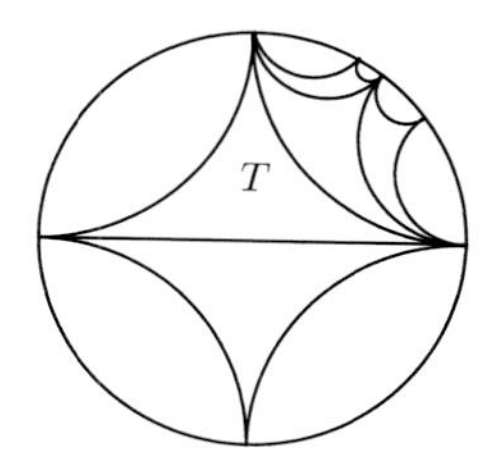

Figure 15.1 Reflections of a circular triangle T.

Let C be the circle with center $1 + i$ and radius 1. Note that C meets the unit circle orthogonally at 1 and at i. Let T be the region in $\mathbb{D}$ bounded by three arcs: the interval $(-1, 1)$, the circular arc $C_1 = C \cap \overline{\mathbb{D}}$ and the reflection of C_1 about the imaginary axis. We call T a **circular triangle** because it is bounded by three circular arcs. See Figure 15.1.

If we reflect T about one of its boundary circles, then by Proposition 15.1 we obtain an adjacent region contained in $\mathbb{D}$ bounded by three orthogonal circles, another circular triangle, with vertices on $\partial\mathbb{D}$. By Carathéodory's theorem, a conformal map of T onto the upper half-plane is a homeomorphism of $\overline{T}$ onto the closure of $\mathbb{H}$ in $\mathbb{C}^*$. See also Exercise 15.2 for a more elementary proof of this fact. Composing with an appropriate LFT, we obtain a conformal map π which maps -1, 1 and i to 0, 1 and ∞, respectively. Each boundary arc of T is mapped to one of the intervals $(-\infty, 0]$, $[0, 1]$ or $[1, \infty)$. By the Schwarz reflection principle, we can extend π to be a conformal map of the reflection of T about any one of its boundary circles onto the lower half-plane. By repeated reflections, we can extend π to be a locally conformal map of the union of all repeated reflections of T and their boundaries onto $\mathbb{C} \setminus \{0, 1\}$. The process of reflection across arcs of $\mathbb{R}$ in Example 14.13 is exactly the same as reflection about these orthogonal arcs in $\mathbb{D}$. In this way we create a new version of the covering surface of $\mathbb{C} \setminus \{0, 1\}$ contained in $\mathbb{D}$.

This gives another proof of Picard's little theorem.

Corollary 15.2 *If g is an entire function which omits the values* 0 *and* 1, *then g is constant.*

Proof Choose z_0 so that $g(z_0) \in \mathbb{H}$. The function $\pi^{-1} \circ g$ can be defined as an analytic function in a neighborhood of z_0. By the Schwarz reflection principle, this function can be continued along all paths in $\mathbb{C}$ beginning at z_0. By the monodromy theorem, it gives an entire function which is bounded and hence constant by Liouville's theorem. Thus g is constant. □

The next proposition may see obvious, but nevertheless we give a proof.

Proposition 15.3 *The repeated reflections of T (and its boundary) fill the unit disk.*

Proof After finitely many reflections of $\overline{T} \cap \mathbb{D}$, we obtain a region in $\mathbb{D}$ which is bounded by finitely many circles orthogonal to $\partial\mathbb{D}$, and hence the complement in $\mathbb{D}$ is the closure of a finite union of disjoint disks intersected with $\mathbb{D}$. Note that reflection of this larger region about any of its bounding arcs is the union of reflections of individual copies of T. If $\Omega \subset \mathbb{D}$ is the region obtained by all possible reflections, then any $p \in \mathbb{D} \setminus \Omega$ is contained in the intersection of a sequence of disks whose boundaries are orthogonal to $\partial\mathbb{D}$. But then there is a maximal disk D with $\overline{D}\cap\mathbb{D}\subset\mathbb{D}\setminus\Omega$ containing p. Since ∂D is orthogonal to $\partial\mathbb{D}$, it must be the limit of orthogonal arcs which come from reflections of ∂T. But reflection about an arc near ∂D will send the origin to a point much closer to $\partial\mathbb{D}$, and hence inside D, which is a contradiction. In fact, by Exercise 15.1(b), the reflection of the origin about a circle which is orthogonal to $\partial\mathbb{D}$ at a and b is the average, $(a+b)/2$. □

It follows that π is an analytic map of $\mathbb{D}$ onto $\mathbb{C}\setminus\{0,1\}$ which is locally one-to-one. As in Example 14.11, we can view the map π as a a one-to-one conformal map of the unit disk $\mathbb{D}$ onto the covering surface of $\mathbb{C}\setminus\{0,1\}$ constructed in Example 14.13, because this surface was constructed by exactly these reflections of $\mathbb{H}$. Thus the unit disk is conformally equivalent to the universal covering surface of $\mathbb{C}\setminus\{0,1\}$. The uniformization theorem will yield the same conclusion for every plane region with at least two points in the complement.

The function $\lambda(z)=\pi((i-z)/(i+z))$, the transplant of π to the upper half-plane, is called the **modular function**. The preimage of T is the region between the vertical lines $\mathrm{Re}z=0$ and $\mathrm{Re}z=1$ and above the circle orthogonal to $\mathbb{R}$ at 0 and 1.

15.2 Green's Function

One of the most important functions on a region or Riemann surface is Green's function. Green's function has many uses, but we will focus in this chapter only on using it to prove the uniformization theorem.

Suppose W is a Riemann surface. Fix $p_0\in W$ and let $z:U\to\mathbb{D}$ be a coordinate function such that $z(p_0)=0$. From now on we assume that we have extended our collection of charts on W to include a local base as described after Definition 14.8 and that each coordinate chart U_α has compact closure $\overline{U}_\alpha\subset W$, by restricting our collection. Moreover, we can assume that the coordinate map z_α is a homeomorphism $\overline{U}_\alpha$ onto $\overline{\mathbb{D}}$.

Let $\mathcal{F}_{p_0}$ be the collection of subharmonic functions v on $W\setminus p_0$ satisfying

$$v=0 \text{ on } W\setminus K, \text{ for some compact } K\subset W \text{ with } K\neq W \tag{15.1a}$$

and

$$\limsup_{p\to p_0}\,(v(p)+\log|z(p)|)<\infty. \tag{15.1b}$$

Note that $v\in\mathcal{F}_{p_0}$ is not assumed to be subharmonic at p_0, and indeed it can tend to $+\infty$ as $p\to p_0$. Set

$$g_W(p,p_0)=\sup\{v(p):v\in\mathcal{F}_{p_0}\}. \tag{15.2}$$

Condition (15.1b) does not depend on the choice of the coordinate function provided it vanishes at p_0. For, if z_α is another coordinate map with $z_\alpha(p_0)=0$, then $\lim_{p\to p_0}|z_\alpha(p)/z(p)|=|(z_\alpha\circ z^{-1})'(0)|\neq 0,\infty$. The collection $\mathcal{F}_{p_0}$ is a Perron family on $W\setminus\{p_0\}$, so one of the following two cases holds by Harnack's theorem:

case 1: $g_W(p,p_0)$ is harmonic in $W\setminus\{p_0\}$, or
case 2: $g_W(p,p_0)=+\infty$ for all $p\in W\setminus\{p_0\}$.

In the first case, $g_W(p, p_0)$ is called **Green's function on** W with pole (or logarithmic singularity) at p_0. In the second case, we say that **Green's function with pole at** p_0 **does not exist on** W.

The primary tool we will use in this section and the rest of this chapter is the maximum principle or Lindelöf's maximum principle, Theorem 7.15, which allows us to deal with an isolated point like p_0. We remark that it is difficult to apply the maximum principle directly to a harmonic or subharmonic function because W is not necessarily contained in a larger Riemann surface so that we cannot consider the "boundary" of W. One possibility is to replace $\lim_{\zeta_n \to \partial W}$ with the requirement that $\{\zeta_n\}$ is eventually outside every compact subset of W. However, there are examples of surfaces W and sequences that are eventually outside of every compact set of W yet g_W does not tend to 0 on these sequences. We avoid these difficulties by considering Perron families of functions which are equal to 0 off a compact set. We remind the reader that our subharmonic functions are assumed to be continuous as maps into $[-\infty, \infty)$, so by continuity the functions are also 0 on the boundary of the appropriate compact set. To bound such a function we need only to use the maximum principle or Lindelöf's maximum principle on the interior of the compact set. Once we obtain a uniform estimate for all members of the family, we can then take the supremum over the family. We shall use this idea many times in this chapter.

Lemma 15.4 *Suppose* $p_0 \in W$ *and suppose* $z : U \to \mathbb{D}$ *is a coordinate function such that* $z(p_0) = 0$. *If* $g_W(p, p_0)$ *exists, then*

$$g_W(p, p_0) > 0 \text{ for } p \in W \setminus \{p_0\}, \text{ and} \tag{15.3}$$

$$g_W(p, p_0) + \log |z(p)| \text{ extends to be harmonic in } U. \tag{15.4}$$

Proof The function

$$v_0(p) = \begin{cases} -\log |z(p)|, & \text{for } p \in U \\ 0, & \text{for } p \in W \setminus U \end{cases}$$

is in $\mathcal{F}_{p_0}$ by Definition 13.1(i). Hence, $g_W(p, p_0) \geq 0$ and $g_W(p, p_0) > 0$ if $p \in U$. By the maximum principle applied to $-g$ in $W \setminus \{p_0\}$, (15.3) holds.

If $v \in \mathcal{F}_{p_0}$ then, by (15.1b) and Lindelöf's maximum principle,

$$\sup_{U \setminus \{p_0\}} (v + \log |z|) = \sup_{\partial U} v \leq \sup_{\partial U} g_W < \infty.$$

Taking the supremum over $v \in \mathcal{F}_{p_0}$, we obtain

$$g_W + \log |z| \leq \sup_{\partial U} g_W < \infty$$

in $U \setminus \{p_0\}$. We also have that

$$g_W + \log |z| \geq v_0 + \log |z| = 0 \tag{15.5}$$

for $p \in U \setminus \{p_0\}$. Thus $g_W + \log |z|$ is bounded and harmonic in $U \setminus \{p_0\}$. Using the Poisson integral formula on $z_\alpha(U)$, we can find a bounded harmonic function on U which agrees with $g_W + \log |z|$ on ∂U. By Lindelöf's maximum principle, $g_W + \log |z|$ extends to be harmonic in U.

□

The Green's function for the unit disk $\mathbb{D}$ is given by

$$g_{\mathbb{D}}(z,a) = \log\left|\frac{1-\bar{a}z}{z-a}\right|.$$

If $g(z) = \log\frac{|1-\bar{a}z|}{|z-a|}$ then, by Lindelöf's maximum principle, each candidate subharmonic function v in the Perron family $\mathcal{F}_a$ is bounded by g. Moreover, $\max(g-\varepsilon,0) \in \mathcal{F}_a$, when $\varepsilon > 0$. Letting $\varepsilon \to 0$, we conclude $g = g_{\mathbb{D}}(z,a)$.

The next theorem gives a large collection of Riemann surfaces for which Green's function exists. See also Exercise 15.3.

Theorem 15.5 *Suppose W_0 is a Riemann surface and suppose U_0 is a coordinate disk whose closure is compact in W_0. Set $W = W_0 \setminus \overline{U_0}$. Then $g_W(p,p_0)$ exists for all $p,p_0 \in W$ with $p \neq p_0$.*

Proof Fix $p_0 \in W$ and let $U \subset W$ be a coordinate disk with compact closure and with coordinate function $z : U \to \mathbb{D}$ and $z(p_0) = 0$. To prove that g_W exists, we show that the family $\mathcal{F}_{p_0}$ is bounded above. Fix r, with $0 < r < 1$, and set $rU = \{p \in W : |z(p)| < r\}$. If $v \in \mathcal{F}_{p_0}$ then, by (15.1b) and Lindelöf's maximum principle,

$$v(p) + \log|z(p)| \le \max_{q\in\partial U}(v(q) + \log|z(q)|) = \max_{q\in\partial U} v(q),$$

for all $p \in U$. Thus

$$\max_{p\in\partial rU} v(p) + \log r \le \max_{p\in\partial U} v(p). \tag{15.6}$$

Equation (15.6) is a growth estimate on v in U. We can obtain another growth estimate which uses the fact that we removed a disk from W_0 by constructing a harmonic function in $W\setminus\overline{rU}$ which is equal to 1 on ∂rU and in some sense equal to 0 on ∂W. More precisely, let $\mathcal{F}$ denote the collection of functions u which are subharmonic on $W\setminus\overline{rU}$ with $u = 0$ on $W_0 \setminus K$ for some compact set $K \subset W_0$, which can depend on u, and such that

$$\limsup_{p\to\zeta} u(p) \le 1 \quad \text{and} \quad \limsup_{p\to\alpha} u(p) \le 0$$

for $\zeta \in \partial rU$ and $\alpha \in \partial U_0$. See Figure 15.2.

Applying the maximum principle on the interior of $K \setminus (U_0 \cup rU)$, we obtain $u \le 1$ on $W\setminus\overline{rU}$. Because $\mathcal{F}$ is a Perron family,

$$\omega(p) = \sup\{u(p) : u \in \mathcal{F}\}$$

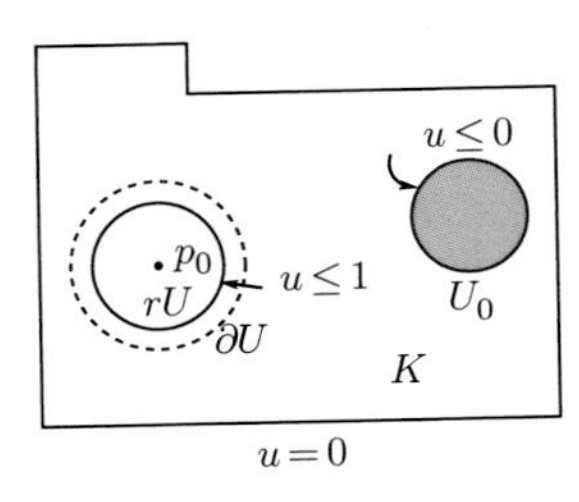

Figure 15.2 Harmonic measure of ∂rU in $W\setminus\overline{rU}$.

is harmonic in $W \setminus \overline{rU}$. We can construct a local barrier at each point of $\partial U_0 \cup \partial rU$ for the region $W \setminus \overline{rU}$ by transporting the problem to a region in $\mathbb{D}$ via a coordinate map. Thus, by exactly the same argument used for the Dirichlet problem, the harmonic function ω extends to be continuous at each point of ∂U_0 and each point of ∂rU so that $\omega(p) = 0$ for $p \in \partial U_0$ and $\omega(p) = 1$ for $p \in \partial rU$. In particular, this implies ω is not constant. Moreover, $0 \leq \omega \leq 1$ because $0 \in \mathcal{F}$ and each candidate $u \in \mathcal{F}$ is bounded by 1. Since ω is not constant, the maximum principle implies $0 < \omega(p) < 1$ for $p \in W \setminus \overline{rU}$. The function ω is called the **harmonic measure** of ∂rU in the region $W \setminus \overline{rU}$.

Returning to our function $v \in \mathcal{F}_{p_0}$, we have that

$$v(p) \leq \left(\max_{\partial rU} v\right) \omega(p),$$

for $p \in W \setminus \overline{rU}$, by the maximum principle applied to $v - (\max_{\partial rU} v)w$. So

$$\max_{\partial U} v \leq \left(\max_{\partial rU} v\right) \max_{p \in \partial U} \omega(p) \leq \left(\max_{\partial rU} v\right)(1 - \delta) \tag{15.7}$$

for some $\delta > 0$. Adding inequalities (15.6) and (15.7) yields

$$\delta \max_{\partial rU} v \leq \log \frac{1}{r}$$

for all $v \in \mathcal{F}$, with δ is independent of v. This implies that case 2 does not hold and hence Green's function exists.

□

Green's function on a Riemann surface can be recovered from Green's function on the universal cover by the following Theorem 15.6. A consequence is Corollary 15.7, which plays a critical role in the proof of the uniformization theorem.

Theorem 15.6 *Suppose W is a Riemann surface for which g_W with pole at $p_0 \in W$ exists. Let W^* be the simply-connected universal covering surface of W, let $\pi : W^* \to W$ be the universal covering map and suppose $\pi(p_0^*) = p_0$. Then g_{W^*} with pole at p_0^* exists and satisfies*

$$g_W(\pi(p^*), \pi(p_0^*)) = \sum_{q^* : \pi(q^*) = \pi(p_0^*)} g_{W^*}(p^*, q^*). \tag{15.8}$$

We interpret the infinite sum on the right-hand side of (15.8) to be the supremum of all sums over finitely many q^*.

Proof Suppose $q_1^*, \ldots, q_n^*$ are distinct points in W^* with $\pi(q_j^*) = p_0 = \pi(p_0^*)$. Suppose $v_j \in \mathcal{F}_{q_j^*}$, the Perron family for the construction of $g_{W^*}(\cdot, q_j^*)$. So $v_j = 0$ off K_j^*, a compact subset of W^* and

$$\limsup_{p^* \to q_j^*} \left(v_j(p^*) + \log |z \circ \pi(p^*)|\right) < \infty,$$

where z is a coordinate chart on W with $z(p_0) = 0$. Recall that $g_W(p, p_0) + \log |z(p)|$ extends to be finite and continuous at p_0, and hence

$$\lim_{p^* \to q_j^*} g_W(\pi(p^*), p_0) + \log |z(\pi(p^*))|$$

exists and is finite because $\pi(q_j^*) = p_0$. Thus

$$\left(\sum_{j=1}^{n} v_j(p^*)\right) - g_W(\pi(p^*), p_0)$$

is bounded in a neighborhood of each q_j^*, for $j = 1, \ldots, n$, and is ≤ 0 off $K^* = \cup_j K_j^*$, by (15.3). By Lindelöf's maximum principle on the interior of $K^* \setminus \{q_1^*, \ldots, q_n^*\}$, it is bounded above by 0. Taking the supremum over all such v_j, we conclude that $g_{W^*}(p^*, q_j^*)$ exists and

$$\left(\sum_{j=1}^{n} g_{W^*}(p^*, q_j^*)\right) - g_W(\pi(p^*), p_0) \leq 0.$$

Taking the supremum over all such finite sums, we have

$$S(p^*) \equiv \sum_{q^*:\pi(q^*)=p_0} g_{W^*}(p^*, q^*) \leq g_W(\pi(p^*), p_0).$$

Moreover, as a supremum of finite sums of positive harmonic functions, $S(p^*)$ is harmonic on $W^* \setminus \pi^{-1}(p_0)$ by Harnack's principle. By (15.4),

$$S(p^*) + \log |z(\pi(p^*))| \tag{15.9}$$

extends to be harmonic in a neighborhood of each $q^* \in \pi^{-1}(p_0)$.

Now take $v \in \mathcal{F}_{p_0}$, the Perron family used to construct $g_W(p, p_0)$. Let U^* be a coordinate disk containing p_0^*, so that $z \circ \pi$ is a coordinate map of U^* onto $\mathbb{D}$, vanishing at p_0^*. Then, by (15.9) and Lindelöf's maximum principle,

$$v(\pi(p^*)) - S(p^*) \leq 0$$

for $p^* \in U^*$. Taking the supremum over all such v, we obtain

$$g_W(\pi(p^*), p_0) \leq S(p^*),$$

and thus (15.8) holds on $U^* \setminus p_0^*$ and therefore on W^*. □

The converse of Theorem 15.6 is not true. See Exercise 15.7.

Corollary 15.7 *Suppose W is a Riemann surface for which Green's function g_W with pole at p exists, for some $p \in W$, and suppose $W^* = \mathbb{D}$. Then g_W with pole at q exists for all $q \in W$ and*

$$g_W(p, q) = g_W(q, p). \tag{15.10}$$

We will remove the hypothesis that $W^* = \mathbb{D}$ in Section 15.3. See Corollary 15.9.

Proof As noted earlier, $g_{\mathbb{D}}(a, b) = -\log |(a - b)/(1 - \overline{b}a)|$. If τ is an LFT of the disk onto the disk, then, by an elementary computation, $g_{\mathbb{D}}(a, \tau(b)) = g_{\mathbb{D}}(\tau^{-1}(a), b)$ for all $a, b \in \mathbb{D}$ with $a \neq \tau(b)$. Let $\mathbb{G}$ denote the group of deck transformations, a group of LFTs τ mapping $\mathbb{D}$ onto $\mathbb{D}$ and satisfying $\pi \circ \tau = \pi$. Moreover, if $\pi(q^*) = \pi(p^*)$ then there is a $\tau \in \mathbb{G}$ such that $\tau(p^*) = q^*$, by Lemma 14.14. Suppose $g(p, p_0)$ exists for some $p_0 \in W$ and all $p \neq p_0$. Choose $p_0^* \in \mathbb{D}$ so that $\pi(p_0^*) = p_0$. By Theorem 15.6 for $p^* \notin \pi^{-1}(p_0)$,

$$g_W(\pi(p^*), \pi(p_0^*)) = \sum_{\tau \in \mathbb{G}} -\log \left| \frac{p^* - \tau(p_0^*)}{1 - \overline{\tau(p_0^*)} p^*} \right| \tag{15.11}$$

$$= \sum_{\tau \in \mathbb{G}} -\log \left| \frac{\tau^{-1}(p^*) - p_0^*}{1 - \overline{\tau^{-1}(p^*)} p_0^*} \right|. \tag{15.12}$$

Fix $p^* \in \mathbb{D} \setminus \pi^{-1}(p_0)$. Each term in the sum

$$S(q^*) = \sum_{\tau \in \mathbb{G}} -\log \left| \frac{\tau^{-1}(p^*) - q^*}{1 - \overline{\tau^{-1}(p^*)} q^*} \right|$$

is a positive harmonic function of $q^* \in \mathbb{D} \setminus \tau^{-1}(p^*)$. Since the sum of these positive harmonic functions converges when $q^* = p_0^*$, the function S is harmonic in $\mathbb{D} \setminus \{\tau^{-1}(p^*) : \tau \in \mathbb{G}\}$, by Harnack's theorem. If $v \in \mathcal{F}_p$, the Perron family for $g_W(q, p)$, where $p = \pi(p^*)$, then, by Lindelöf's maximum principle, $v \leq S \circ \pi^{-1}$. Taking the supremum over all $v \in \mathcal{F}_p$, we conclude that $g_W(q, p)$ exists and $g_W(q, p) \leq S \circ \pi^{-1}(q)$, for all $q \neq p$. Thus

$$g_W(\pi(p_0^*), \pi(p^*)) \leq S(p_0^*) = g_W(\pi(p^*), \pi(p_0^*)).$$

Reversing the roles of p_0^* and p^* proves (15.10) for $q = p_0$ and all $p \neq p_0$. Because the Green's function g_W with pole at p then exists for every $p \in W$, (15.10) must hold for all p and q. □

15.3 Simply-Connected Riemann Surfaces

We will now consider simply-connected Riemann surfaces and prove the uniformization theorem in case 1.

Theorem 15.8 (uniformization, case 1) *If W is a simply-connected Riemann surface then the following are equivalent:*

$$g_W(p, p_0) \textit{ exists for some } p_0 \in W, \tag{15.13}$$

$$g_W(p, p_0) \textit{ exists for all } p_0 \in W, \tag{15.14}$$

$$\textit{there is a one-to-one analytic map } \varphi \textit{ of } W \textit{ onto } \mathbb{D}. \tag{15.15}$$

Moreover, if g_W exists, then

$$g_W(p_1, p_0) = g_W(p_0, p_1), \tag{15.16}$$

and $g_W(p, p_0) = -\log |\varphi(p)|$, where $\varphi(p_0) = 0$.

Proof Suppose there is a one-to-one analytic map φ of W onto $\mathbb{D}$ and let $p_0 \in W$. By composing φ with an LFT, we can assume that $\varphi(p_0) = 0$. If $v \in \mathcal{F}_{p_0}$ then $v = 0$ off a compact set K, so, by (15.1b) and Lindelöf's maximum principle applied on the interior of $K \setminus \{p_0\}$,

$$v + \log|\varphi| \leq 0$$

on W. Taking the supremum over all such v shows that $g_W(p, p_0) < \infty$ and therefore (15.14) holds. Clearly, (15.14) implies (15.13).

Now suppose (15.13) holds. By (15.4) there is an analytic function f defined on a coordinate disk U containing p_0 so that

$$\text{Re} f(p) = g_W(p, p_0) + \log|z(p)|$$

for $p \in U$. Hence the function

$$\varphi(p) = ze^{-f(p)}$$

is analytic in U and satisfies $|\varphi(p)| = e^{-g_W(p,p_0)}$ and $\varphi(p_0) = 0$. On any coordinate disk U_α with $p_0 \notin U_\alpha$, $g_W(p, p_0)$ is the real part of an analytic function because it is harmonic. The difference of two analytic functions with the same real part is constant, so that φ on U can be analytically continued along all curves in W beginning at p_0. By the monodromy theorem, there is a function φ, analytic on W, such that

$$|\varphi(p)| = e^{-g_W(p,p_0)} < 1.$$

We claim that φ is one-to-one. If $\varphi(p) = \varphi(p_0) = 0$, then $p = p_0$ because $g_W(p, p_0)$ is finite for $p \neq p_0$. Let $p_1 \in W$, with $p_1 \neq p_0$. Then, by (15.3), $|\varphi(p_1)| < 1$ and

$$\varphi_1 \equiv \frac{\varphi - \varphi(p_1)}{1 - \overline{\varphi(p_1)}\varphi}$$

is analytic on W and $|\varphi_1| < 1$. If $v \in \mathcal{F}_{p_1}$, then, by (15.1) and Lindelöf's maximum principle, as argued above,

$$v + \log|\varphi_1| \leq 0.$$

Taking the supremum over all such v, we conclude that $g_W(p, p_1)$ exists and that

$$g_W(p, p_1) + \log|\varphi_1| \leq 0. \tag{15.17}$$

Setting $p = p_0$ in (15.17) gives

$$g_W(p_0, p_1) \leq -\log|\varphi_1(p_0)| = -\log|\varphi(p_1)| = g_W(p_1, p_0).$$

Switching the roles of p_0 and p_1 gives (15.16). Moreover, equality holds in (15.17) at $p = p_0$ so that, by the maximum principle, $g_W(p, p_1) = -\log|\varphi_1(p)|$ for all $p \in W \setminus \{p_1\}$. Now, if $\varphi(p_2) = \varphi(p_1)$, then, by the definition of φ_1, $\varphi_1(p_2) = 0$. Thus $g_W(p_2, p_1) = \infty$ and $p_2 = p_1$. Therefore φ is one-to-one.

The image $\varphi(W) \subset \mathbb{D}$ is simply-connected, for if $\gamma \subset \varphi(W)$ is a closed curve then $\varphi^{-1}(\gamma) \subset W$ is closed and therefore homotopic to a constant curve. Applying the map φ to the homotopy gives a homotopy in $\varphi(W)$ of γ to a constant curve. If $\varphi(W) \neq \mathbb{D}$ then, by the Riemann mapping theorem, we can find a one-to-one analytic map ψ of $\varphi(W)$ onto $\mathbb{D}$ with $\psi(0) = 0$. The map $\psi \circ \varphi$ is then a one-to-one analytic map of W onto $\mathbb{D}$, with $\psi \circ \varphi(p_0) = 0$, proving (15.15). □

We remark that the map φ_1 in the proof of Theorem 15.8 is actually onto, as can be seen by applying the "onto" argument in the proof of the Riemann mapping theorem. See Exercise 15.5.

As promised, we now remove the hypothesis that $W^* = \mathbb{D}$ in Corollary 15.7.

Corollary 15.9 *Suppose W is a Riemann surface for which Green's function g_W with pole at p exists, for some $p \in W$. Then g_W with pole at q exists for all $q \in W$ and*

$$g_W(p,q) = g_W(q,p). \tag{15.18}$$

Proof If W is a Riemann surface such that g_W with pole at some $p \in W$ exists, then g_{W^*} exists by Theorem 15.6. But then, by Theorem 15.8, W^* is conformally equivalent to $\mathbb{D}$. Applying Exercise 15.4(a) and Corollary 15.7 yields (15.18). □

Theorem 15.10 (uniformization, case 2) *Suppose W is a simply-connected Riemann surface for which Green's function does not exist. If W is compact, then there is a one-to-one analytic map of W onto $\mathbb{C}^*$. If W is not compact, then there is a one-to-one analytic map of W onto $\mathbb{C}$.*

As seen above, Green's function for the disk with pole at 0 is given by $G = -\log|z|$. There is no Green's function on the sphere or the plane, but this same function G plays a similar role. Instead of one pole, or logarithmic singularity, G has two poles on the sphere, with opposite signs. We will call it a **dipole Green's function**, because it is the two-dimensional analog of an electric dipole, two charges with opposite signs. The next lemma says that a dipole Green's function exists for every Riemann surface. For a simply-connected Riemann surface without Green's function, the dipole Green's function will be used to construct a conformal map to the sphere or the plane in much the same way as Green's function was used in case 1.

Lemma 15.11 *Suppose W is a Riemann surface and, for $j = 1, 2$, suppose that $z_j : U_j \to \mathbb{D}$ are coordinate functions with coordinate disks U_j satisfying $\overline{U_1} \cap \overline{U_2} = \emptyset$, and $z_j(p_j) = 0$. Then there is a function $G(p) \equiv G(p, p_1, p_2)$, harmonic in $p \in W \setminus \{p_1, p_2\}$, such that*

$$G + \log|z_1| \text{ extends to be harmonic in } U_1, \tag{15.19}$$

$$G - \log|z_2| \text{ extends to be harmonic in } U_2 \tag{15.20}$$

and

$$\sup_{p \in W \setminus (U_1 \cup U_2)} |G(p)| < \infty. \tag{15.21}$$

Before proving Lemma 15.11, we will use it to prove the uniformization theorem in case 2, since the proof is similar to the proof in case 1.

Proof of Theorem 15.10 By Theorem 15.8, we may suppose that $g_W(p, p_1)$ does not exist for all $p, p_1 \in W$. Because W is simply-connected, we can apply the monodromy theorem, as in the proof of Theorem 15.8, to obtain a meromorphic function φ_1 defined on W such that

$$|\varphi_1(p)| = e^{-G(p,p_1,p_2)},$$

where G is the dipole Green's function from Lemma 15.11. Note that φ_1 has a simple zero at p_1, a simple pole at p_2 and no other zeros or poles.

Let us prove φ_1 is one-to-one. If $p_0 \in W \setminus \{p_1, p_2\}$, then $\varphi_1(p_0) \neq 0, \infty$. Let φ_0 be the meromorphic function on W such that

$$|\varphi_0(p)| = e^{-G(p,p_0,p_2)}$$

and consider the function

$$H(p) = \frac{\varphi_1(p) - \varphi_1(p_0)}{\varphi_0(p)}.$$

Then H is analytic on W because its poles at p_2 cancel and because φ_0 has a simple zero at p_0. By (15.21) and the analyticity of H, $|H|$ is bounded on W. But if $v \in \mathcal{F}_{p_1}$, the Perron family used to construct $g_W(p, p_1)$, then, by Lindelöf's maximum principle,

$$v(p) + \log \left| \frac{H(p) - H(p_1)}{2 \sup_W |H|} \right| \leq 0.$$

Because $g_W(p, p_1)$ does not exist, $\sup\{v(p) : v \in \mathcal{F}_{p_1}\} \equiv +\infty$ for every $p \in W \setminus \{p_1\}$, and therefore

$$H(p) \equiv H(p_1) = -\varphi_1(p_0)/\varphi_0(p_1) \neq 0, \infty.$$

Since $H \neq 0$, we conclude that $\varphi_1(p) \neq \varphi_1(p_0)$, unless $\varphi_0(p) = 0$. But if $\varphi_0(p) = 0$, then $p = p_0$. Thus φ_1 is one-to-one on $W \setminus \{p_1, p_2\}$. But the only zero of φ_1 is p_1 and the only pole of φ_1 is p_2, so that φ_1 is one-to-one on W.

We have shown that φ_1 is a one-to-one analytic map from W to a simply-connected region $\varphi_1(W) \subset \mathbb{C}^*$. If $\mathbb{C}^* \setminus \varphi_1(W)$ contains more than one point, then, by the Riemann mapping theorem, there is a one-to-one analytic map of $\varphi_1(W)$, and hence of W, onto $\mathbb{D}$. Since we assumed that g_W does not exist, this contradicts Theorem 15.8. Thus $\mathbb{C}^* \setminus \varphi_1(W)$ contains at most one point, and the last two statements of Theorem 15.10 are now obvious. □

We now complete the proof of the uniformization theorem by proving Lemma 15.11.

Proof of Lemma 15.11 If g_W exists, then $G(p) = g(p, p_1) - g(p, p_2)$ satisfies (15.19), (15.20) and (15.21). The idea of the proof when there is no Green's function is to remove a small disk from W, giving a surface with Green's function by Theorem 15.5 and therefore a corresponding dipole Green's function, then let the radius of the removed disk decrease to zero. However, we need some bound on the difference to be able to pass to a limiting dipole Green's function. Because an approximating G is a difference, we cannot use the maximum principle on the difference of two candidates for the corresponding Perron families. So we reduce the size of each Green's function by using $g(p, p_1) - g(p_2, p_1)$ and $g(p, p_2) - g(p_1, p_2)$. By Corollary 15.9, this does not affect the difference, but we can estimate each on $\partial U_1 \cup \partial U_2$.

Suppose z_0 is a coordinate function with coordinate chart U_0 such that $\overline{U_0} \cap \overline{U_j} = \emptyset$ for $j = 1, 2$. Let p_0 be the point in U_0 such that $z_0(p_0) = 0$. Set $tU_0 = \{p \in W : |z_0(p)| < t\}$ and set $W_t = W \setminus tU_0$. See Figure 15.3.

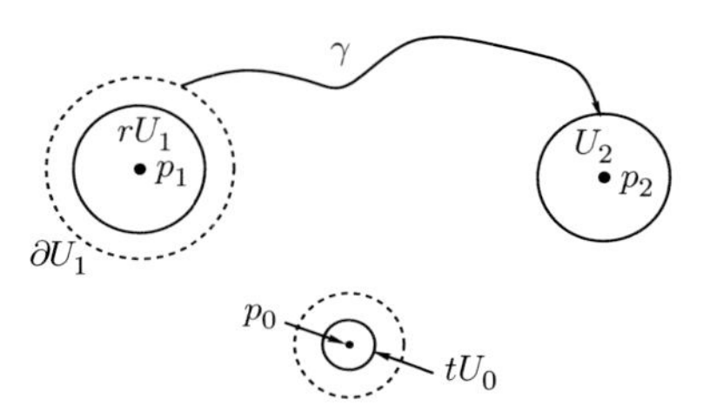

Figure 15.3 The surface W_t.

By Theorem 15.5, $g_{W_t}(p, p_1)$ exists for all $p, p_1 \in W_t$ with $p \neq p_1$. Fix r, $0 < r < 1$, and set $rU_1 = \{p \in W : |z_1(p)| < r\}$. By the maximum principle,

$$g_{W_t}(p, p_1) \leq M_1(t) \equiv \max_{q \in \partial rU_1} g_{W_t}(q, p_1), \tag{15.22}$$

for all $p \in W_t \setminus rU_1$, because the same bound holds for all candidates in the Perron family defining g_{W_t}. The growth estimate (15.6) shows that

$$M_1(t) \leq \max_{p \in \partial U_1} g_{W_t}(p, p_1) + \log \frac{1}{r}. \tag{15.23}$$

By (15.22), $u_t(p) \equiv M_1(t) - g_{W_t}(p, p_1)$ is a positive harmonic function in $W_t \setminus rU_1$, and by (15.23) there exists $q \in \partial U_1$ with $u_t(q) \leq \log \frac{1}{r}$.

Riemann surfaces are pathwise connected so let γ be a curve in $W \setminus (U_1 \cup U_2)$ connecting ∂U_1 to ∂U_2 which does not pass through p_0. Then, for $t \leq t_0$, $K = \partial U_1 \cup \overline{U_2} \cup \gamma \subset W_t \setminus \overline{rU_1}$ is compact and connected. By Harnack's inequality, there is a constant $C < \infty$ depending on K and r but not on t so that, for all $p \in K$ and $t \leq t_0$,

$$0 \leq u_t(p) \leq C,$$

and

$$|g_{W_t}(p, p_1) - g_{W_t}(p_2, p_1)| = |u_t(p_2) - u_t(p)| \leq 2C.$$

Likewise, if $K' = \partial U_2 \cup \overline{U_1} \cup \gamma$ there is a constant $C < \infty$ so that

$$|g_{W_t}(p, p_2) - g_{W_t}(p_1, p_2)| \leq C,$$

for all $p \in K'$ and $t \leq t_0$.

By Corollary 15.9, $g_{W_t}(p_1, p_2) = g_{W_t}(p_2, p_1)$ and so the function

$$\begin{aligned} G_t(p, p_1, p_2) &\equiv g_{W_t}(p, p_1) - g_{W_t}(p, p_2) \\ &= (g_{W_t}(p, p_1) - g_{W_t}(p_2, p_1)) - (g_{W_t}(p, p_2) - g_{W_t}(p_1, p_2)) \end{aligned}$$

is harmonic in $W_t \setminus \{p_1, p_2\}$ and satisfies

$$|G_t(p, p_1, p_2)| \leq C,$$

for all $p \in K \cap K' \supset \partial U_1 \cup \partial U_2$ and some finite C independent of t.

If $v \in \mathcal{F}_{p_1}$, the Perron family for $g_{W_t}(p, p_1)$, then $v = 0$ off a compact subset of W_t and $g_{W_t} > 0$ so that, by the maximum principle,

$$\sup_{W_t \setminus U_1} \left[v(p) - g_{W_t}(p, p_2) \right] \le \max \left(0, \sup_{\partial U_1} \left[v(p) - g_{W_t}(p, p_2) \right] \right)$$
$$\le \max \left(0, \sup_{\partial U_1} \left[g_{W_t}(p, p_1) - g_{W_t}(p, p_2) \right] \right) \le C.$$

Taking the supremum over all such v yields

$$\sup_{p \in W_t \setminus U_1} G_t(p, p_1, p_2) \le C.$$

Similarly,

$$\inf_{p \in W_t \setminus U_2} G_t(p, p_1, p_2) = - \sup_{p \in W_t \setminus U_2} -G_t(p, p_1, p_2) \ge -C,$$

and so

$$|G_t(p, p_1, p_2)| \le C$$

for all $p \in W_t \setminus \{U_1 \cup U_2\}$. The function $G_t + \log |z_1|$ extends to be harmonic in U_1, so, by the maximum principle, we have that

$$\sup_{U_1} |G_t + \log |z_1|| = \sup_{\partial U_1} |G_t + \log |z_1|| = \sup_{\partial U_1} |G_t| \le C.$$

Similarly,

$$\sup_{U_2} |G_t - \log |z_2|| = \sup_{\partial U_2} |G_t - \log |z_2|| = \sup_{\partial U_2} |G_t| \le C.$$

By normal families, there exists a sequence $t_n \to 0$ so that G_{t_n} converges uniformly on compact subsets of $W \setminus \{p_0, p_1, p_2\}$ to a function $G(p, p_1, p_2)$ harmonic on $W \setminus \{p_0, p_1, p_2\}$ satisifying (15.19), (15.20) and (15.21). The function $G(p, p_1, p_2)$ extends to be harmonic at p_0 because it is bounded in a punctured neighborhood of p_0. Indeed, we can transfer this problem to the unit disk via the coordinate map, then use the Poisson integral formula to create a bounded harmonic function on the disk with the same values on $\partial \mathbb{D}$. By Lindelöf's maximum principle, this is the harmonic extension. □

Combining Theorems 15.8 and 15.10, we have proved Koebe's uniformization theorem.

Theorem 15.12 (uniformization) *Suppose W is a simply-connected Riemann surface.*

(i) *If Green's function exists for W, then there is a one-to-one analytic map of W onto $\mathbb{D}$.*
(ii) *If W is compact, then there is a one-to-one analytic map of W onto $\mathbb{C}^*$.*
(iii) *If W is not compact and if Green's function does not exist for W, then there is a one-to-one analytic map of W onto $\mathbb{C}$.*

Since each of the spaces $\mathbb{D}, \mathbb{C}$ and $\mathbb{C}^*$ is second countable, we have the following consequence.

Corollary 15.13 (Rado) *Every Riemann surface satisfies the second axiom of countability.*

Proof The universal covering map π sends a countable base on the universal covering surface W to a countable base on W. □

We remark that the normal families argument used in the proof of Lemma 15.11 does not require second countability. If we find a sequence t_n so that G_{t_n} converges on a disk B then, on any compact set K containing B, the sequence G_{t_n} has only one subsequential limit by the uniqueness theorem. Thus the full sequence G_{t_n} converges on K. Alternatively, we can deduce Rado's Corollary 15.13 using only Theorem 15.5, 15.6 and 15.8, avoiding the construction of dipole Green's function entirely.

15.4 Classification of All Riemann Surfaces

We can now give a classification of all Riemann surfaces up to one-to-one analytic maps (conformal equivalence). By the uniformization theorem, the universal covering surface W^* of a Riemann surface W is, up to conformal equivalence, the disk, the plane or the sphere. By Corollary 14.16, the deck transformations form a properly discontinuous group $\mathbb{G}$ of one-to-one analytic maps of W^* onto W^*, such that W is conformally equivalent to $W^*/\mathbb{G}$. But, by Exercise 14.9, the deck transformations are LFTs. See also Exercise 14.8.

Theorem 15.14 *If $U = \mathbb{C}^*$, $\mathbb{C}$ or $\mathbb{D}$, and if $\mathbb{G}$ is a properly discontinuous group of LFTs of U onto U, then $U/\mathbb{G}$ is a Riemann surface. A function f is analytic, meromorphic, harmonic or subharmonic on $U/\mathbb{G}$ if and only if there is a function h defined on U which is (respectively) analytic, meromorphic, harmonic or subharmonic on U satisfying $h \circ \tau = h$ for all $\tau \in \mathbb{G}$ and $h = f \circ \pi$, where $\pi : U \to U/\mathbb{G}$ is the quotient map. Every Riemann surface is conformally equivalent to $U/\mathbb{G}$ for some such U and $\mathbb{G}$.*

Test your understanding of this result by checking all remaining details.

In Chapter 16, we will show that, up to conformal equivalance, the only Riemann surface covered by the $\mathbb{C}^*$ is $\mathbb{C}^*$, and that the only surfaces covered by $\mathbb{C}$ are $\mathbb{C}$, $\mathbb{C} \setminus \{0\}$ and tori. Any Riemann surface not conformally equivalent to one of these is covered by the disk $\mathbb{D}$.

15.5 Exercises

A

15.1 (a) Prove that an LFT M of $\mathbb{D}$ onto $\mathbb{D}$ is given by two reflections. Hint: If $M = \lambda(z-c)/(1-\bar{c}z)$, then $|M(0)| = |M^{-1}(0)|$. Let C be the circle orthogonal to $\partial\mathbb{D}$ centered at $1/\overline{M(0)}$ and let L be the diameter of $\mathbb{D}$ which bisects the angle determined by the three points $M(0), 0, M^{-1}(0)$. Reflect about L then reflect about C.

(b) If $a, b \in \partial\mathbb{D}$ give an explicit formula for reflection R about the circle orthogonal to $\partial\mathbb{D}$ through a and b, show that $R(0) = (a+b)/2$.

15.2 In Section 15.1, we used Carathéodory's theorem to conclude that a conformal map of the circular triangle T onto the upper half-plane extends to be a homeomorphism of the closure T onto the closure of the upper half-plane in $\mathbb{C}^*$. Give a more elementary proof by using Corollary 8.18. Hint: Near a vertex v of T, open up the region by applying $C/(z-v)$ then e^z to map a neighborhood of v in T to a neighborhood of 0 in $\mathbb{H}$.

15.3 If E_0 is a compact subset of a Riemann surface W_0 such that there exists a local barrier in $W = W_0 \setminus E_0$ for some $\zeta_0 \in E_0$, prove g_W exists. Then prove that $\lim_{p \in W \to \zeta_0} g_W(p, p_0) = 0$.

15.4 (a) Prove that if φ is a one-to-one analytic map of a Riemann surface W_1 onto a Riemann surface W_2, then Green's function on W_1 exists if and only if Green's function on W_2 exists. Moreover, show that $g_{W_2}(\varphi(p), \varphi(p_0)) = g_{W_1}(p, p_0)$.

(b) Prove that if $W_1 \subset W_2$ are Riemann surfaces, and if g_{W_2} exists, then g_{W_1} exists.

The Riemann mapping theorem is used in two places in the proof of the uniformization theorem. It is possible to avoid both by incorporating parts of the proof of the Riemann mapping theorem, so that the Riemann mapping theorem becomes a consequence of the uniformization theorem. The next two exercises outline the ideas. Fill in the details.

15.5 At the end of the proof of Theorem 15.8, we obtain a conformal map φ of W onto a possibly proper subset of $\mathbb{D}$. If it is proper, apply the "onto" portion of the proof of Theorem 10.11 to obtain a map σ mapping $\varphi(W)$ into $\mathbb{D}$ such that

$$|\sigma \circ \varphi(z)| > |\varphi(z)|. \tag{15.24}$$

By the argument used to establish (15.14) and (15.16), $g_W + \log |\sigma \circ \varphi| \leq 0$. But $g_W(p, p_0) = -\log |\varphi(p)|$, so that $\log |\sigma \circ \varphi| \leq \log |\varphi|$, contradicting (15.24).

15.6 At the conclusion of the proof of Theorem 15.10, we obtained a conformal map of W into $\mathbb{C}^*$. If two points were omitted, we obtained a conformal map φ of W onto $\mathbb{D}$, then obtained a contradiction. Another way to proceed is as follows: if there are at least two points $a, b \in \mathbb{C}^* \setminus \varphi_1(W)$ then use a construction from the proof of Theorem 8.11 to map $\varphi_1(W)$ into $\mathbb{D}$, then apply Exercise 15.4(b) to obtain a contradiction.

B

15.7 Prove that if $W = \mathbb{C} \setminus \{0, 1\}$ then Green's function g_W does not exist. Prove also that g_{W^*} does exist.

15.8 (a) Prove that a positive harmonic function on a Riemann surface without a Green's function is constant. Hint: Follow the proof of Theorem 15.5, except instead of requiring $\limsup_{p \to \alpha} u(p) \leq 0$ for $\alpha \in \partial U_0$, require u to be 0 off some compact set of W. Show $\omega(p) = \sup_u u(p)$ is not identically equal to 1, as otherwise the positive harmonic function will achieve its minimum on W at a point of $\partial r U$.

(b) Prove that Green's function with pole at p_0 exists on a Riemann surface W if and only if there exists a non-constant positive harmonic function on $W \setminus \{p_0\}$. Hint: Use the proof of (a) and Lindelöf's maximum principle.

15.9 Prove Montel's theorem using the modular function. Some care is needed if a sequence composed with the inverse of the modular function converges to a constant with absolute value 1 since there are copies of $\mathbb{C} \setminus \{0, 1\}$ arbitrarily close to some points on the unit circle.

15.10 (a) Suppose Ω is a bounded region in $\mathbb{C}$ bounded by finitely many analytic Jordan curves. Paste two copies of Ω together along the boundaries. Give the resulting surface a natural structure as a Riemann surface (called the "double" of Ω) by describing the coordinate charts and maps.

(b) With Ω as in part (a), suppose J is a finite union of open arcs on $\partial\Omega$ with pairwise disjoint closures. Use a similar construction as in (a), attaching the two copies of Ω along J to solve the mixed Dirichlet–Neumann problem: $\Delta u = 0$ in Ω, $u = f_1$ on $\partial\Omega \setminus J$ and $\nabla u \cdot n = 0$ on J, where $n = n(\zeta)$ is the unit normal to $\partial\Omega$ at ζ, and f_1 is continuous. Hint: See Exercise 8.9.

15.11 Cut out a finite number of equilateral triangles from a sheet of paper and paste them together along their edges. The resulting surface in $\mathbb{R}^3$ may not sit in the plane if there are fewer or more than six triangles meeting at a point. Construct a Riemann surface on the union of the triangles which agrees on the interiors of each triangle with the topology it inherits as a flat Euclidean triangle. If the surface is topologically a sphere then describe the conformal image on the sphere: what kind of curves are the edges, and how do they meet at a vertex?

15.12 Prove that there is no finite properly discontinuous group of LFTs mapping $\mathbb{D}$ onto $\mathbb{D}$, except the group containing just the identity map. Do the same for maps of $\mathbb{C}$ onto $\mathbb{C}$. Hint: See Exercise 15.1.

C

15.13 Suppose $D_1, \ldots, D_n$ are disjoint open disks whose boundaries are orthogonal to $\mathbb{R}$ and contained in $\{z : \mathrm{Re} z > 0\}$. Let $\Omega = \{z : \mathrm{Re} z > 0 \text{ and } \mathrm{Im} z > 0\} \setminus \cup_{j=1}^{n} \overline{D_j}$, let φ be a conformal map of Ω onto the upper half-plane $\mathbb{H}$ which fixes ∞, and let $E = \varphi(\overline{\Omega} \cap \mathbb{R})$. Reflect Ω about each bounding circle (including the imaginary axis), creating a larger region bounded by circles orthogonal to $\mathbb{R}$. Then continue reflecting about the newly created circles, and continue doing so as much as possible.

(a) Show that these reflections fill the upper half-plane.
(b) Show that $\mathbb{H}$ is conformally equivalent to the universal covering surface of $\mathbb{C} \setminus E$ using the Schwarz reflection principle.
(c) Prove that the generators of the group of deck transformations are given by reflection about the imaginary axis followed by reflection about ∂D_j for $j = 1, \ldots, n$.
(d) If E is a compact subset of $\mathbb{R}$ then $\mathbb{R} \setminus E = \cup_{j=1}^{\infty} I_j$, where I_j are disjoint open intervals. It is known that there is then a conformal map ψ of $\mathbb{H}$ onto $\mathbb{H} \setminus \cup D_j$, where D_j are closed disks orthogonal to $\mathbb{R}$ such that $\varphi(I_j) = D_j \cap H$. Use this fact to construct the universal cover and covering map of $\mathbb{C} \setminus E$.

16 Meromorphic Functions on a Riemann Surface

In this chapter we will study properly discontinuous groups of linear fractional transformations on the plane and on the disk, and give an introduction to the corresponding function theory on surfaces covered by the plane and disk.

16.1 Existence of Meromorphic Functions

Not all Riemann surfaces admit non-constant analytic or harmonic functions. If W is a compact Riemann surface then there are no non-constant analytic or harmonic functions on W by the maximum principle. However, Theorem 16.1 shows that there are plenty of meromorphic functions on a Riemann surface.

Theorem 16.1 *If W is a Riemann surface with distinct points $p_1, p_2, p_3 \in W$ then there is a meromorphic function f on W with $f(p_1) = 0$, $f(p_2) = 1$ and $f(p_3) = \infty$.*

Proof Let $G(p, p_1, p_2)$ and $G(p, p_3, p_2)$ be dipole Green's functions on W as in Lemma 15.11. If $z_\alpha : U_\alpha \to \mathbb{D}$ is a coordinate map, let $u(\zeta) = G(z_\alpha^{-1}(\zeta), p_1, p_2)$. Then $u_x - iu_y$ is analytic on $D_1 = z_\alpha(U_\alpha \setminus \{p_1, p_2\})$ with a simple pole at $z_\alpha(p_1)$ and residue -1, if $p_1 \in U_\alpha$, because $u(\zeta) + \log|\zeta - z_\alpha(p_1)|$ is harmonic in a neighborhood of $z_\alpha(p_1)$. Similarly $u_x - iu_y$ has a simple pole at $z_\alpha(p_2)$ with residue 1, if $p_2 \in U_\alpha$. Likewise, if $v(\zeta) = G(z_\alpha^{-1}(\zeta), p_3, p_2)$ then $v_x - iv_y$ is analytic on $D_2 = z_\alpha(U_\alpha \setminus \{p_3, p_2\})$ with a simple pole at $z_\alpha(p_3)$ and residue -1, if $p_3 \in U_\alpha$, and a simple pole at $z_\alpha(p_2)$ and residue 1, if $p_2 \in U_\alpha$. Then the function

$$F = \frac{v_x - iv_y}{u_x - iu_y}$$

is analytic on $z_\alpha(U_\alpha \setminus \{p_1, p_2, p_3\})$ with a zero of order at least 1 at $z_\alpha(p_1)$, a pole of order at least 1 at $z_\alpha(p_3)$, and a removable singularity at $z_\alpha(p_2)$. In fact, the value at $z_\alpha(p_2)$ is 1 because the residues of $u_x - iu_y$ and $v_x - iv_y$ are both 1 at $z_\alpha(p_2)$. Set $f = F \circ z_\alpha$ on U_α. Observe that the definition of F does not depend on the choice of the chart map z_α by the chain rule because the transition functions are analytic. Thus f is a well-defined meromorphic function on all of W. □

16.2 Properly Discontinuous Groups on $\mathbb{C}^*$ and $\mathbb{C}$

As defined in Chapter 14, a group $\mathbb{G}$ of LFTs on $W^* = \mathbb{C}^*, \mathbb{C}$ or $\mathbb{D}$ is properly discontinuous if each $z \in W^*$ has a neighborhood B^* so that $\tau(B^*) \cap B^* = \emptyset$ for all $\tau \in \mathbb{G} \setminus \mathbb{G}_0$, where

$\mathbb{G}_0$ contains only the identity map. By Theorem 15.14, every Riemann surface is given by a properly discontinuous group of LFTs on $\mathbb{C}^*$, $\mathbb{C}$ or $\mathbb{D}$.

Proposition 16.2 *The only properly discontinuous group of LFTs on $\mathbb{C}^*$ is the group $\mathbb{G}_0$ with one element, the identity map.*

Proof If $\sigma(z) = (az+b)/(cz+d)$ then there is a solution to $\sigma(z) = z$ in $\mathbb{C}^*$. But if the group is properly discontinuous then any element which is not the identity cannot have a fixed point. Thus $\sigma(z) \equiv z$, and Proposition 16.2 follows. □

For the remainder of this chapter we will assume $\mathbb{G}$ is a properly discontinuous group of LFTs on $\mathbb{C}$. Let $\mathbb{Z}$ denote the integers.

Theorem 16.3 *The properly discontinuous groups $\mathbb{G}$ on $\mathbb{C}$ are*

(i) *the group with one element $\mathbb{G}_0 = \{z\}$;*
(ii) *$\mathbb{G}_b = \{z + nb : n \in \mathbb{Z}\}$, where $b \in \mathbb{C}$, $b \neq 0$;*
(iii) *$\mathbb{G}_{b,c} = \{z + nb + mc : n, m \in \mathbb{Z}\}$, where $b, c \in \mathbb{C}$ with b/c not real.*

Proof If $\sigma(z) = (az + b)/(cz + d)$ is a map of $\mathbb{C}$ onto $\mathbb{C}$ then it cannot have a pole and so we can write $\sigma(z) = az + b$ for some constants a, b. But if $a \neq 1$ then σ has a fixed point in $\mathbb{C}$. Since the group is properly discontinuous, we must have $a = 1$. So each $\sigma(z) = z + b$ is determined by its value at 0, since $b = \sigma(0)$.

Suppose $\mathbb{G} \neq \mathbb{G}_0$. Set $r = \inf_{\sigma \in \mathbb{G}\setminus\mathbb{G}_0} |\sigma(0)|$. If $|\sigma_n(0)| \to r$, then $\sigma_n(0)$ has a cluster point. The values of elements of $\mathbb{G}$ at 0 form an additive group, so that the difference of two values at 0 is in this additive group. Because the group is properly discontinuous, this difference cannot be arbitrarily small. Thus there is a $b \in \mathbb{C}$ so that $z + b \in \mathbb{G}$ and

$$|b| = r = \inf_{\sigma \in \mathbb{G}\setminus\mathbb{G}_0} |\sigma(0)|. \tag{16.1}$$

If $\mathbb{G} \neq \mathbb{G}_b = \{z + nb : n \in \mathbb{Z}\}$, then choose $\tau = z + c \in \mathbb{G}$ with

$$|c| = |\tau(0)| = \inf\{|\alpha(0)| : \alpha \in \mathbb{G} \setminus \mathbb{G}_b\}. \tag{16.2}$$

If $\sigma(z) = z + b$ and $\tau(z) = z + c$ are chosen to satisfy (16.1) and (16.2), and if $c/b \in \mathbb{R}$, then $|c/b - n| \leq 1/2$ for some integer n. But then $|c - nb| \leq |b|/2$ and so $\delta = \tau - n\sigma \in \mathbb{G}$ satisfies $|\delta(0)| < |\sigma(0)|$. By our choice of σ, we must have $c - nb = 0$, contradicting our choice of c. We conclude c/b is not real.

Now suppose $\alpha \in \mathbb{G}$, $\alpha(z) = z + d$. Since c/b is not real, we can find real numbers t_1, t_2 so that $d = t_1 b + t_2 c$. Choose integers m, n so that $|t_1 - m| \leq \frac{1}{2}$ and $|t_2 - n| \leq \frac{1}{2}$. Set $a_1 = d - (mb + nc)$. Then

$$|a_1| = |(t_1 - m)b + (t_2 - n)c| \leq \frac{1}{2}|b| + \frac{1}{2}|c| \leq |c|. \tag{16.3}$$

But by (16.2), if a_1 is not an integer multiple of b then $|c| \leq |a_1|$, so each of the inequalities in (16.3) must be equalities. But then $|b| = |c|$, $|t_1 - m| = \frac{1}{2}$ and $|t_2 - n| = \frac{1}{2}$. If $t_1 - m = \frac{1}{2}$ and $t_2 - n = \frac{1}{2}$ then a_1 is the midpoint of the line segment between b and c, and hence $|a_1| < |c|$. By our choice of b and c, a_1 must be an integer multiple of b, in which case $d \in \mathbb{G}_{b,c}$. The same conclusion holds if $t_1 - m$ or $t_2 - n$ is negative since then a_1 is the midpoint of the segment from either b or $-b$ to either c or $-c$. We conclude that (iii) holds. This proves Theorem 16.3. □

If $\mathbb{G} = \{z + nb : n \in \mathbb{Z}\}$, let

$$\mathbb{S} = \{z : |z| < |z - nb| \text{ for all non-zero integers } n\}.$$

If $\mathbb{S}_1 = \{z \in \mathbb{C}; |\mathrm{Re} z| < 1/2\}$, then $\mathbb{S}$ is the rotated and dilated strip $b\mathbb{S}_1$. The analytic function

$$\pi(z) = e^{2\pi i z/b}$$

maps the plane onto the punctured plane $\mathbb{C} \setminus \{0\}$. It is a conformal map of $\mathbb{S}$ onto the slit plane $\mathbb{C} \setminus (-\infty, 0]$. The map π identifies the two edges of $\mathbb{S}$, mapping each of them to $(-\infty, 0]$. Moreover, $\pi(z) = \pi(w)$ if and only if $(z - w)(2\pi/b) = 2\pi n$ if and only if $z = \tau(w)$ for some $\tau \in \mathbb{G}$. The function π maps each translate of $\mathbb{S}$ by an integer multiple of b onto $\mathbb{C} \setminus (-\infty, 0]$, and maps the boundary of such translates onto $(-\infty, 0]$. A closed curve $\gamma \subset \mathbb{C} \setminus \{0\}$ can be lifted, using π^{-1} defined locally, to a curve $\gamma^* \in \mathbb{C}$. The curve γ^* will be closed if and only if γ is homotopic to a constant curve. Thus $\mathbb{C}$ is conformally equivalent to the universal covering surface of $\mathbb{C} \setminus \{0\}$ and $\mathbb{C}/\mathbb{G}$ is conformally equivalent to $\mathbb{C} \setminus \{0\}$. The map π is the projection map of the covering surface $\mathbb{C}$ onto the Riemann surface $\mathbb{C} \setminus \{0\}$. In this context, Theorem 15.14 says that g is entire and satisfies $g(z + b) = g(z)$ for all $z \in \mathbb{C}$ if and only if $g(z) = f(e^{2\pi i z/b})$ for some f which is analytic on $\mathbb{C} \setminus \{0\}$. A similar statement holds for meromorphic, harmonic and subharmonic functions.

Suppose now that $\mathbb{G}_{b,c} = \{z + nb + mc : n, m \in \mathbb{Z}\}$, where $b, c \in \mathbb{C}$ with b/c not real, such that (16.1) and (16.2) hold. Let $\mathbb{P}$ be the parallelogram with vertices 0, b, c and $b + c$. The elements of the group $\mathbb{G}$ translate $\overline{\mathbb{P}}$ to cover the plane. The Riemann surface $\mathbb{C}/\mathbb{G}_{b,c}$ is topologically a torus obtained by identifying the opposite sides of $\mathbb{P}$, and $\mathbb{C}$ is conformally equivalent to the universal covering surface of $\mathbb{C}/\mathbb{G}_{b,c}$. See Exercise 14.8.

Theorems 16.3 and 15.14 yield the following.

Corollary 16.4 *The plane is a universal covering surface of the plane, the punctured plane and tori. The sphere is a universal covering surface of the sphere. Each Riemann surface not conformally equivalent to one of these has the unit disk as a universal covering surface.*

16.3 Elliptic Functions

In this section we will explore function theory on a torus. Fix a properly discontinuous group $\mathbb{G}$ on $\mathbb{C}$ generated by $\sigma(z) = z + b$ and $\tau(z) = z + c$ with $c/b \notin \mathbb{R}$, such that (16.1) and (16.2) hold. The parallelogram $\mathbb{P}$ with vertices 0, b, c and $b + c$ is called a **fundamental domain** for the group $\mathbb{G}$. For convenience in the following, we let $\mathbb{P}_0$ denote $\mathbb{P}$ together with two adjacent edges:

$$\mathbb{P}_0 = \{tb + sc : 0 \le t < 1 \text{ and } 0 \le s < 1\}.$$

Thus every point in $\mathbb{C}$ is equivalent in a point in $\mathbb{P}_0$, and no two points in $\mathbb{P}_0$ are equivalent.

If g is meromorphic in $\mathbb{C}$ and $g(z + b) = g(z + c) = g(z)$, then g is called an **elliptic function**. By Theorem 15.14, g is an elliptic function if and only if $g \circ \pi^{-1}$ is a meromorphic function on the torus $\mathbb{C}/\mathbb{G}$, where π is the quotient map. The zeros and poles of an elliptic function are isolated, so there can be at most finitely many zeros and poles in $\mathbb{P}_0$.

Lemma 16.5 *If g is elliptic and has no poles, then g is constant.*

Proof Every point in $\mathbb{C}$ is equivalent to a point in the compact set $\overline{\mathbb{P}}$. But g has a maximum on $\overline{\mathbb{P}}$ and therefore in $\mathbb{C}$. By the maximum principle, g is constant. □

Lemma 16.6 *If g is elliptic then the number of zeros of g in $\mathbb{P}_0$ equals the number of poles of g in $\mathbb{P}_0$, counting multiplicity. Thus $g \circ \pi^{-1}$ is an n-to-one (counting multiplicity) map of $\mathbb{C}/\mathbb{G}$ onto $\mathbb{C}^*$, where n is the number of poles of g in $\mathbb{P}$.*

Proof Let $\mathbb{P}_\varepsilon = \mathbb{P} - \varepsilon(b+c)$. Then, for $\varepsilon > 0$ sufficiently small, g has no zeros or poles on $\partial\mathbb{P}_\varepsilon$. So $\mathbb{P}_\varepsilon$ contains exactly one point from each equivalence class of zeros and poles. Because $g(z+b) = g(z+c) = g(z)$, it is also true that $g'(z+b) = g'(z+c) = g'(z)$. So

$$\int_{\partial\mathbb{P}_\varepsilon} \frac{g'(z)}{g(z)} dz = 0,$$

because the integrals on opposite edges cancel. By the argument principle, the number of zeros of g in $\mathbb{P}_\varepsilon$ equals the number of poles of g in $\mathbb{P}_\varepsilon$. By the definition of $\mathbb{P}_0$, this statement holds for $\mathbb{P}_0$ as well.

Observe that if $c \in \mathbb{C}$ then the poles of $g - c$ are exactly the poles of g, and hence the number of zeros of $g - c$ in $\mathbb{P}_0$ equals the number of poles of g in $\mathbb{P}_0$. This proves the last statement of Lemma 16.6. □

Exercise 14.9 is the analog of Lemma 16.6 for analytic functions mapping the sphere into the sphere.

Lemma 16.7 *Suppose that g is elliptic, that $z_1, \ldots, z_n$ are the zeros of g in $\mathbb{P}_0$ and that $p_1, \ldots, p_n$ are the poles of g in $\mathbb{P}_0$, where zeros and poles are repeated in the list according to multiplicity. Then*

$$\sum_{j=1}^{n} (z_j - p_j) = nb + mc \in \{\sigma(0) : \sigma \in \mathbb{G}\},$$

for some $n, m \in \mathbb{Z}$.

Proof Let $\mathbb{P}_\varepsilon$ be the translate of $\mathbb{P}$ in the proof of Lemma 16.6. By the residue theorem,

$$\int_{\partial\mathbb{P}_\varepsilon} \frac{zg'(z)}{g(z)} dz = 2\pi i \sum_{j=1}^{n} (z_j - p_j). \tag{16.4}$$

Note that if I is an edge of $\partial\mathbb{P}_\varepsilon$ then $g(I)$ is a closed curve with winding number $n = \int_I g'(z)/g(z)dz/(2\pi i)$. But $(z+b)g'(z+b)/g(z+b) = zg'(z)/g(z) + bg'(z)/g(z)$. Similarly $(z+c)g'(z+c)/g(z+c) = zg'(z)/g(z) + cg'(z)/g(z)$. Thus (16.4) equals

$$-b\int_I \frac{g'(z)}{g(z)}dz - c\int_J \frac{g'(z)}{g(z)}dz = b\,2\pi in + c\,2\pi im,$$

where I and J are adjacent edges of $\partial\mathbb{P}_\varepsilon$ and $n, m \in \mathbb{Z}$. Lemma 16.7 follows. □

For example, if b/c is not real and $G = \{nb+mc : n, m \in \mathbb{Z}\}$, then Weierstrass's $\mathcal{P}$ function,

$$\mathcal{P}(z) = \frac{1}{z^2} + \sum_{\omega\in G\setminus\{0\}} \frac{1}{(z-\omega)^2} - \frac{1}{\omega^2},$$

satisfies $\mathcal{P}(z+b) = \mathcal{P}(z+c) = \mathcal{P}(z)$ and thus is elliptic. See Example 11.5. This function has a double pole at the equivalents of 0 and no other poles in $\mathbb{P}_0$. By Lemma 16.6, $\mathcal{P}$ a is two-to-one map of $\mathbb{P}_0$ onto $\mathbb{C}^*$, counting multiplicity.

Theorem 16.8 *Suppose b/c is not real and $G = \{nb+mc : n, m \in \mathbb{Z}\}$. Then*

$$\sigma(z) = z\prod_{\omega\in G\setminus\{0\}} \left(1 - \frac{z}{\omega}\right)e^{\frac{z}{\omega}+\frac{1}{2}\frac{z^2}{\omega^2}}$$

converges and is entire. Moreover, there exist $t_1, t_2 \in \mathbb{C}$ such that

$$\sigma(z+b) = -\sigma(z)e^{t_1(z+\frac{b}{2})} \text{ and } \sigma(z+c) = -\sigma(z)e^{t_2(z+\frac{c}{2})}. \tag{16.5}$$

Proof In Example 11.5 we proved that there are at most Ck points $\omega \in G$ satisfying $k \le |\omega| < k+1$. Thus

$$\sum_{\omega\in G\setminus\{0\}} \frac{1}{|\omega|^3} < \infty.$$

By Theorem 11.11, the infinite product in Theorem 16.8 converges to an entire function. Moreover,

$$\frac{\sigma'(z)}{\sigma(z)} = \frac{1}{z} + \sum_{\omega\in G\setminus\{0\}} \left(\frac{1}{z-\omega} + \frac{1}{\omega} + \frac{z}{\omega^2}\right)$$

converges. But then $-(\frac{\sigma'}{\sigma})' = \mathcal{P}$ by Example 11.5. By the invariance of $\mathcal{P}$,

$$\left(\frac{\sigma'}{\sigma}\right)'(z+b) = \left(\frac{\sigma'}{\sigma}\right)'(z+c) = \left(\frac{\sigma'}{\sigma}\right)'(z).$$

By an integration, it follows that

$$\frac{\sigma'}{\sigma}(z+b) = \frac{\sigma'}{\sigma}(z) + t_1,$$

for some constant t_1, and hence

$$\sigma(z+b) = \sigma(z)de^{t_1 z},$$

for some constant d. We are allowed to rearrange the terms in the infinite product representation of σ because it converges absolutely. Multiplying the terms involving ω and $-\omega$, if $\omega \neq 0$, gives even functions. Thus σ is odd, and $\sigma(b/2) = \sigma(-b/2)de^{-t_1 b/2} = -\sigma(b/2)de^{-t_1 b/2}$. Thus $d = -e^{t_1 b/2}$. This proves the first equality in (16.5), and the second follows similarly. □

The next theorem gives a complete characterization of elliptic functions.

Theorem 16.9 *Suppose b/c is not real. Choose $z_1, \ldots, z_n, p_1, \ldots, p_n \in \mathbb{C}$ satisfying $\sum_{j=1}^{n}(z_j - p_j) = 0$. Let D be a constant, then*

$$g(z) = D\prod_{j=1}^{n}\frac{\sigma(z-z_j)}{\sigma(z-p_j)} \tag{16.6}$$

is an elliptic function with poles $\{p_j + kb + mc : k, m \in \mathbb{Z}\}$ and zeros $\{z_j + kb + mc : k, m \in \mathbb{Z}\}$, $j = 1, \ldots, n$, and $g(z+b) = g(z+c) = g(z)$, for all $z \in \mathbb{C}$. Conversely, every elliptic function satisfying $g(z+b) = g(z+c) = g(z)$ for all $z \in \mathbb{C}$ can be written in this way.

Proof Using the notation of Theorem 16.8,

$$g(z+b) = D\prod_{j=1}^{n}\frac{-\sigma(z-z_j)e^{t_1(z-z_j+\frac{b}{2})}}{-\sigma(z-p_j)e^{t_1(z-p_j+\frac{b}{2})}} = g(z)e^{t_1\sum(p_j-z_j)} = g(z).$$

Similarly $g(z+c) = g(z)$.

Now suppose h is elliptic with $h(z+b) = h(z+c) = h(z)$. If the zeros of h in $\mathbb{P}_0$ are $z_1, \ldots, z_n$, and poles in $\mathbb{P}_0$ are $p_1, \ldots, p_n$, repeated according to multiplicity, then, by Lemma 16.7, $\sum(z_j - p_j) = kb + mc$ for some $k, m \in \mathbb{Z}$. If we replace z_1 by $z_1 - kb - mc$, then $\sum(z_j - p_j) = 0$. Now form g as in (16.6). Then g has exactly the same zeros and poles as h so that g/h is elliptic without zeros or poles. By Lemma 16.5, $h = Dg$ for some constant D. □

The Weierstrass function $\mathcal{P}$ is even so there exists $\alpha \in \mathbb{P}_0$ with $\mathcal{P}(\alpha) = 0$ and $\mathcal{P}(b+c-\alpha) = \mathcal{P}(-\alpha) = 0$. It follows from Theorem 16.9 that the Weierstrass $\mathcal{P}$ function can be written in the form

$$\mathcal{P}(z) = D\frac{\sigma(z-\alpha)\sigma(z+\alpha)}{\sigma(z)^2},$$

where σ is the infinite product in the statement of Theorem 16.8, and D is a constant. Since $z^2\mathcal{P}(z) \to 1$ and $\sigma(z)/z \to 1$ as $z \to 0$, we must have $1 = D\sigma(-\alpha)\sigma(\alpha)$. The first explicit formula for α was found in 1981.

16.4 Fuchsian Groups

A properly discontinuous group of LFTs of $\mathbb{D}$ onto $\mathbb{D}$ is called a **Fuchsian group**. The metric used for function theory on $\mathbb{D}$ is called the **pseudohyperbolic metric**, and it is given by

$$\rho(z, w) = \left|\frac{z-w}{1-\overline{w}z}\right|.$$

The invariant form of Schwarz's lemma says that an analytic function f mapping $\mathbb{D}$ into $\mathbb{D}$ is a contraction in this metric: $\rho(f(z), f(w)) \le \rho(z, w)$, with equality if and only if f is an LFT of $\mathbb{D}$ onto $\mathbb{D}$. See Exercises 3.9 and 6.8 for more information on this metric.

If $\mathbb{G}$ is a Fuchsian group then the **normal fundamental domain** for $\mathbb{G}$ is

$$\mathcal{F} = \{z \in \mathbb{D} : \rho(z, 0) < \rho(z, \sigma(0)) \text{ for all } \sigma \in \mathbb{G} \setminus \mathbb{G}_0\},$$

where $\mathbb{G}_0$ is the identity map. Every $w \in \mathbb{D}$ can be written as $w = \sigma(z)$ for some $z \in \overline{\mathcal{F}}$ and $\sigma \in \mathbb{G}$ because $\rho(w, \sigma(0)) = \rho(\sigma^{-1}(w), 0)$ by Exercise 3.9(a). If $\sigma(0)$ is a closest equivalent of 0 to w, then $z = \sigma^{-1}(w) \in \overline{\mathcal{F}}$. Since $\mathbb{G}$ is properly discontinuous, $\mathcal{F}$ contains a neighborhood of 0. If $\rho(z, 0) = \rho(z, \sigma(0))$, then, by Exercise 3.9(b),

$$|1 - \overline{\sigma(0)}z|^2 = 1 - |\sigma(0)|^2,$$

so that

$$\left|\frac{1}{\overline{\sigma(0)}} - z\right|^2 = \frac{1}{|\sigma(0)|^2} - 1.$$

This means that z lies on the circle centered at $1/\overline{\sigma(0)}$, the reflection of $\sigma(0)$ about $\partial\mathbb{D}$, with radius $\sqrt{\frac{1}{|\sigma(0)|^2} - 1}$. Thus the points equidistant between 0 and $\sigma(0)$ in the pseudohyperbolic metric lie on a circular arc C_σ which is orthogonal to $\partial\mathbb{D}$. See Exercise 16.3 and Figure 16.1.

If the closed disk bounded by this circle is denoted by D_σ then

$$\mathcal{F} = \mathbb{D} \setminus \cup_{\sigma \in \mathbb{G} \setminus \mathbb{G}_0} D_\sigma.$$

If $\sigma(z) = \lambda(\frac{z-a}{1-\bar{a}z})$, with $|\lambda| = 1$ and $|a| < 1$, then $\sigma(0) = -\lambda a$ and $\sigma^{-1}(0) = a$ so that $|\sigma(0)| = |\sigma^{-1}(0)|$. Thus the circles associated with $\sigma(0)$ and $\sigma^{-1}(0)$ have the same radius and their centers have the same absolute value.

Note also $\rho(z, 0) = \rho(z, \sigma^{-1}(0))$ if and only if $\rho(\sigma(z), \sigma(0)) = \rho(\sigma(z), 0)$ by Exercise 3.9(a). Thus σ maps the circular arc $C_{\sigma^{-1}}$ onto the circular arc C_σ. By Exercise 15.1, the map σ is then given by reflection about the diameter, which lies halfway between $\sigma(0)$ and $\sigma^{-1}(0)$, followed by reflection about the circular arc C_σ. This implies that a point of intersection in $\mathbb{D}$ of the two circles is a fixed point of the map σ. Properly discontinuous maps have no fixed points so the two circles cannot intersect in $\mathbb{D}$. Circular arcs associated with $\sigma, \tau \in \mathbb{G}$, with $\sigma \neq \tau^{-1}$, however, can intersect in $\mathbb{D}$. The boundary of $\mathcal{F}$ then consists of a subset of $\partial\mathbb{D}$ together with subsets of circles orthogonal to $\partial\mathbb{D}$. Because the group is properly discontinuous, there can be only finitely many $\sigma(0)$ in $|z| < r < 1$, and hence the subsets of these circles are subarcs, with at most finitely many arcs in $|z| < r$ for each $r < 1$. The arcs are identified in pairs by elements of $\mathbb{G}$.

Note that the equalities $\sigma(0) = -\lambda a$ and $\sigma^{-1}(0) = a$ show that we can recover the LFT σ from subarcs of C_σ and $C_{\sigma^{-1}}$ by geometrically locating the centers of the corresponding circles.

We have established the following result.

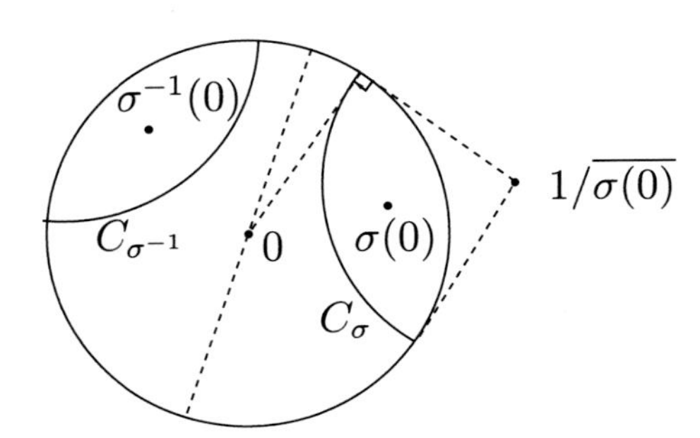

Figure 16.1 On C_σ, $\rho(z, 0) = \rho(z, \sigma(0))$.

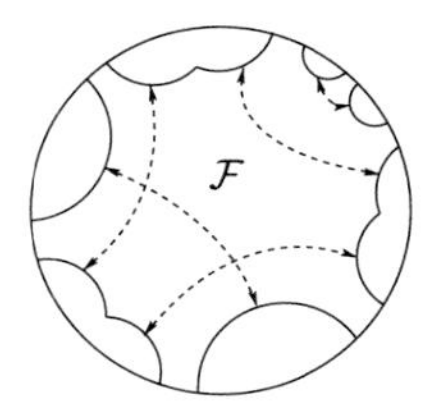

Figure 16.2 Normal fundamental domain.

Theorem 16.10 *Suppose $\mathbb{G}$ is a Fuchsian group with normal fundamental domain $\mathcal{F}$. Then*

$$\mathbb{D} = \bigcup_{\tau \in \mathbb{G}} \tau(\overline{\mathcal{F}} \cap \mathbb{D}).$$

Moreover, $\partial\mathcal{F} \cap \mathbb{D}$ consists of arcs of circles orthogonal to $\partial\mathbb{D}$ which are identified in pairs $C_\sigma \cap \partial\mathcal{F}$, $C_{\sigma^{-1}} \cap \partial\mathcal{F}$ by elements $\sigma \in \mathbb{G}$. The Riemann surface $\mathbb{D}\backslash\mathbb{G}$ is conformally equivalent to the Riemann surface obtained from $\mathcal{F}$ by this identification. See Figure 16.2.

If

$$G_1 = \{\sigma \in \mathbb{G} : \mathbb{D} \cap C_\sigma \cap \partial\mathcal{F} \neq \emptyset\},$$

then the subgroup of $\mathbb{G}$ generated by G_1 is still properly discontinuous and has the same normal fundamental domain and thus corresponds to the same Riemann surface. Thus $\mathbb{G}$ is the smallest group containing G_1. The normal fundamental domain gives us a geometric way of visualizing the corresponding Riemann surface and the associated fundamental group (see Figure 16.2).

16.5 Blaschke Products and Convergence Type

Theorem 15.14 shows that function theory on a Riemann surface W covered by $\mathbb{D}$ can be transferred to function theory on $\mathbb{D}$. In this section we will illustrate this principle by constructing a Green's function, solving the Dirichlet problem and building analytic functions using the corresponding Fuchsian group on $\mathbb{D}$.

Not all Riemann surfaces covered by the unit disk have a Green's function. For example, $W = \mathbb{C}\backslash\{0, 1\}$ does not have a Green's function. Indeed, each v in the Perron family for $g(z, p_0)$ is bounded in $|z| < r$ by $\sup_{\{|z|=r\}} g(z, p_0)$ for r sufficiently small by the maximum principle. Thus $g_W(z, p_0)$ extends harmonically to a neighborhood of 0. Similarly, $g(z, p_0)$ extends harmonically to a neighborhood of 1 and ∞, and hence $-g(z, p_0)$ would have a maximum on $\mathbb{C}^* \setminus \{p_0\}$, which is impossible.

However, it is possible to tell if a Green's function for a Riemann surface W exists directly from the corresponding Fuchsian group.

Theorem 16.11 *Suppose W is a Riemann surface with universal covering surface conformally equivalent to $\mathbb{D}$, and suppose $\pi : \mathbb{D} \to W$ is a universal covering map. Let $\mathbb{G}$ be the corresponding Fuchsian group. Then Green's function g_W exists if and only if*

$$\sum_{\sigma \in \mathbb{G}} 1 - |\sigma(0)|^2 < \infty. \tag{16.7}$$

If (16.7) *holds then the Blaschke product*

$$B(z) = \prod_{\sigma \in \mathbb{G}} \frac{|\sigma(0)|}{-\sigma(0)} \left(\frac{z - \sigma(0)}{1 - \overline{\sigma(0)}z} \right) = \prod_{\sigma^{-1} \in \mathbb{G}} \frac{|\sigma^{-1}(0)|}{\sigma^{-1}(0)} \sigma^{-1}(z)$$

converges and

$$g_W(\pi(z), \pi(0)) = -\log |B(z)|.$$

If (16.7) holds, then we say that the Fuchsian group $\mathbb{G}$ is of **convergence type**. Otherwise, we say that $\mathbb{G}$ is of **divergence type**.

Proof If g_W exists, then, by Theorem 15.6,

$$g_W(\pi(z), \pi(0)) = \sum_{\sigma \in \mathbb{G}} -\log \left| \frac{z - \sigma(0)}{1 - \overline{\sigma(0)}z} \right|. \tag{16.8}$$

Because $\lim_{x \uparrow 1} \frac{-\log x}{1 - x^2} = \frac{1}{2}$, the sum in (16.8) converges if and only if

$$\sum_{\sigma \in \mathbb{G}} 1 - \left| \frac{z - \sigma(0)}{1 - \overline{\sigma(0)}z} \right|^2 < \infty. \tag{16.9}$$

Setting $z = 0$, we obtain (16.7). By Theorem 11.19, the Blaschke product $B(z)$ in the statement of Theorem 16.11 converges, and $g_W(\pi(z), \pi(0)) = -\log |B(z)|$.

Conversely, if (16.7) holds, then, by Theorem 11.19, the Blaschke product B converges uniformly and absolutely on compact subsets of $\mathbb{D}$ and vanishes precisely at $\{\sigma(0) : \sigma \in \mathbb{G}\}$. Thus $G(z) = -\log |B(z)|$ is harmonic in $\mathbb{D} \setminus \{\sigma(0) : \sigma \in \mathbb{G}\}$. Moreover, $B \circ \sigma$ has exactly the same zero set as B so that $|B \circ \sigma| = |B|$ by Theorem 11.20. Thus $G \circ \sigma = G$ for all $\sigma \in \mathbb{G}$. Then $H = G \circ \pi^{-1}$ is a well-defined function which is harmonic and positive on $W \setminus \{\pi(0)\}$ with the property that $H(w) + \log |w - \pi(0)|$ is bounded in a neighborhood of $\pi(0)$. So if v is a subharmonic function in the Perron family for $g_W(w, \pi(0))$ then $v \leq H$. By Theorem 13.3, g_W exists. □

We remark that, by Theorem 11.20, $B \circ \sigma = c_\sigma B$, where c_σ is a constant of absolute value 1. But because the composition will have different convergence factors for the linear transformations in the product, and we cannot necessarily rearrange these convergence factors, the constants c_σ are not necessarily equal to 1.

For the second illustration we give an explicit solution to the Dirichlet problem for Riemann surfaces contained in a larger Riemann surface. Theorem 16.12 below remains true under less restrictive hypotheses, but we will be content with the statement below to illustrate the idea. If $\tau(z) = \lambda(z - a)/(1 - \overline{a}z)$ then

$$\frac{\tau'(z)}{\tau(z)} = \frac{1 - |a|^2}{(z - a)(1 - \overline{a}z)}.$$

If $z = e^{it}$, then

$$\frac{z\tau'(z)}{\tau(z)} = \frac{1-|a|^2}{|e^{it}-a|^2} = P_a(e^{it}),$$

where P_a is the Poisson kernel on $\partial\mathbb{D}$ for the point $a = \tau^{-1}(0) \in \mathbb{D}$. We can use this to give a solution to the Dirichlet problem on nicely bounded Riemann surfaces.

Theorem 16.12 *Suppose W is a Riemann surface such that $\overline{W}$ is a compact subset of a Riemann surface W_1 with $W_1 \setminus \overline{W} \neq \emptyset$. Then $\mathbb{D}$ is a universal covering surface of W. Let π be a universal covering map $\pi : \mathbb{D} \to W$ and let $\mathbb{G}$ be the corresponding Fuchsian group. Suppose that the boundary of the normal fundamental domain $\mathcal{F}$ consists of finitely many arcs on circles orthogonal to $\partial\mathbb{D}$ and finitely many arcs on $\partial\mathbb{D}$, none of which are a single point. Suppose also that π extends to be a continuous map on $\overline{\mathcal{F}}$. If f is continuous on ∂W then set*

$$u(z) = \int_{\partial\mathbb{D}} P_z(e^{it}) f(\pi(e^{it}))\frac{dt}{2\pi}. \tag{16.10}$$

Then $u \circ \tau = u$ for all $\tau \in \mathbb{G}$ and $h(w) = u(\pi^{-1}(w))$ is a well-defined harmonic function on W which is continuous on $\overline{W}$ and satisfies $h = f$ on ∂W. Moreover, we can write

$$u(z) = \int_{\mathcal{F}_b}\left(\sum_{\sigma\in\mathbb{G}} P_{\sigma(z)}(e^{it})\right) f(\pi(e^{it}))\frac{dt}{2\pi} \tag{16.11}$$

$$= \sum_{\sigma\in\mathbb{G}}\int_{\mathcal{F}_b} P_{\sigma(z)}(e^{it}) f(\pi(e^{it}))\frac{dt}{2\pi}, \tag{16.12}$$

where $\mathcal{F}_b = \partial\mathcal{F} \cap \partial\mathbb{D}$.

Equality (16.12) says that $u(z)$ is the sum of the harmonic extensions of $\chi_{\mathcal{F}_b}(f\circ\pi)$ at the equivalents of z, where $\chi_{\mathcal{F}_b}$ is the characteristic function of $\mathcal{F}_b$. If $W \subset \mathbb{C}^*$ such that $\mathbb{C}^* \setminus W$ consists of finitely many compact sets, each of which contains more than one point, then, by Exercise 14.11, the corresponding Fuchsian group is finitely generated and satisfies the hypotheses of Theorem 16.12. See Exercise 16.4.

Proof Because W cannot be the sphere, the plane, the punctured plane or a torus, W must be covered by the disk, by Corollary 16.4. A short computation using Exercise 3.9(b) shows that

$$P_z(\tau(e^{it}))\frac{e^{it}\tau'(e^{it})}{\tau(e^{it})} = P_{\tau^{-1}(z)}(e^{it}).$$

Changing variables $e^{it} = \tau(e^{i\theta})$ in (16.10) yields

$$u(z) = \int_{\partial\mathbb{D}} P_z(\tau(e^{i\theta})) f(\pi(e^{i\theta}))\frac{e^{i\theta}\tau'(e^{i\theta})}{\tau(e^{i\theta})}\frac{d\theta}{2\pi}$$

$$= \int_{\partial\mathbb{D}} P_{\tau^{-1}(z)}(e^{i\theta}) f(\pi(e^{i\theta}))\frac{d\theta}{2\pi} = u(\tau^{-1}(z)).$$

Thus there is a harmonic function h on W so that $h(\pi(z)) = u(z)$. As shown in the proof of Theorem 16.10, each $\sigma \in \mathbb{G}$ is given by reflection about the diameter equidistant between

$\sigma(0)$ and $\sigma^{-1}(0)$, followed by reflection about the orthogonal circle C_σ. Since F_b consists of finitely many arcs, none of which is a single point, the set $E = \cup_{\sigma\in\mathbb{G}}\sigma(\mathcal{F}_b)$ contains a neighborhood of $\mathcal{F}_b$ in $\partial\mathbb{D}$. This implies that $f\circ\pi$ is continuous in a neighborhood of $\mathcal{F}_b$. By Schwarz's theorem, u extends to be continuous and equal to $f\circ\pi$ on $\mathcal{F}_b$. Since $\pi(\mathcal{F}_b) = \partial W$, h extends to be continuous on $\overline{W}$ with $h = f$ on ∂W.

We claim that the set E has length 2π. Note that if $\zeta\in\sigma(\mathcal{F}_b)\cap\tau(\mathcal{F}_b)$ with $\sigma,\tau\in\mathbb{G}$, then $\zeta_1 = \sigma^{-1}(\zeta)\in\mathcal{F}_b$ and $\zeta_2 = \tau^{-1}(\zeta)\in\mathcal{F}_b$, with $\alpha(\zeta_1) = \zeta_2$, where $\alpha = \tau^{-1}\circ\sigma\in\mathbb{G}$. But this only occurs if ζ_1 is an endpoint of one of the finitely many intervals in $\mathcal{F}_b$. Set

$$\omega(z) = \frac{1}{2\pi}\int_E P_z(e^{it})dt.$$

Then $\omega(\tau(z)) = \omega(z)$ for all $z\in\mathbb{D}$ by the change of variables argument used above. Moreover, by Schwarz's theorem, $\omega = 1$ on a neighborhood of $\mathcal{F}_b$ in $\partial\mathbb{D}$, and hence $\omega\circ\pi^{-1}(w)$ is a well-defined harmonic function on W which equals 1 on ∂W. By the maximum principle, $\omega\circ\pi^{-1}(w) = 1$ for all $w\in W$. In particular, $\omega(0) = \int_E \frac{dt}{2\pi} = 1$. This proves that E has length 2π. In particular, $\int_{\partial\mathbb{D}\setminus E} P_z(e^{it})f(\pi(e^{it}))dt = 0$. Thus (16.11) holds, and (16.12) follows by the change of variables argument presented above. □

It is harder to represent analytic functions on a Riemann surface because of the difficulty dealing with harmonic conjugates. However, we have the following. Differentiating the Blaschke product B in Theorem 16.11 logarithmically,

$$\frac{B'(z)}{B(z)} = \sum_{\tau\in\mathbb{G}}\frac{\tau'(z)}{\tau(z)} = \sum_{\tau\in\mathbb{G}}\frac{1-|\tau(0)|^2}{(z-\tau(0))(1-\overline{\tau(0)}z)}. \tag{16.13}$$

Locally on $\mathbb{D}\setminus\{\sigma(0) : \sigma\in\mathbb{G}\}$ we can find an analytic function F such that $\mathrm{Re}F = g_W(\pi(z),\pi(0))$. Equation (16.13) is an explicit expression for $F'(z)$ which is meromorphic in $\mathbb{D}$ with simple poles at the equivalents of 0, and residue 1. If W satisfies the hypotheses in Theorem 16.12, then the zeros of B, which are the equivalents of 0, do not accumulate on $\mathcal{F}_b$, so that B and B' extend to be analytic across a neighborhood of $\mathcal{F}_b$ in $\partial\mathbb{D}$. See Exercise 11.4. Moreover,

$$\left.\frac{zB'(z)}{B(z)}\right|_{z=e^{it}} = \sum_{\tau\in\mathbb{G}} P_{\tau(0)}(e^{it}) > 0$$

on $\mathcal{F}_b$.

Theorem 16.13 *Suppose W, π and $\mathbb{G}$ satisfy the hypotheses in Theorem 16.12. If f is bounded and analytic on $\mathbb{D}$, then set*

$$E(f)(z) = \sum_{\tau\in\mathbb{G}} f\circ\tau(z)\tau'(z)\frac{B(z)}{\tau(z)}\frac{1}{B'(z)}. \tag{16.14}$$

Then $E(f)$ is analytic on $\mathbb{D}\setminus\{z : B'(z) = 0\}$ and satisfies $E(f)\circ\tau = E(f)$ for all $\tau\in\mathbb{G}$. The function $H\equiv E(f)\circ\pi^{-1}$ is a well-defined meromorphic function on W with poles only at the critical points of $g_W(z,\pi(0))$. Moreover, if h is meromorphic on W then $E(h\circ\pi\cdot f) = h\circ\pi\cdot E(f)$, and, since $E(1) = 1$, $E(h\circ\pi) = h\circ\pi$. Finally,

$$\limsup_{w\to\partial W}|H(w)| \le \limsup_{z\to\partial\mathbb{D}}|f(z)|.$$

Theorem 16.13 says that the map $f \to E(f) \circ \pi^{-1}$ is a projection from the bounded analytic functions on $\mathbb{D}$ onto the meromorphic functions on W bounded near ∂W with poles only at the critical points of g_W.

Proof The analyticity of $E(f)$ follows from the infinite product representation of B given in Theorem 16.11. The Blaschke product B is not necessarily invariant under the Fuchsian group $\mathbb{G}$, but the zeros of the Blaschke product B are invariant. So

$$B \circ \tau = c_\tau B, \tag{16.15}$$

where c_τ is a constant with absolute value 1. A short computation then proves that $E(f) \circ \tau = E(f)$ for all $\tau \in \mathbb{G}$. The critical points of $g_W(w, \pi(0))$ are the places where its gradient vanishes. By the Cauchy–Riemann equations and Theorem 16.11, the critical points correspond precisely to the zeros of B'. The identity $E(h \circ \pi \cdot f) = h \circ \pi E(f)$ follows from the chain rule and (16.15) because $\pi \circ \tau = \pi$. The identity $E(1) = 1$ follows from (16.13). To prove the final statement of Theorem 16.13, let $w_n \in W \to \zeta \in \partial W$. Choose $z_n \in \overline{\mathcal{F}}$ with $\pi(z_n) = w_n$. Then $z_n \to \alpha \in \mathcal{F}_b$. But the sum in (16.14) converges uniformly on $\mathcal{F}_b$ since B' and B are analytic across $\mathcal{F}_b$. Moreover, $z\tau'(z)/\tau(z) > 0$ and $B(z)/(zB'(z)) > 0$ on a neighborhood of $\mathcal{F}_b$ in $\partial\mathbb{D}$. Thus

$$\limsup_{w \to \partial W} H(w) \le \limsup_{z \to \partial\mathbb{D}} |f(z)| \cdot E(1) = \limsup_{z \to \partial\mathbb{D}} |f(z)|. \qquad \square$$

Note that the zeros of B' in Theorem 16.13 do not accumulate on $\mathcal{F}_b$ and hence g_W has only finitely many critical points in W. The function g_W is not necessarily differentiable on $\partial W \subset W_1$, but this comes from the possible non-differentiability of π, not B, under our assumptions on W.

16.6 Exercises

A

16.1 Set $G = \{z = nb + mc : n, m \in \mathbb{Z}\}$. Prove b, c satisfy (16.1) and (16.2) if and only if $-\frac{1}{2} \le \mathrm{Re}(c/b) \le \frac{1}{2}$ and $|c/b| \ge 1$. Draw a picture of the region in which c lies, given b.

16.2 Find a normal fundamental domain for a torus with one point removed.

16.3 Prove that the circle C_σ in Section 16.3 is orthogonal to $\partial\mathbb{D}$. Hint: Use the converse of Pythagoras's theorem.

B

16.4 If $W \subset \mathbb{C}^* $ such that $\mathbb{C}^* \setminus W$ consists of finitely many pairwise disjoint compact connected sets, each of which contains more than one point, then, by Exercise 14.11, the corresponding Fuchsian group is finitely generated. Prove that W satisfies the hypotheses of Theorem 16.12. Hint: First find a conformal map of W onto a region bounded by finitely many analytic Jordan curves, by applying a finite sequence of conformal maps of simply-connected regions onto the disk.

16.5 (a) Suppose that b/c is not real and that $G = \{nb + mc : n, m \in \mathbb{Z}\}$. Prove that the Weierstrass $\mathcal{P}$ function satisfies $\mathcal{P}(z) = \frac{1}{z^2} + Az^2 + Bz^4 + \ldots$ near 0, where

$$A = 3 \sum_{\omega \in G\setminus\{0\}} \omega^{-4} \text{ and } B = 5 \sum_{\omega \in G\setminus\{0\}} \omega^{-6}.$$

(b) Prove that $(\mathcal{P}')^2 - 4\mathcal{P}^3 + 20A\mathcal{P} = -28B$. An equation of the form $y^2 = x^3 + ax + b$ is called an **elliptic curve**. Hint: Use (a) to show that the left-hand side is analytic and elliptic.

(c) Use (b) to show that $\mathcal{P}$ is the inverse of an elliptic integral:

$$z - z_0 = \int_{\mathcal{P}(z_0)}^{\mathcal{P}(z)} \frac{d\zeta}{\sqrt{4\zeta^3 - 20A\zeta - 28B}}.$$

The square root and path of integration from $\mathcal{P}(z_0)$ to $\mathcal{P}(z)$ must be properly chosen.

(d) Find the Schwarz–Christoffel integral for a conformal map of the upper half-plane onto a rectangle such that ∞ is mapped to one of the vertices. Compare your answer with (c).

16.6 (a) Prove that two tori are conformally equivalent if and only if there is a linear map which maps one fundamental domain onto the other.

(b) Give a list of fundamental domains, exactly one in each conformal equivalence class.

16.7 If W is a torus as in Section 16.2, find the dipole Green's function in terms of the function σ given in Theorem 16.8. Hint: Use σ to build a function with the right logarithmic singularities then add $ax + by$ for some a, b with $z = x + iy$.

16.8 If $\mathbb{G}$ is a group of deck transformations associated with a torus W as in Section 16.2, set $\mathcal{F} = \{z : |z| < |z - \sigma(0)| : \sigma \in \mathbb{G}, \ \sigma(z) \not\equiv z\}$.

(a) Prove that if $w \in \mathbb{C}$, then there exists $z \in \overline{\mathcal{F}}$ such that $w = \sigma(z)$, for some $\sigma \in \mathbb{G}$.

(b) Show that $\mathcal{F}$ is bounded by four or six line segments.

(c) Prove that the Riemann surface obtained by identifying opposite edges of $\mathcal{F}$ is conformally equivalent to W.

Remark: $\mathcal{F}$ is the analog in $\mathbb{C}$ of the normal fundamental domains in $\mathbb{D}$.

16.9 Suppose W satisfies the hypotheses of Theorem 16.13 and suppose f is bounded and analytic on $\mathbb{D}$ with the property that it is invariant at the zeros of B': if $B'(z) = 0$, then $f \circ \tau(z) = f(z)$ for all $\tau \in \mathbb{G}$. Prove that $E(f)$ is bounded and analytic on $\mathbb{D}$ and that $H(w) = E(f) \circ \pi^{-1}(w)$ is a well-defined bounded analytic function on W.

C

16.10 Suppose $\Omega \subset \mathbb{C}$ is a region bounded by $n < \infty$ analytic Jordan curves.

(a) Prove that

$$u(z) = \int_{\partial\Omega} \frac{\partial g_W(\zeta, z)}{\partial \eta_\zeta} f(\zeta) \frac{|d\zeta|}{2\pi}$$

is the harmonic extension of $f \in C(\partial\Omega)$ to Ω, where η_ζ is the unit inner normal at $\zeta \in \partial\Omega$ and $g_W(z,b)$ is Green's function with pole at b. Hint: Use Green's theorem or use $\partial g_W/\partial s = 0$, where s is the arc-length on $\partial\Omega$, $\partial g_W/\partial\eta > 0$ and $zB'(z)/B(z) > 0$ on $\mathcal{F}_b$.

(b) Prove that, for each $b \in W$, $\nabla g_W(\zeta, b) = (0,0)$ has at most $n-1$ solutions in Ω.

(c) Prove that

$$\int_{\partial\Omega} \frac{\partial g_W(\zeta,b)}{\partial\eta_\zeta} \log \frac{\partial g_W(\zeta,b)}{\partial\eta_\zeta} \frac{|d\zeta|}{2\pi} = \gamma_b + \sum_{\zeta:\nabla g_w=0} g_w(\zeta,b),$$

where η_ζ is the unit inner normal at $\zeta \in \partial\Omega$ and $g_W(z,b)$ is Green's function with pole at b. The constant γ_b comes from the expansion of g near b: $g(z,b) = -\log|z-b| + \gamma_b + O(|z-b|)$, and is sometimes called the **Robin constant**.

16.11 Suppose $\mathbb{G}$ is a group of LFTs τ of $\mathbb{D}$ onto $\mathbb{D}$ with no fixed points: if $\tau \in \mathbb{G}$ and $\tau(a) = a$ for some $a \in \mathbb{D}$, then $\tau(z) = z$ for all $z \in \mathbb{D}$. Is $\mathbb{G}$ necessarily properly discontinuous? (Prove or give a counter-example.)

Appendix

A.1 Fifteen Conditions Equivalent to Analytic

In the text and in the exercises we encountered several ways to determine when a function is analytic. As a review, we list some here. For simplicity, we assume the region is the unit disk $\mathbb{D}$. These statements can be transferred to other disks using linear maps of the form $az + b$, $a, b \in \mathbb{C}$.

If f is continuous on $\mathbb{D}$ then the following are equivalent:

1. For each $z_0 \in \mathbb{D}$, f has a power series expansion centered at z_0, valid in a neighborhood of z_0. (Definition 2.6)
2. f has a power series expansion centered at 0 with radius of convergence at least 1. (Theorem 2.7 and Corollary 4.15)
3. $f'(z)$ exists for each $z \in \mathbb{D}$ and f' is continuous on $\mathbb{D}$. (Corollary 4.15)
4. $f'(z)$ exists for each $z \in \mathbb{D}$. (Goursat's theorem, Exercise 4.12)
5. $\int_\gamma f(z)dz = 0$ for all curves $\gamma \subset \mathbb{D}$. (Cauchy's Theorem 5.1 and Morera's Theorem 4.19)
6. $\int_{\partial R} f(z)dz = 0$ for all closed rectangles $R \subset \mathbb{D}$, with sides parallel to the axes. (Local Cauchy's theorem, Corollary 4.18, and Morera's Theorem 4.19)
7. $\int_{\mathbb{D}} \frac{\partial \varphi}{\partial \bar{z}} f dxdy = 0$ for all continuously differentiable functions φ with compact support contained in $\mathbb{D}$. (Weyl's lemma, Exercise 7.13)
8. There exist rational functions r_n with poles in $\{z : |z| \geq 1\}$ such that r_n converges uniformly on compact subsets of $\mathbb{D}$ to f. (Runge's Corollary 4.28 and Weierstrass's Theorem 4.29)
9. There exist polynomials p_n such that p_n converges uniformly on compact subsets of $\mathbb{D}$ to f. (Runge's Theorem 4.27 and Weierstrass's Theorem 4.29)
10. $f = u + iv$, where u, v are continuously differentiable real-valued functions satisfying $u_x = v_y$ and $u_y = -v_x$. (Cauchy–Riemann equations, Theorem 7.9)
11. $f = g'$ for some g analytic on $\mathbb{D}$. (Theorem 2.12 and Corollary 2.15)
12. $f = u_x - iu_y$ for some u harmonic on $\mathbb{D}$. (Theorem 7.10 and Corollary 2.15)
13. f is continuously differentiable (with respect to x and y) and f preserves angles, including direction, between curves except at isolated points. (Section 3.2, Corollary 7.13 and Corollary 5.10)

With a little more work, this list can be extended to include:

14. $\int_{\partial S} f(z)dz = 0$ for all closed squares $S \subset \mathbb{D}$, with sides parallel to the axes.

    Hint: If $R \subset \mathbb{D}$ is a closed rectangle with rational side ratio, subdivide $\mathbb{R}$ into squares. If the side ratio is not rational, write $\partial R = \partial R_1 + \partial R_2$, where R_1 and R_2 are rectangles, R_1

has rational side ratio and ∂R_2 has two sides with length $< \varepsilon$. By uniform continuity, the integrals on the two long sides of R_2 almost cancel.

15. For each $r < 1$ there exists $M = M(r) < \infty$ so that, for each triple z_1, z_2, z_3 with $|z_j| < r$, $j = 1, 2, 3$, there exists a polynomial p with $p(z_j) = f(z_j), j = 1, 2, 3$, and $|p(z)| \leq M$ on $|z| < r$.
Hint: One direction follows from Exercise 4.13. For the other direction, find three polynomials p_j so that $p_j(z_k) = 0$ if $j \neq k$, and $p_j(z_j) = 1$. Truncate the power series for f to get a polynomial p_0 which is uniformly close to f on $|z| \leq r$. Then use $p_0 + \sum c_j p_j$ with $|c_j|$ small.

Analyticity is a local condition. But there are related global consequences. If f is analytic on a region Ω then

1. $\int_\gamma f(z)dz = 0$ for all curves $\gamma \sim 0$ (homologous to 0). (Cauchy's Theorem 5.1)
2. $\int_\gamma f(z)dz = 0$ for all curves $\gamma \approx \{\gamma(0)\}$ (homotopic to a constant curve). (Corollary 14.4)
3. There exist rational functions r_n with poles in $\partial\Omega$ converging to f uniformly on compact subsets of Ω. (Runge's Corollary 4.28)
4. If $z_0 \in \Omega$ then f has a power series expansion centered at z_0 with radius of convergence at least equal to the distance from z_0 to $\partial\Omega$. (Corollary 4.15)

Exercises

There are many ways to show a region is simply-connected. Locate results in the text for each item below or give a proof. Suppose $\Omega \subset \mathbb{C}$ is a region, then the following are equivalent to simple connectivity.

1. $\mathbb{C}^* \setminus \Omega$ is connected.
2. The boundary of Ω in $\mathbb{C}^*$ is connected.
3. $\gamma \sim 0$ for all closed curves $\gamma \subset \Omega$.
4. $\gamma \sim 0$ for all closed polygonal curves $\gamma \subset \Omega$.
5. $\gamma \approx \{\gamma(0)\}$ for all closed curves $\gamma : [0, 1] \to \Omega$.
6. $\gamma \approx \{\gamma(0)\}$ for all closed polygonal curves $\gamma : [0, 1] \to \Omega$.
7. $\Omega = \mathbb{C}$ or there is a conformal map f of Ω onto $\mathbb{D}$.
8. Every f analytic on Ω can be written as $f = g'$ for some g analytic on Ω.
9. Every non-vanishing f analytic on Ω can be written as $f = e^g$ for some g analytic on Ω.
10. Every non-vanishing f analytic on Ω can be written as $f = g^2$ for some g analytic on Ω.
11. Every u harmonic on Ω can be written as $u = \text{Re}f$ for some f analytic on Ω.

A.2 Program for Color Pictures

In this section we give a sample Matlab program for drawing the color pictures from Chapter 6. A number of examples are also given to illustrate the Matlab notation. The 3D option gives a surface colored according to $\arg f(z)$ with height $\log|f(z)|$ above each point z. The axes can be rotated using a mouse. The center and size of the domain can also be changed with a mouse. For more advanced code, and better pictures, see [27] and http://www.visual.wegert.com

```
function fcn(action)
% Choices: (alternatives follow %)
   domain=0; %plane=0, disk=1, ext(disk)=2, UHP=-1, RHP=-2
   range=0; % plane=0, disk=1, ext(disk)=2, UHP=-1, RHP=-2
   dimension=2; % 2d plot=2 3D plot =3
   Size=8; %width of plot (domain)
   resol=0.001; %relative grid size of domain (resolution)
   modlines=12; % number of modulus lines -1.
   mmod=1.8; %mod lines are in range:-mmod \le log|f| \le mmod
   arglines=64; % number of argument lines
   xcenter=0; % x coordinate of center of plot
   ycenter=0 ; %y coordinate of center of plot
% end choices, except for the function, given below.
% The program runs faster with a bigger resolution size, but
% may result in jagged lines if it is too big.
% Warning: the function contour doesn't work well with discont
% functions. Use NaN to remove pts where not cont. See below.
global Size
global xcenter
global ycenter
global resol
global range
global domain
global dimension
global mmod
global modlines
global arglines
if nargin < 1
   fname = mfilename;
   callbackStr=[fname '(''zoomin'')'];
   uicontrol(...
         'Style','pushbutton',...
         'Units','normalized',...
         'Position', [.70 .02 .1 .03],...
         'Background','white',...
         'Foreground','black',...
         'String',' zoomin',...
         'tag','zoomin',...
         'Callback',callbackStr);
   callbackStr=[fname '(''zoomout'')'];
   uicontrol(...
         'Style','pushbutton',...
         'Units','normalized',...
         'Position', [.80 .02 .1 .03],...
         'Background','white',...
```

```
          'Foreground','black',...
          'String',' zoomout',...
          'tag','zoomout',...
          'Callback',callbackStr);
    callbackStr=[fname '(''quit'')'];
    uicontrol(...
          'Style','pushbutton',...
          'Units','normalized',...
          'Position', [.90 .02 .1 .03],...
          'Background','white',...
          'Foreground','black',...
          'String',' quit',...
          'tag','quit',...
          'Callback',callbackStr);
    callbackStr=[fname '(''2D/3D'')'];
    uicontrol(...
          'Style','pushbutton',...
          'Units','normalized',...
          'Position', [.60 .02 .1 .03],...
          'Background','white',...
          'Foreground','black',...
          'String',' 2D/3D',...
          'tag','2D/3D',...
          'Callback',callbackStr);
elseif strcmp(action,'zoomin')
   [xcenter,ycenter]=ginput(1);
   Size=Size/2;
elseif strcmp(action,'zoomout')
   Size=Size*2;
elseif strcmp(action,'quit')
   close;
   return
elseif strcmp(action,'2D/3D')
   dimension=5-dimension;
end
   deltax=resol*Size;
   X = -Size/2.+xcenter:deltax:Size/2.+xcenter;
   Y=X-xcenter+ycenter;
   [x,y]=meshgrid(X,Y);
   z = x+1i*y;
%
%FUNCTION CHOICE:
   w=(z-3).^2./(z.^2+4);
% Some other functions to try: (put % in front of w in line
%       above then remove a % on a line below)
```

```
%       you can also change your choice for domain, size, etc.
% w=z;
% w=(z-1i)./(z+1i); %Cayley Transform
% w=(z-1).*(z+2).*(z-i);
% w=exp(z);
%    w=log(z);
%   w=log(z)+2*pi*1i;
%    w=z.^1.5
%  w=z.^4;
%     w=z.^(1/4);
%     w=-z.^3;
%    w=exp(z.^3);
%    w=tan(z);
% w=(z+1./z)/2;
%    w=exp((z+1./z)/2);
%   w=.5*z.*(1+sqrt(1-1./z.^2));%
%     w=1i*(1i*z).^0.5-.5*1i;
%   w=exp(z+1./z);
%       w=(exp(z)+exp(-z))/2;
%      w=(w+1i)./(w-1i);
%  w=(besselj(0,z).^2-besselj(1,z).^2).*z./sin(2*z);
%   w=(besselj(1,z).^2-besselj(3,z).^2).*z./sin(2*z);
%  w=1./z-1./sin(z);
%      w=besselj(0,z);
%Joukovski
%  w=z.*(1+sqrt(1-1./z.^2));
%   w=1./z;
%    w=z+1i*sqrt(1-z.^2);% add NaN statement below
%    w=z-1i*sqrt(1-z.^2);% add NaN statement below
%     w(imag(z) < 0)=NaN;
%   w(imag(z)==0 & real(z).^2 > 1)=NaN;
%         ---Kill where mod is not cont.
%  NOTE: if extraneous contours occur, decrease resol.
%   Same map as sequence of compositions,
%      w=-1i*sqrt((1-z)./(1+z));w=(1+w)./(1-w);
%      w(imag(z)==0 & real(z).^2 > 1)=NaN;
%          ---Kill where mod is not cont.
% in addition to get arccos using J map:
%     w=-1i*log(w);
%    w=exp(15./z);
%    w=(z-1i/3)./(1+1i*z/3);
%   w=sin(z)-z;
%  w=cos(z);
%     wz=z./(z-(1+1i));
%    w=((z.^4-1)./(z.^4+1));
```

```
%   w=sqrt(z.^2+1);
%     w=z./(1-z);
%      w=(z.*sqrt(1+1./z.^2));
%      w=-1i*log(z+1i*sqrt(1-z.^2));
% w=exp((z+1)./(z-1));
%  M.C. Escher:
% the push forward of a polar grid
% by the map exp((1+i)z/2) is the same as the
%pull back of the polar grid by the inverse map
%(1-i)log(w). The push forward of a rectangular grid
% is the same as the pull back of the polar grid by the map
% exp((1-i)logw)= w^{1-i).
%    w=exp((1-1i)*log(z));
% w=z.^(1-1i);w(imag(z)==0 & real(z) < 0 )=NaN;
%       ----Kills pts where mod is not continuous.
%If we color the plane with a rectangular grid instead
% of polar, then w would have the coloring of:
% w=exp(z);
%   w=sqrt(1i*z+.25)-0.5;%exterior of parabola
%    w=1i*w;
%    w=exp(z);%strip to half disk
%     w(abs(w)>1)=NaN;
%     w(imag(z)>pi)=NaN;
%     w(imag(z)<0)=NaN;
%       w=z;w(imag(z)<0)=NaN;
% use formula only if z in domain, w in range
   if range==1
     w(abs(w)>1)=NaN;
   elseif range==2
     w(abs(w)<1)=NaN;
   elseif range==-1
     w(imag(w)<0)=NaN;
   elseif range==-2
     w(real(w)<0)=NaN;
   end
   if domain==1
       w(abs(z)>1)=NaN;
   elseif domain==2
       w(abs(z)<1)=NaN;
   elseif domain==-1
       w(imag(z)<0)=NaN;
   elseif domain==-2
       w(real(z) < 0)=NaN;
   end
   ww=angle(w)/pi;
```

```
%       fix red color at discontinuity:
   ww(imag(w)==0 & real(w) < 0)=-1;
% NEED: axis equal before imagesc; axis square after imagesc
   axis equal
  if dimension==2
    ww(isnan(ww))=2; %NaNs are white
    imagesc(X,Y,ww);
    view(0,90)
    rotate3d off
  elseif dimension==3
    surf(x,y,log(abs(w)),ww); %NaNs are automatically white
    rotate3d on
    view(-37.5,30)
  end
  shading interp
  set(gca,'YDir','normal');
  caxis([-1,2*modlines/arglines+1])%color=[-1,1] gray=[1,last]
  axis square
  colormap([hsv(arglines);gray(modlines)])
  hold on
  ww2=log(abs(w));
%flip order so small=white:
  ww3=(mmod-ww2)*modlines/(mmod*arglines)+1;
  if dimension==2
    contour(x,y,ww3,1:2/arglines:2*modlines/arglines+1,...
            'LineWidth',1);
    contour(x,y,abs(w),[1 1],'-k','LineWidth',2)
  elseif dimension==3
    contour3(x,y,ww2,[0 0],'-k')
    xlabel('real'), ylabel('imaginary'), zlabel('log-modulus')
  end
  hold off
end
```

Bibliography

[1] Lars V. Ahlfors. *Complex Analysis: An Introduction to the Theory of Analytic Functions of One Complex Variable*, 3rd edn. McGraw-Hill Book Co., New York, 1979.

[2] Lars V. Ahlfors. *Conformal Invariants: Topics in Geometric Function Theory*. McGraw-Hill Series in Higher Mathematics. McGraw-Hill Book Co., New York, 1973.

[3] Robert B. Burckel. *An Introduction to Classical Complex Analysis, Vol. 1*. Pure and Applied Mathematics, vol. 82. Academic Press, New York, 1979.

[4] Henri Cartan. *Elementary Theory of Analytic Functions of One or Several Complex Variables*. Éditions Scientifiques, Hermann, Paris; Addison-Wesley Publishing Co., Reading, 1963.

[5] James Clunie, Alexandre Eremenko and John Rossi. On equilibrium points of logarithmic and Newtonian potentials. *J. London Math. Soc.*, **s2-47**(2):309–320, 1993.

[6] John B. Conway. *Functions of One Complex Variable*, 2nd edn. Graduate Texts in Mathematics, vol. 11. Springer, New York, 1978.

[7] Bart de Smit, Hendrik W. Lenstra, Jr., Douglas Dunham and Reza Sarhangi. Artful mathematics: the heritage of M. C. Escher. *Notices Amer. Math. Soc.*, **50**(4):446–457, 2003.

[8] Alexandre Eremenko, Jim Langley and John Rossi. On the zeros of meromorphic functions of the form $f(z) = \sum_{k=1}^{\infty} a_k/(z - z_k)$. *J. Anal. Math.*, **62**:271–286, 1994.

[9] Theodore W. Gamelin. *Complex Analysis*. Undergraduate Texts in Mathematics. Springer-Verlag, New York, 2001.

[10] John B. Garnett and Donald E. Marshall. *Harmonic Measure*. New Mathematical Monographs, vol. 2. Cambridge University Press, 2008. (Reprint of the 2005 original.)

[11] Jeremy Gray. On the history of the Riemann mapping theorem. *Rend. Circ. Mat. Palermo (2) Suppl.*, **34**:47–94, 1994.

[12] Robert E. Greene and Steven G. Krantz. *Function Theory of One Complex Variable*, 3rd edn. Graduate Studies in Mathematics, vol. 40. American Mathematical Society, Providence, 2006.

[13] Einar Hille. *Analytic Function Theory, Vol. 1*. Introductions to Higher Mathematics. Ginn and Co., Boston, 1959.

[14] Einar Hille. *Analytic Function Theory, Vol. II*. Introductions to Higher Mathematics. Ginn and Co., Boston, 1962.

[15] Otto Hölder. Ueber den Casus Irreducibilis bei der Gleichung dritten Grades. *Math. Ann.*, **38**(2):307–312, 1891.

[16] Morris Marden. *The Geometry of the Zeros of a Polynomial in a Complex Variable*. Mathematical Surveys, no. 3. American Mathematical Society, New York, 1949.

[17] Tristan Needham. *Visual Complex Analysis*. Oxford University Press, New York, 1997.

[18] University of Washington Mathematics Department. *Complex Analysis Preliminary Examinations 1953–2017*. University of Washington, Seattle, 2017.

[19] Roger Penrose and Wolfgang Rindler. *Spinors and Space-Time, Vol. 1: Two-Spinor Calculus and Relativistic Fields*. Cambridge Monographs on Mathematical Physics. Cambridge University Press, 1984.

[20] Christian Pommerenke. *Univalent Functions. With a Chapter on Quadratic Differentials by Gerd Jensen. (Studia mathematica Band XXV.)* Vandenhoeck & Ruprecht, Göttingen, 1975.

[21] Thomas Ransford. *Potential Theory in the Complex Plane*. London Mathematical Society Student Texts, vol. 28. Cambridge University Press, 1995.

[22] Walter Rudin. *Principles of Mathematical Analysis*, 3rd edn. International Series in Pure and Applied Mathematics. McGraw-Hill Book Co., New York, 1976.

[23] Walter Rudin. *Real and Complex Analysis*, 3rd edn. McGraw-Hill Book Co., New York, 1987.

[24] Donald Sarason. *Complex Function Theory*, 2nd edn. American Mathematical Society, Providence, 2007.

[25] Steve Smale. The fundamental theorem of algebra and complexity theory. *Bull. Amer. Math. Soc. (N.S.)*, **4**(1):1–36, 1981.

[26] Kenneth Stephenson. *Introduction to Circle Packing: The Theory of Discrete Analytic Functions.* Cambridge University Press, 2005.

[27] Elias Wegert. *Visual Complex Functions: An Introduction with Phase Portraits.* Birkhauser/Springer Basel AG, Basel, 2012.

Index

Printed in the United States
by Baker & Taylor Publisher Services